Teubner Skripten zur
Numerik

Stefan Vandewalle
Parallel Multigrid Waveform Relaxation for
Parabolic Problems

Teubner Skripten zur Numerik

Herausgegeben von
Prof. Dr. rer. nat. Hans Georg Bock, Universität Heidelberg
Prof. Dr. rer. nat. Wolfgang Hackbusch, Universität Kiel
Prof. Dr. phil. nat. Rolf Rannacher, Universität Heidelberg

Die Reihe soll ein Forum für Einzel- sowie Sammelbeiträge zu aktuellen Themen der Numerischen Mathematik und ihrer Anwendungen in Naturwissenschaften und Technik sein. Das Programm der Reihe reicht von der Behandlung klassischer Themen aus neuen Blickwinkeln bis hin zur Beschreibung neuartiger noch nicht etablierter Verfahrensansätze. Es umfaßt insbesondere die mathematische Fundierung moderner numerischer Methoden sowie deren Aufbereitung für praxisrelevante Anwendungen. Dabei wird bewußt eine gewisse Vorläufigkeit und Unvollständigkeit der Stoffauswahl und Darstellung in Kauf genommen, um den Leser schnell mit aktuellen Entwicklungen auf dem Gebiet der Numerik vertraut zu machen. Dadurch soll in den Texten die Lebendigkeit und Originalität von Vorlesungen und Forschungsseminaren erhalten bleiben. Hauptziel ist es, in knapper aber fundierter Weise über aktuelle Entwicklungen zu informieren und damit weitergehende Studien anzuregen und zu erleichtern.

Parallel Multigrid Waveform Relaxation for Parabolic Problems

by Dr. Stefan Vandewalle
Katholieke Universiteit Leuven

B. G. Teubner Stuttgart 1993

Dr. Stefan Vandewalle

born in 1962 in Brugge, Belgium. From 1980 till 1985, engineering studies in computerscience, numerical analysis and applied mathematics at the Katholieke Universiteit Leuven in Belgium. From 1985, research assistant and, later on, senior research assistant of the Belgian National Fund for Scientific Research (N.F.W.O.). Special graduate student at the California Institute of Technology, Pasadena, in 1987. Doctoral thesis on parallel algorithms for parabolic problems in 1992. Currently, post-doctoral researcher at the Katholieke Universiteit Leuven.

Die Deutsche Bibliothek – CIP-Einheitsaufnahme

Vandewalle, Stefan:
Parallel multigrid waveform relaxation for parabolic problems /
by Stefan Vandewalle. – Stuttgart : Teubner, 1993
(Teubner-Skripten zur Numerik)
ISBN 978-3-519-02717-1   ISBN 978-3-322-94761-1 (eBook)
DOI 10.1007/978-3-322-94761-1

Herstellung: Druckhaus Beltz, Hemsbach/Bergstraße

Preface

The numerical solution of a parabolic partial differential equation is usually calculated by using a time-stepping method. This precludes the efficient use of parallelism and vectorization, unless the problem to be solved at each time-level is very large. This monograph investigates the use of an algorithm that overcomes the limitations of the standard schemes by calculating the solution at many time-levels, or along a continuous time-window simultaneously. The algorithm is based on *waveform relaxation*, a highly parallel technique for solving very large systems of ordinary differential equations, and *multigrid*, a very fast method for solving elliptic partial differential equations. The resulting *multigrid waveform relaxation method* is applicable to both initial boundary value and time-periodic parabolic problems.

We analyse in this book theoretical and practical aspects of the multigrid waveform relaxation algorithm. Its implementation on a distributed memory message-passing computer and its computational complexity (arithmetic complexity, communication complexity and potential for vectorization) are studied. The method has been implemented and extensively tested on a hypercube multiprocessor with vector nodes. Results of numerical experiments are given, which illustrate a severalfold performance gain when compared to parallel implementations of a variety of standard initial boundary value and time-periodic solvers.

This monograph is based on my PhD-thesis, obtained at the Katholieke Universiteit Leuven in Belgium. It is a pleasure to express my gratitude to the many people that have played vital roles in all stages of this work.

First of all, I wish to express my gratitude to Prof. R. Piessens for giving me the opportunity to do research in the very active field of parallel computing and applied mathematics. I wish to thank Prof. D. Roose, Prof. P. Dierckx, Prof. R. Piessens, Prof. H. Van de Vel and Prof. W. Hackbusch as members of the reading committee and thesis jury for the careful and critical reading of the manuscript.

This work has greatly benefited from numerous discussions held at conferences, and from an intense e-mail correspondence with people working in closely related areas.

These discussions have influenced, defined and sometimes altered the course of my research. I would like to thank my colleagues in Leuven and abroad for many fruitful discussions and for creating the cordial atmosphere in which this work could grow.

At the beginning of my research I was given the opportunity to be a special graduate student at the California Institute of Technology. For this, I sincerely wish to thank Prof. G. Fox and dr. Eric Van de Velde. It is with great pleasure that I remember those L.A.-days. Part of the computing experiments discussed in this work were obtained during a stay at the Gesellschaft für Mathematik und Datenverarbeitung mbH, Sankt Augustin, Germany. I gratefully acknowledge dr. K. Stüben and dr. R. Hempel for making this possible.

I recognize the financial support of the Belgian National Science Foundation (Nationaal Fonds voor Wetenschappelijk Onderzoek). I was on their payroll, initially as a research assistant, and later on as a senior research assistant.

Special thanks go to Nele Geurden, who carefully corrected and anglicized the language of the preliminary version of this text. Any errors that remain are certainly to blame on myself. Finally, I wish to dedicate this dissertation to my wife, Ann, for her great love, her patience and understanding, displayed throughout the duration of this work.

Stefan Vandewalle
Leuven, November 1992

Contents

Preface   **5**

Table of contents   **7**

Notations   **11**

1 Introduction   **15**
   1.1 Numerical simulation and parallel processing 15
   1.2 The simulation of time-dependent processes 17
       1.2.1 The serial bottleneck of time-marching 18
       1.2.2 Accelerating the time-marching process 19
   1.3 Outline 21

2 Waveform Relaxation Methods   **23**
   2.1 Introduction 23
   2.2 Waveform relaxation: basic ideas 25
   2.3 A classification of waveform methods 28
   2.4 General convergence results 30
       2.4.1 The contraction mapping principle 30
       2.4.2 Waveform relaxation as a contraction map 31
       2.4.3 On the order of accuracy 33
   2.5 Convergence analysis for linear systems 35
       2.5.1 Functional analysis preliminaries 35
       2.5.2 Waveform relaxation for linear systems 37
       2.5.3 Continuous-time convergence results 38
       2.5.4 Discrete-time convergence results 40
   2.6 Waveform relaxation acceleration techniques 45
   2.7 Some concluding remarks 48

3 Waveform Relaxation Methods for Initial Boundary Value Problems   **49**
   3.1 Introduction and notations 49
   3.2 Standard waveform relaxation 50
       3.2.1 A model problem: the two-dimensional heat equation 51
       3.2.2 Standard waveform relaxation methods 51
       3.2.3 Some convergence results 52
       3.2.4 Numerical experiments 54

	3.2.5	Further remarks	59
3.3		Linear multigrid acceleration	59
	3.3.1	The multigrid principle	59
	3.3.2	Linear multigrid waveform relaxation	64
3.4		Convergence analysis	68
	3.4.1	Continuous-time convergence analysis	68
	3.4.2	Discrete-time convergence analysis	70
3.5		Experimental results	73
3.6		Nonlinear multigrid waveform relaxation	78
3.7		A multigrid method on a space-time grid	80
3.8		Concluding remarks	82

4 Waveform Relaxation for Solving Time-Periodic Problems   83
4.1		Introduction	83
4.2		Standard time-periodic PDE solvers	85
4.3		Time-periodic waveform relaxation	88
4.4		Analysis of the continuous-time iteration	90
	4.4.1	The linear model problem	90
	4.4.2	The time-periodic waveform iteration	92
	4.4.3	The time-periodic integral operator	93
	4.4.4	Convergence of the time-periodic waveform relaxation	95
	4.4.5	The existence of convergent and divergent splittings	99
4.5		Analysis of the discrete-time iteration	100
	4.5.1	Discretization and solution of the model problem	100
	4.5.2	Reformulation as a linear algebra problem	102
	4.5.3	The discrete-time time-periodic waveform iteration	104
	4.5.4	Convergence of the discrete-time iteration	105
	4.5.5	The spectral picture	107
4.6		Multigrid acceleration	110
	4.6.1	Introduction	110
	4.6.2	Time-periodic multigrid waveform relaxation	110
	4.6.3	Analysis of the continuous-time iteration	112
	4.6.4	Analysis of the discrete-time iteration	114
	4.6.5	Numerical example	115
4.7		Autonomous time-periodic problems	116
	4.7.1	Introduction	116
	4.7.2	Shooting with coarse grid Jacobian approximation	118
	4.7.3	The use of waveform relaxation within shooting	120
	4.7.4	A numerical example: the Brusselator	120

5 A Short Introduction to Parallel Computers and Parallel Computing 123
5.1		Introduction	123
5.2		Classification of parallel computers	124
5.3		The hypercube topology	126
	5.3.1	Definition and properties	126
	5.3.2	Binary reflected Gray codes	128

    5.3.3 Topology embedding onto the hypercube 128
  5.4 The Intel iPSC/2 hypercube multiprocessor 130
    5.4.1 System overview 130
    5.4.2 Some computation benchmarks 131
  5.5 Parallel performance parameters 132

6 Parallel Implementation of Standard Parabolic Marching Schemes 135
  6.1 Introduction 135
  6.2 Problem class and discretization 136
    6.2.1 Problem class 136
    6.2.2 Discretization 137
    6.2.3 The numerical kernels 139
  6.3 Parallel implementation: preliminaries 140
    6.3.1 The parallel computer model 140
    6.3.2 The grid partitioning approach 141
  6.4 The explicit update step 142
  6.5 The multigrid solver 144
    6.5.1 Introduction 144
    6.5.2 Parallelizing the multigrid components 146
    6.5.3 Agglomeration/deagglomeration strategies 150
  6.6 The tridiagonal systems solver 154
    6.6.1 Introduction 154
    6.6.2 Substructured Gaussian elimination 155
    6.6.3 Solution of the intermediate system 157
  6.7 Timing results on the Intel hypercube 165
    6.7.1 The explicit update step 165
    6.7.2 The multigrid solver 166
    6.7.3 The tridiagonal system solver 170
  6.8 Numerical examples 174
    6.8.1 Introduction 174
    6.8.2 Test problem one 175
    6.8.3 Test problem two 176
    6.8.4 Test problem three 178
  6.9 Concluding remarks 178

7 Computational Complexity of Multigrid Waveform Relaxation 183
  7.1 Introduction 183
  7.2 Arithmetic complexity 184
    7.2.1 Initial boundary value problems 184
    7.2.2 Time-periodic problems 187
  7.3 Parallel implementation 190
    7.3.1 Grid partitioning 190
    7.3.2 Communication complexity 191
  7.4 Vectorization 193
  7.5 Concluding remarks 194

8 Case Studies   **195**
   8.1   Introduction   195
   8.2   Programming considerations   196
   8.3   Representation of the results   197
   8.4   Linear initial boundary value problems   198
         8.4.1   Example 1   198
         8.4.2   Example 2   204
   8.5   Nonlinear initial boundary value problems   207
         8.5.1   Example 3   207
         8.5.2   Example 4   210
   8.6   Linear time-periodic problems   214
         8.6.1   Example 5   214
         8.6.2   Example 6   217
   8.7   Example 7: a nonlinear periodic system   220
   8.8   Further remarks, limits of applicability   222

9 Concluding Remarks and Suggestions for Future Research   **225**

A Discretization and Stencils   **231**

Bibliography   **237**

Index   **249**

Notations

He examined the engraved characters...
They looked like the footprints of a spider
that had had one too many of whatever it is
that spiders have on a night out.

—D. Adams *"The restaurant at the end of the Universe"*
(ISBN 0-671-66494-8, p. 176)

Below we present lists of symbols and abbreviations used in the text, together with a very brief explanation of their meaning. We have added the number of the page on which they are first used, and numbers of pages where a more detailed explanation is given. The symbol list is restricted to symbols that are used more or less frequently, e.g., in different sections and on different pages. Symbols with a generally well-known meaning are excluded from the list, e.g.: number spaces $(I\!N, Z\!\!\!Z, I\!R, C\!\!\!\!\,)$, differentiation operators $(d/dt, \partial/\partial x, \partial^2/\partial x^2, \ldots)$, relational operators $(\in, \subset, \gg, \leq, \ldots)$, various abbreviations $(\min, \inf, \sup, \lim, \det, \bmod)$, etc.

List of symbols

$\star$	:	convolution operator, p. 38
$*$	:	don't care symbol, p. 128
$\parallel$	:	bit concatenation, p. 129
$(.)^h$	:	function, operator defined on Ω^h (e.g. x^h), p. 17
$(.)_\tau$	:	a sequence or vector (e.g. x_τ), p. 41
$(\tilde{.})_n$	:	Fourier series coefficient (e.g. $\tilde{x}_n$), p. 94
$(\hat{.})_m$	:	discrete Fourier series coefficient (e.g. $\hat{x}_m$), p. 100
$(\bar{.})$	:	bit complement (e.g. $\bar{x}$), p. 127
$(.)^{(\nu)}$	:	ν'th iterate (e.g. $x^{(\nu)}$), p. 25
$\{.\}_{n=0}^N$	:	sequence with $N+1$ elements (e.g. $\{x_n\}_{n=0}^N$), p. 41
$[.]^t$	:	transposition operator, p. 25
$(\ldots\ldots)_2$	:	binary representation, p. 127
$\lvert . \rvert$	:	modulus of complex number, p. 39

$\| \cdot \|$	:	vector- and induced matrix-norm in $I\!R^d, C^d$, p. 30
$\| \cdot \|_p$	:	p-norm in L_p, p. 35, and in l_p, p. 36
$\| \cdot \|_b$	:	exponentially scaled norm, p. 31
$\| \cdot \|_T$	:	maximum-norm, p. 30
$\| \cdot \|_X$	:	general Banach space norm, p. 30
$a(\xi)$	:	first characteristic polynomial of linear multistep method, p. 41
arg z	:	argument of complex number z, p. 72
$A\ (=D\text{-}L\text{-}U)$	:	matrix (diagonal, lower- and upper-triangular parts), p. 37
A_c	:	accuracy order, p. 33
$b(\xi)$	:	second characteristic polynomial of linear multistep method, p. 41
$\bar{C}$	:	set of complex numbers with point at infinity, p. 41
$C, C[0,T]$	:	space of continuous functions on $[0,T]$, p. 30
$C_V, C_P, C_R, ..$	:	multigrid cycle and operator cost factors, p. 185
$e^{(\nu)}, e_T^{(\nu)}, e_k^{(\nu)}$	:	error of ν-th iterate (continuous, discrete), component at t_k, p. 55
E_p	:	parallel efficiency, p. 133
$F(t,u,v)$	:	waveform iteration function, p. 28
$\mathcal{F}$	:	waveform relaxation operator, p. 28
G	:	static iteration matrix, p. 37
h, h_i, H	:	mesh sizes, p. 17, p. 59, p. 61
I	:	unit matrix, p. 39
$I_h^H, \bar{I}_h^H, I_{h_i}^{h_{i-1}}$	:	restriction operators, p. 60, p. 79
$I_H^h, I_{h_{i-1}}^{h_i}, \bar{I}_{h_{i-1}}^{h_i}$	:	prolongation operators, p. 60, p. 61
$J_{.k}$	:	k'th column of matrix J, p. 119
$K(z)$	:	symbol of convolution operator $\mathcal{K}$, p. 39
$\mathcal{K}, k$	:	continuous-time initial value WR operator, kernel, p. 38
$\mathcal{K}_\tau, k_\tau$	:	discrete-time initial value WR operator, kernel, p. 43
$\bar{\mathcal{K}}, \bar{k}$	:	continuous-time time-periodic WR operator, kernel, p. 92
$\bar{\mathcal{K}}_\tau, \bar{k}_\tau$	:	discrete-time time-periodic WR operator, kernel, p. 105
L_i	:	total message length in one cycle on Ω^{h_i}, p. 191
L^h, L^H, L^{h_i}	:	discretized elliptic operator, p. 17, p. 50, p. 61
$L_p, L_p(0,T)$	:	space of p'th power Lebesgue integrable functions, p. 35
$l_p, l_p(0..N)$	:	space of p-summable sequences, p. 36
$\mathcal{L}$	:	elliptic PDE operator, p. 17, p. 49

$M(z)$:   symbol of multigrid convolution operator, p. 68

M_i :   number of messages in one cycle on Ω^{h_i}, p. 191

$\mathcal{M}, m$:   contin.-time initial value multigrid WR operator, kernel, p. 68

$\bar{\mathcal{M}}, \bar{m}$:   contin.-time time-periodic multigrid WR operator, kernel, p. 112

$\mathcal{M}_\tau, m_\tau$:   discrete-time initial value multigrid WR operator, kernel, p. 71

$\bar{\mathcal{M}}_\tau, \bar{m}_\tau$:   discrete-time time-periodic multigrid WR operator, kernel, p. 114

n_i :   number of IBVPs on grid Ω^{h_i}, p. 189

$n_{t,i}$:   number of time-steps, p. 190

n_x, n_y, n_t :   number of unknown in x-, y- and t-direction, p. 197, p. 235

O :   order of magnitude, p. 61, big-"O"-symbol, p. 33

(P, Q) :   matrix splitting $(A = P - Q)$, p. 37

$Re(z)$:   real part of complex number z, p. 39

$S\ (\partial S, \mathrm{int}S)$:   stability region (boundary, interior), p. 41

S_p :   parallel speedup, p. 133

t_{send} :   communication-time parameter, p. 131

$t_{startup}$:   message startup time, p. 131

T :   length of time interval, period of periodic solution, p. 25, p. 83

TL_i :   total message length for solving PDE on Ω^{h_i}, p. 191

TM_i :   total number of messages for solving PDE on Ω^{h_i}, p. 191

$u, u(t, x)$:   solution of a partial differential equation, p. 17, p. 49

$u^h, u^h(t), u^h_{ij}(t)$:   solution of numerical method of lines ODE, p. 17, p. 50

$u^h_\tau, u^h_k, u^h_{k,i,j}$:   fully discrete solution, p. 50

$\mathcal{U}$:   general linear operator, p. 30

W_i :   arithmetic complexity of solving parabolic PDE on Ω^{h_i}, p. 189

$\mathcal{X}$:   general Banach space, p. 30

$\alpha\ (\alpha_M, \alpha_E, ..)$:   number of multigrid cycles, p. 186

$\alpha_0, \alpha_1, .., \alpha_k$:   linear multistep coefficients, p. 40

$\beta_0, \beta_1, .., \beta_k$:   linear multistep coefficients, p. 40

Γ :   root locus curve, p. 41

δ :   diagonal element of constant diagonal matrix $(D = \delta I)$, p. 40

Δt_k :   time-increment $(= t_k - t_{k-1})$, p. 18

Δ :   correction, e.g. $\Delta u^h = u^{(\nu)} - u^{(\nu-1)}$, p. 118

$\varepsilon_{\tau,m}$:   basic N-periodic sequence, p. 100

λ	:	spectral element, eigenvalue, p. 36
ν_1, ν_2	:	number of pre- and post-smoothing steps, p. 61
ρ	:	spectral radius (e.g. in $\rho(\mathcal{U})$, $\rho(A)$), p. 36
$\rho^{(\nu)}$	:	iteration convergence factor, p. 55
$\rho_k^{(\nu)}$	:	time-level convergence factor, p. 56
$\bar{\rho}$	:	averaged convergence factor, p. 55
σ	:	spectrum (e.g. in $\sigma(\mathcal{U})$, $\sigma(A)$), p. 36
τ	:	fixed time-increment, p. 40
$\omega, (\omega_{opt})$	:	(optimal) overrelaxation parameter, p. 37
ω	:	pulsation ($=2\pi/T$), p. 91
Ω	:	PDE domain, p. 17, p. 49
$\Omega^h, \Omega^{h_i}, \Omega^H$	:	spatial grid with mesh size h, h_i, or H, p. 17, p. 61

List of abbreviations

a.e.	:	almost everywhere (see p. 35)
BDF	:	backward differentiation formula
CN	:	Crank-Nicolson
DFS	:	discrete Fourier series
flop	:	floating point operation
FMG	:	full multigrid
GS	:	Gauss-Seidel
IBVP	:	initial boundary value problem
JAC	:	Jacobi
JOR	:	Jacobi overrelaxation
ODE	:	ordinary differential equation
PDE	:	partial differential equation
SOR	:	successive overrelaxation
TGET	:	transpose - Gaussian elimination - transpose
TPP	:	time-periodic problem
WR	:	waveform relaxation

Chapter 1

Introduction

The successful solution of a realistic problem in applied mathematics
requires the fusion of four distinct ingredients:

1. knowledge of the subject area of the problem

2. knowledge of the relevant mathematics

3. knowledge of the relevant computer science

4. a talent for selecting just what part of all this knowledge will actually
solve the problem, and ignoring the rest

—George E. Forsythe, *in [36]*

We emphasize the increasing importance of parallel supercomputers for the solution
of large-scale scientific and engineering problems. A sequential bottleneck which limits
the obtainable parallelism and performance when simulating time-dependent processes
with standard time-marching schemes is identified. Some of the approaches that have
been suggested in the literature for eliminating or alleviating this fundamental problem
are reviewed. Finally, we present an overview of the book.

1.1 Numerical simulation and parallel processing

Over the years the roles of mathematical modelling, numerical simulation and prediction have dramatically increased in both science and engineering. In science, more and
more, conventional experimental testing is being replaced by simulation on computers.
Theoretical analysis is being complemented with computer generated data. In the aircraft and automobile industry complicated fluid flow problems are numerically solved
in order to optimize the design of wings and engines. The construction of very high
density integrated circuits has only become possible thanks to availability of simulation
tools that allow the verification of the circuit before its actual fabrication. Geophysical
exploration is guided by computer analysis of seismic data combined with extensive oil

reservoir simulation. Numerical simulation allows the study of nuclear power plants and weapons in operating conditions which exceed the safety limits for the practical experiment.

More accurate predictions require a more accurate modelling of the physics involved. This necessitates, for instance, the incorporation of more species and interreacting chemical substances, and the use of finer discrete meshes with a larger number of sampling points. All of this leads to a corresponding explosion in the number of equations and unknowns that are to be solved for. As such, the computational complexity often renders the modelling and simulation task very difficult and extremely time-consuming. Therefore, many of the recent developments have only become feasible thanks to advances in both computer architectures and numerical algorithms.

The development of faster electronic switching devices has made current-day personal computers comparable in speed to the large main-frames of a decade ago. The introduction of powerful vector processors has led to severalfold increases in speed over conventional serial architectures. Yet, a great number of problems remains beyond the reach of these so-called *mono-processor vector-supercomputers*. Furthermore, additional orders of magnitude increase in computational power is mandatory for addressing some of the future "Grand-Challenges"-projects. For instance, unprecedented computational problems are faced by the scientists involved in Global Climate Change research. Equally formidable are the goals of the so-called Human Genome project, which aims at deciphering the nucleotide sequence of human DNA. Because of physical and engineering limitations the increase in single-processor speed as required by these problems is no longer feasible. It is only the use of systems consisting of many cooperating processors which is expected to match the computational needs.

Several architectures of such machines have emerged. Some consist of a small number of powerful processors cooperating through the use of a vast shared memory (e.g., Cray, IBM, Sequent). Others provide a large number of processors each with a local memory, communicating and exchanging data over an interconnection network (e.g., Intel iPSC, Ncube, Meiko Computing Surface). Still others consist of a massive number of very simple processing elements operating in a lock-step mode, i.e., executing an identical sequence of instructions on their local data (e.g., Distributed Array Processor, Connection Machine, Masspar).

Since about one decade, we have witnessed this evolution towards parallelization. Only recently however have massively parallel systems proven to outdo the vectorprocessors on non-academic problems. This delay is to some extent due to the difficulty of programming such machines. In particular, sequential algorithms have to be analysed and possibly reformulated in such a way that as many operations as possible can be executed concurrently. This is to be done with the additional goal of minimizing the amount of data that is to be shared by different processors, in order to reduce message exchange or common memory access costs. New algorithms have to be devised when the parallel execution of standard serial approaches is inefficient. Very often this has to be done with the hardware characteristics of the target architecture in mind. All of this constitutes an additional burden for the algorithm developer who can no longer be solely concerned with the mathematics of the problem.

It is fair to say that most standard serial algorithms have nowadays been analysed for their parallel potential, and implemented on parallel machines. We refer to [31, 94] for an overview of (some of) the literature. Many standard sequential techniques have found efficient parallel counterparts, especially in the fields of linear algebra and partial differential equations. More and more, parallel computing research is nowadays specializing towards the implementation of algorithms for complicated real-life problems. Such problems raise important new issues related to adaptivity and load-balancing. Considerable effort is also focussed towards the construction of parallel software development tools, such as automatic code parallelizers, load-balancing and monitoring tools. Finally, a lot of people continue to work on the parallelization of the inherently sequential algorithms that have withstood the first wave of parallel computing research. One example of such a problem, to be discussed in the next section, is the simulation of time-dependent processes, modelled by ordinary or partial differential equations.

1.2 The simulation of time-dependent processes

This book deals with the numerical simulation of time-evolving physical processes modelled by *parabolic partial differential equations* (parabolic *PDEs*). Such problems are frequently encountered in the study of diffusion, convection and reaction phenomena, and arise in such diverse fields like thermodynamics, chemistry, hydrodynamics, aerodynamics etc. They are generally modelled by one or more equations of the form

$$\frac{\partial u}{\partial t} = \mathcal{L}(u) + f. \tag{1.1}$$

The unknown function u is a function of the spatial variable x and of the time-coordinate t, i.e., $u = u(t, x)$, and is defined on some compact spatial domain, $x \in \Omega$, and some interval of time, $t_0 \leq t \leq t_f$. The same holds for the right-hand side forcing function f. $\mathcal{L}(\ .\ )$ is a general time-dependent uniformly elliptic operator comprising diffusion, reaction and convection terms. Equation (1.1) is usually complemented with suitable boundary conditions, specifying for instance a given temperature or particle flux along the boundary of Ω. Finally the equations are completed with a "boundary condition" in the time-dimension, like a given initial condition on t_0, or a demand for time-periodicity, $u(t_0, x) = u(t_f, x)$.

The solution of the partial differential equation is approximated by constructing a discrete counterpart, obtained by finite differences, finite elements or finite volumes. The solution is then found by solving a system with a finite number of unknowns, which, for instance, approximate the continuous solution in a discrete set of points laying on a grid that covers the domain. A commonly used approach, which will be followed in a large part of this book, is the so-called *numerical method of lines*. The spatial derivatives in $\mathcal{L}$ are replaced by discrete analogues obtained by using finite differences on a spatial mesh. This transforms (1.1) into a large system of ordinary differential equations (*ODEs*),

$$\frac{du^h}{dt} = L^h(u^h) + f^h, \tag{1.2}$$

where u^h is a vector of functions which approximate u at the grid points x_i. Standard techniques for solving ordinary differential equations can then be applied.

1.2.1   The serial bottleneck of time-marching

Standard numerical methods for solving the equations (1.1) and (1.2) are based on the *time-stepping* or *time-marching* idea. Starting from a given solution at t_0 the solution is advanced time-step per time-step with a time-increment Δt_k between successive time-levels t_{k-1} and t_k. On each time-level a discrete set of variables which approximate $u(t_k, x_i)$ is solved for. This may be done by *explicit* methods, in which each variable can be calculated independently, or by *implicit* methods, which require the solution of a large (non)linear system of equations in each time-step. The time-marching process is graphically illustrated for a rectangular two-dimensional domain in figure 1.1.

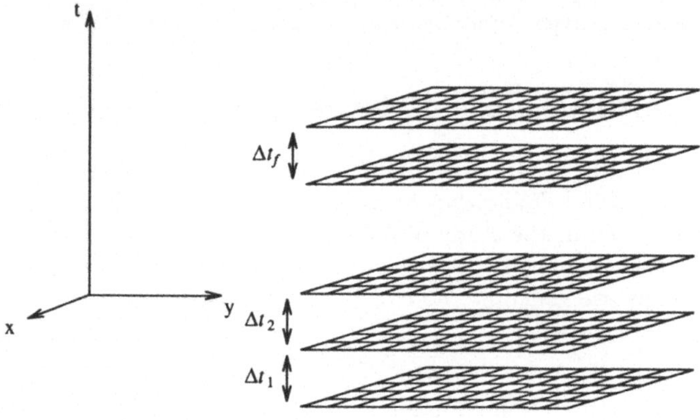

Figure 1.1: Standard time-marching methods calculate the solution variables in a sequential manner, time-level after time-level.

The standard approach to parallelizing the time-marching process consists of applying a *spatial grid partitioning* on each time-level. By that, each processor is assigned to the grid points in a subset of the computational domain, for which it gets the *update right*. In each time-step each processor has to calculate the new values at its grid points. In a distributed memory machine this requires a cooperation with the other processors through message passing. When the number of spatial grid points per processor is *sufficiently large*, both explicit and implicit methods can be parallelized with high efficiency, i.e., the losses due to parallel overheads are negligible. When the number of grid points per processor is *relatively small*, due to the use of coarse computational meshes, or, more importantly, due to the availability of a large number of processors, only explicit methods retain their parallel efficiency. However, they suffer from a severe stability constraint which forces the use of very small time-steps. This makes them unattractive for solving problems on fine meshes. For relatively small problems, the performance of the implicit methods is also very unsatisfactory. This is especially so when very fast multigrid solvers are applied for solving the system in each time-step. The computation at each time-level is then easily dominated by parallel overheads due to communication, processor idling or load imbalance.

The serial nature of classic time-marching schemes imposes severe restrictions on the obtainable parallel performance. The parallelism is limited to the parallelism inherent in the relatively small problems defined on the time-levels, even though the total number of variables and the corresponding arithmetic complexity may be formidable. Furthermore, the comparison of explicit to implicit methods shows that parallel efficiency must usually be traded off against numerical efficiency.

1.2.2 Accelerating the time-marching process

Several approaches have been suggested for accelerating the parallel simulation of time-dependent processes modelled by parabolic partial differential equations. In the following overview we discern basically three classes: methods that start from the PDE formulation (1.1), methods that start from the ODE formulation (1.2), and, finally, methods that entirely circumvent the mathematical PDE and ODE formulation, but directly simulate the underlying physics instead. Note however that this classification is to be taken rather loosely as some methods do fit in more than one class.

Methods based on the PDE formulation

If one retains the standard time and space discretization principles, two obvious research directions arise. One may start from standard explicit techniques which parallelize well, and try to improve their numerical quality. Alternatively, one may wish to keep the numerical quality of the implicit methods, and try to improve their parallel performance.

The number of time-steps required by an explicit technique can be reduced if its stability boundary is enlarged. For instance, Rodrigue and Wolitzer have devised fully explicit methods which require about the same amount of computational work as the explicit Euler method, and yet have a stability interval several times as large. Their method, *preconditioned time-differencing*, also called the *predictor corrector iterative* method, is based on the use of a classical explicit corrector to yield an initial guess, and the use of a fixed number of iterations of a standard relaxation scheme applied to an implicit corrector, [105]. Evans and coworkers have developed stable explicit schemes, the *alternating group explicit* methods, in which the unknowns are explicitly computed in groups of two, four or eight, [28]. In [122], Sommeijer shows how to increase the stability boundary by combining extrapolation techniques for bridging the largest part of a time-step, and simple explicit techniques to perform the remaining part. The author illustrates a trade-off between stability and accuracy. This seems to be a general characteristic for methods of this type.

Others aim at improving the parallel performance of implicit methods by applying iterative techniques on several time-levels simultaneously, instead of by iterating until convergence over each time-step before moving on to the next. In the *windowed block relaxation* methods of Saltz and Naik, each processor iterates in its own subdomain over a number of time-steps, a so-called *window*. This is shown to allow an efficient overlap of computation with communication, and to lead to a reduction in the number of messages, [110]. In Womble's *parallel time-stepping* method processors that otherwise would be idle are used to improve the initial guess at time-levels $t_{k+1}, t_{k+2}, \ldots, t_{k+N}$, while the solution is being computed at time-level t_k, [149]. He illustrates his algorithm with timing results obtained on a 1024-processor Ncube. In 1984 Hackbusch proposed the *parabolic multigrid method*, a multigrid technique with typical multigrid convergence rates, which calculates the unknowns on several time-levels simultaneously, [43]. A *parallel smoothing* variant allows most calculations on different time levels to be performed independently. The method was first implemented on a parallel processor by Bastian, Burmeister and Horton, [5] and later on extensively studied by Horton, Burmeister and Knirsch [14, 59, 60, 58].

Finally, some techniques were suggested that are based on the *domain decomposition* principle, which is well-known as a parallel method for solving elliptic partial differential equations. The computational domain is then split into several possibly overlapping subdomains on which computations may proceed independently. Of course, special precautions are to be taken along subdomain interfaces. In [26], Dawson, Du and Dupont present non-overlapping domain decomposition procedures for parabolic problems in which the interfacial degrees of freedom are advanced explicitly. The subdomains are advanced implicitly and independently. A domain splitting scheme that is locally implicit on slightly overlapping subdomains is analysed by Blum, Lisky and Rannacher in [7]. The local subdomain boundary data is propagated by a simple explicit process. In [114], Scroggs studies a certain class of convection diffusion equations. An asymptotic analysis allows him to identify regions where simplified versions of the equation may be solved. In such a way he arrives at a physically motivated domain decomposition which allows different regions to be treated simultaneously.

Methods based on the ODE formulation

A second class of methods starts from the equations derived by the method of lines. They apply parallel ODE techniques such as time-discretization schemes with high inherent degree of parallelism, extrapolation methods, or waveform relaxation techniques. Gallopoulos and Saad analyse in [33, 34] the use of *rational Chebyshev and Padé expansions* to the matrix exponential which arises in the analytical solution of (1.2). (They deal with the constant coefficient case.) They derive formulae of very high degree the computations of which can be parallelized through the use of a partial fraction expansion. In [60], Horton and Knirsch combine a parallel time-stepping method with *extrapolation* techniques in order to solve the Navier-Stokes equations. Solutions obtained with different magnitudes of the step-size are calculated concurrently.

The *waveform relaxation* method is conceptually very similar to the Gauss-Seidel and Jacobi relaxation methods for iteratively solving systems of (non)linear equations. Instead of successively solving equations in one unknown value, one successively solves (non)linear ODEs in one unknown function. The use of this method for solving certain semi-discretized parabolic PDEs was first considered by Miekkala and Nevanlinna in [84]. A multigrid acceleration was analysed by Lubich and Ostermann, [78].

Methods that circumvent the PDE and ODE formulation

Partial differential equations of reaction-convection-diffusion type often represent a mathematical idealization of a large population of particles, cells, organisms, etc., which are interacting locally while diffusing and drifting in space. Whereas the PDE formulation is ideally suited for symbolic manipulation and theoretical analysis, several researchers have expressed doubts whether it should be kept as the starting point for numerical simulation. Instead, they have suggested a direct simulation of simple physical models which closely approximate the microscopic dynamics of the studied phenomena. Such *moving point* and *particle* methods, or *cellular automata* techniques are discussed, e.g., by Hebert [50], and by Rees and Morton [103]. Simulations of this type require large numbers of simple local computations and are therefore excellently suited for implementation on massively parallel systems.

1.3 Outline

We analyse and illustrate in this monograph the use of waveform relaxation methods for solving parabolic partial differential equations on parallel computers. The central algorithm is the *multigrid waveform relaxation method*. This technique combines the very fast convergence of multigrid, a method originally developed for solving elliptic partial differential equations, with the high parallel efficiency of waveform relaxation, an algorithm for solving very large systems of ordinary differential equations.

The multigrid waveform relaxation algorithm

The multigrid waveform relaxation algorithm for solving linear problems was first published by Lubich and Ostermann, in an article which appeared in 1987, [78]. These authors pointed out the potential for parallelism, and illustrated their theoretical results with a sequential implementation for a one-dimensional model problem.

Independently of the previous authors, the current author arrived at the same algorithm which he first documented in [129]. A nonlinear variant based on the multigrid full approximation scheme was presented in [140]. The method was implemented on a parallel machine, [133, 141], and also vectorized on a vector-multiprocessor [134]. From the start substantial interest was focused on comparing waveform relaxation with standard parabolic solvers. To this end a careful implementation was made of a large number of standard time-stepping techniques, [142, 143, 144]. These implementations were mainly based on a program library for solving linear second order elliptic partial differential equations on two-dimensional, rectangular domains, [131, 132]. The comparisons established the real performance of the waveform relaxation method, as they showed a severalfold reduction in computing time over the parallel implementation of the standard techniques. The numerical evidence is largely surveyed in [136, 137].

Apart from initial boundary value problems also *time-periodic* parabolic partial differential equations were considered, for which a new *time-periodic multigrid waveform relaxation method* was developed. To the author's initial surprise an experimental comparison with standard time-periodic solvers showed that the new method outdid the best standard technique with a factor 2.5 *on a sequential machine*, [130, 135]. As such, the quest for better parallel algorithms had led to a more efficient sequential algorithm. Later on, a theoretical framework was developed for the study of the time-periodic waveform relaxation algorithms, [139], and results were proven similar to the ones for the initial value iteration developed by Miekkala and Nevanlinna, [84, 85] and by. Lubich and Ostermann, [78]. Finally, in cooperation with D. Roose a waveform relaxation based technique was studied for calculating the time-periodic solutions to autonomous, nonlinear systems of parabolic partial differential equations, [106].

Overview of the text

In chapter 2 we present an overview of standard waveform relaxation results and we direct the interested reader to further references. A qualitative analysis of convergence results, mainly found in the electrical engineering literature, is presented first. A quantitative analysis for linear constant coefficient ordinary differential equations is given next. We also enumerate a number of techniques that have been suggested for accelerating the waveform iteration.

Chapter 3 deals with waveform relaxation methods for initial boundary value problems. The convergence behaviour and the deficiencies of standard waveform relaxation methods are analysed in detail. The linear and nonlinear multigrid waveform relaxation methods are introduced and illustrated by numerical examples.

In chapter 4 the applicability of the waveform technique is extended to time-periodic problems. We theoretically analyse the time-periodic waveform relaxation method and prove a number of convergence results for the continuous-time and discrete-time iteration. It is shown that the convergence is intimately related to the convergence of the initial value iteration, as discussed by Miekkala and Nevanlinna, and by Lubich and Ostermann. We also present a modified shooting method for solving autonomous time-periodic problems.

Chapter 5 provides a short introduction to parallel computers and parallel computing. We discuss one architecture and one machine in particular: the Intel iPSC/2 hypercube multiprocessor, which is the machine used in our experiments.

The parallel implementation and the parallel performance of standard parabolic time-stepping schemes form the subject of chapter 6. We discuss various explicit, implicit and line-implicit time-discretization schemes. In particular, we detail the parallel implementation of an explicit update step, the solution of a system of equations by a multigrid method, and the parallel solution of tridiagonal linear systems.

The parallel computational complexity of the waveform relaxation method is discussed in chapter 7. We analyse the arithmetic complexity and the communication complexity, and we discuss the vectorization of waveform relaxation methods.

A large number of case studies is presented in chapter 8. For each example we compare the performance of the best standard method with that of the appropriate waveform relaxation variant. It is shown that for many problems multigrid waveform relaxation leads to a many-fold reduction in computing time.

Finally, in chapter 9 some general conclusions are given and some possible directions for future research are suggested.

Chapter 2

Waveform Relaxation Methods

—Sir Arthur Conan Doyle, *Silver Blaze*
(borrowed from [51, p.740])

We survey standard waveform relaxation results, mainly for future reference. The method is introduced and its use for solving ordinary differential equations is illustrated. The convergence theorems based on a contraction mapping argument are recalled and some alternative proofs are presented. A detailed analysis for linear constant coefficient ordinary differential equations is given. Finally, we enumerate techniques which have been proposed for accelerating the computational process.

2.1   Introduction

Standard numerical solvers for stiff ordinary differential equations usually implement a three-stage process that is executed in every consecutive time-step.

- To start with, the differential equations are *discretized* with an implicit and stable time-discretization scheme. This leads to a nonlinear system of equations in a set of variables which approximate the ODE solution at a particular time-level.

- This nonlinear system is *linearized* by the Newton-Raphson procedure or by a related technique. This results in a large linear system of equations.

- This linear system is *solved* by using the well-known direct Gaussian elimination algorithm or any of its many variants.

(Note that the latter two stages may have to be performed a number of times in order to ensure convergence.)

While this procedure has proven to be adequate for solving systems consisting of a relatively small number of equations, say of the order of a hundred, it has shown to fail for systems consisting of thousands of equations. The reasons are twofold. First, the time-stepping nature forces each variable to be discretized on an identical set of time points. As such, the time-step is restricted to one which is fine enough to resolve the most rapidly changing component. Very often though, large systems of differential equations have variables which change at very different rates. The resolution of the slowly changing variables is then unnecessarily accurate, and computation time is wasted. Secondly, the cost of the Gaussian elimination at each time-level rapidly increases with an increasing number of variables. Even if sophisticated sparse matrix solvers are used in order to avoid substantial matrix fill-in, this elimination step soon becomes prohibitively expensive.

These shortcomings are strongly felt by researchers and engineers involved in the numerical simulation of very large-scale integrated circuits. Such electrical circuits often comprise hundreds or thousands of components like transistors and capacitors. Their numerical simulation gives rise to very large nonlinear systems of ordinary differential or differential-algebraic equations. These systems are characterized by a very loose coupling of the variables, which reflects the limited and local coupling between the electrical components. Moreover, very often only a small number of the components are active, while most other components are *latent*, i.e., not changing state. By consequence, in the electrical engineering community substantial effort has been focussed towards the development of numerical techniques that overcome the deficiencies of standard time-stepping schemes. Various methods have been proposed which effectively exploit the typical *latency* and *multi-rate* behaviour of such electrical circuits. Most of these techniques are iterative techniques, and based on *relaxation*, either at the level of the linear systems (*linear relaxation*), or at the level of the nonlinear systems (*nonlinear relaxation*), or at the differential equation level (*waveform relaxation*). We refer to a paper of Newton and Sangiovanni, [92], and a book by White and Sangiovanni, [147], for an extensive overview of the various relaxation-based electrical simulation techniques and for a large number of pointers into the relevant literature.

The waveform relaxation method as a computational procedure was first proposed by Lelarasmee in his PhD-thesis, [74], which appeared in 1982. Convergence results for fairly general nonlinear systems were published in the electrical engineering literature in articles by Lelarasmee, Ruehli and Sangiovanni, [75], and by White et al., [148]. The convergence for linear ordinary differential equations was extensively studied by Miekkala and Nevanlinna, [84, 85], and by Nevanlinna, [88, 89, 90, 91]. The latter author has pointed out the correspondence of the waveform relaxation method to a technique introduced a century ago by Picard and Lindelöf in studies to prove the existence of solutions of differential equations. The waveform method is correspondingly called the *Picard-Lindelöf iteration*, see [88]. Miekkala and Nevanlinna also introduced the alternative name *dynamic iteration*. The method is then opposed to standard relaxation methods, which they named *static iteration* methods.

We explain the basic principles behind the waveform relaxation method and introduce the Gauss-Seidel and Jacobi algorithms in section 2.2. A classification of waveform methods is given in section 2.3. We survey a number of general convergence results for waveform relaxation applied to nonlinear systems of ordinary differential equations in section 2.4. The theoretical framework for linear systems as developed by Miekkala and Nevanlinna is overviewed in some detail in section 2.5. Acceleration techniques for improving the convergence and for speeding up the waveform computations are reported in section 2.6. We end this chapter in section 2.7 weighing the waveform relaxation pros and cons.

2.2 Waveform relaxation: basic ideas

Consider a general nonlinear system of d ordinary differential equations with associated initial conditions,

$$\frac{d}{dt}y = f(t,y) , \quad y(0) = y_0 , \quad t \in [0,T], \tag{2.1}$$

where $T > 0$, $f : [0,T] \times I\!\!R^d \to I\!\!R^d$, $y_0 = [y_{1,0}\, y_{2,0}\, \ldots\, y_{d,0}]^t \in I\!\!R^d$ is a vector which contains the initial values, and $y(t) = [y_1(t)\, y_2(t)\, \ldots\, y_d(t)]^t \in I\!\!R^d$ is the solution vector. Componentwise, this system is written as follows,

$$\begin{cases} \dfrac{d}{dt}y_1 = f_1(t,y_1,y_2,y_3,\ldots,y_d) \\[2mm] \dfrac{d}{dt}y_2 = f_2(t,y_1,y_2,y_3,\ldots,y_d) \\[1mm] \quad\vdots \\[1mm] \dfrac{d}{dt}y_d = f_d(t,y_1,y_2,y_3,\ldots,y_d) \end{cases} \text{with} \quad \begin{array}{l} y_1(0) = y_{1,0} \\[2mm] y_2(0) = y_{2,0} \\[1mm] \quad\vdots \\[1mm] y_d(0) = y_{d,0} \end{array}$$

The waveform relaxation method for solving (2.1) is a *continuous-time* iterative method. That is to say, given a *function* which approximates the solution, it calculates a new approximation along the *whole time-interval of interest*. Obviously it differs from most standard iterative techniques in that its iterates are functions in time (i.e., *waveforms*) instead of scalar values. The iteration formula is chosen in such a way that one avoids having to solve a large system of differential equations. A particularly simple, but often very effective iteration scheme is written below. It maps the "old" iterate $y^{(\nu-1)}$ into the "new" iterate $y^{(\nu)}$.

$$\begin{cases} \dfrac{d}{dt}y_1^{(\nu)} = f_1(t,y_1^{(\nu)},y_2^{(\nu-1)},y_3^{(\nu-1)},\ldots,y_d^{(\nu-1)}) \\[2mm] \dfrac{d}{dt}y_2^{(\nu)} = f_2(t,y_1^{(\nu)},y_2^{(\nu)},y_3^{(\nu-1)},\ldots,y_d^{(\nu-1)}) \\[1mm] \quad\vdots \\[1mm] \dfrac{d}{dt}y_d^{(\nu)} = f_d(t,y_1^{(\nu)},y_2^{(\nu)},y_3^{(\nu)},\ldots,y_d^{(\nu)}) \end{cases} \text{with} \quad \begin{array}{l} y_1^{(\nu)}(0) = y_{1,0} \\[2mm] y_2^{(\nu)}(0) = y_{2,0} \\[1mm] \quad\vdots \\[1mm] y_d^{(\nu)}(0) = y_{d,0} \end{array} \tag{2.2}$$

It is called the *Gauss-Seidel waveform relaxation* scheme because of its obvious resemblance to the Gauss-Seidel method for iteratively solving linear and nonlinear systems of algebraic equations. It converts the task of solving a differential equation in d variables into the task of solving a sequence of differential equations in a single variable.

In an actual computer implementation these differential equations are solved by using
a standard numerical ODE integration method. The time-step of the numerical integration scheme may then be chosen to reflect the behaviour of the local variable, as in
a genuine *multi-rate* time-integration method. A closely related iteration is the *Jacobi
waveform relaxation* scheme, the iteration formula of which is given by,

$$
\begin{cases}
\dfrac{d}{dt}y_1^{(\nu)} = f_1(t, y_1^{(\nu)}, y_2^{(\nu-1)}, y_3^{(\nu-1)}, \ldots, y_d^{(\nu-1)}) & & y_1^{(\nu)}(0) = y_{1,0} \\[2mm]
\dfrac{d}{dt}y_2^{(\nu)} = f_2(t, y_1^{(\nu-1)}, y_2^{(\nu)}, y_3^{(\nu-1)}, \ldots, y_d^{(\nu-1)}) & \text{with} & y_2^{(\nu)}(0) = y_{2,0} \\[2mm]
\;\;\vdots & & \;\;\vdots \\[2mm]
\dfrac{d}{dt}y_d^{(\nu)} = f_d(t, y_1^{(\nu-1)}, y_2^{(\nu-1)}, y_3^{(\nu-1)}, \ldots, y_d^{(\nu)}) & & y_d^{(\nu)}(0) = y_{d,0}
\end{cases}
\tag{2.3}
$$

Both iterations are started with an initial approximation $y^{(0)}(t)$ defined along the
whole time-interval. A natural choice is to take this approximation constant and equal
to the values specified by the initial condition,

$$
y_i^{(0)}(t) = y_{i,0} , \quad t \in [0, T] , \quad i = 1, \ldots, d .
\tag{2.4}
$$

Alternatively, one could start from the solution of a related problem, for instance, one
obtained with a slightly different (possibly simplified) right-hand side, or one calculated
with slightly different initial conditions[1].

The Gauss-Seidel waveform relaxation algorithm is formulated in alg. 2.1. The
Jacobi algorithm is very similar. Note that the Gauss-Seidel process is typically sequential. The equations are solved the one after the other. The Jacobi algorithm on
the contrary is fully parallel. The equations can be solved simultaneously. Finally,
the waveform relaxation methods are not necessarily tied to finite time-intervals $[0, T]$.
The principle is equally applicable to ODEs defined on the infinite interval $[0, \infty)$.

Algorithm 2.1: Gauss-Seidel Waveform Relaxation

$\nu := 0$; choose $\;y_i^{(0)}(t)$ for $t \in [0, T], \;\; i = 1, \ldots, d$

repeat

    $\nu := \nu + 1$

    for $\;i = 1, \ldots, d$

        solve $\quad \dfrac{d}{dt}y_i^{(\nu)} = f_i(t, y_1^{(\nu)}, \ldots, y_{i-1}^{(\nu)}, y_i^{(\nu)}, y_{i+1}^{(\nu-1)}, \ldots, y_d^{(\nu-1)})$

        with $\quad y_i^{(\nu)}(0) = y_{i,0}$

    endfor

until convergence

____

[1] An example of the latter will be given in section 4.7 where we discuss the use of waveform
relaxation inside a shooting procedure.

Example 2.2.1 We shall apply the Gauss-Seidel waveform relaxation method to the following system of two differential equations, which we solve for $t \geq 0$,

$$\begin{cases} \frac{d}{dt}y_1 = y_2 \\ \frac{d}{dt}y_2 = -y_1 \end{cases} \quad \text{with} \quad \begin{matrix} y_1(0) = 0 \\ y_2(0) = 1 \end{matrix} \tag{2.5}$$

The reader will certainly recognise (2.5) as the equations defining the "sine" and "cosine" functions. The iterates $y_1^{(\nu)}$, $y_2^{(\nu)}$, satisfying $y_1^{(\nu)}(0)=0$ and $y_2^{(\nu)}=1$ are calculated as below. But first, the iteration is started by choosing an initial approximation.

- initialization: $\begin{cases} y_1^{(0)}(t) = 0 \\ y_2^{(0)}(t) = 1 \end{cases}$

- iteration 1: $\begin{cases} \frac{d}{dt}y_1^{(1)} = y_2^{(0)} \\ \frac{d}{dt}y_2^{(1)} = -y_1^{(1)} \end{cases} \implies \begin{cases} y_1^{(1)}(t) = t \\ y_2^{(1)}(t) = 1 - \frac{t^2}{2} \end{cases}$

- iteration 2: $\begin{cases} \frac{d}{dt}y_1^{(2)} = y_2^{(1)} \\ \frac{d}{dt}y_2^{(2)} = -y_1^{(2)} \end{cases} \implies \begin{cases} y_1^{(2)}(t) = t - \frac{t^3}{3!} \\ y_2^{(2)}(t) = 1 - \frac{t^2}{2} + \frac{t^4}{4!} \end{cases}$

- iteration 3: $\begin{cases} \frac{d}{dt}y_1^{(3)} = y_2^{(2)} \\ \frac{d}{dt}y_2^{(3)} = -y_1^{(3)} \end{cases} \implies \begin{cases} y_1^{(3)}(t) = t - \frac{t^3}{3!} + \frac{t^5}{5!} \\ y_2^{(3)}(t) = 1 - \frac{t^2}{2} + \frac{t^4}{4!} - \frac{t^6}{6!} \end{cases}$

- $\quad \vdots$

- iteration ν: $\begin{cases} \frac{d}{dt}y_1^{(\nu)} = y_2^{(\nu-1)} \\ \frac{d}{dt}y_2^{(\nu)} = -y_1^{(\nu)} \end{cases} \implies \begin{cases} y_1^{(\nu)}(t) = \sum_{i=0}^{\nu-1}(-1)^i \frac{t^{2i+1}}{(2i+1)!} \\ y_2^{(\nu)}(t) = \sum_{i=0}^{\nu}(-1)^i \frac{t^{2i}}{(2i)!} \end{cases}$

The iteration obviously converges to $y_1^{(\infty)} = \sin(t)$ and $y_2^{(\infty)} = \cos(t)$ as each iterate picks up one additional term of the Taylor expansion. Independent of the length of the time-interval there is convergence. The successive iterates $y_1^{(\nu)}(t)$, $y_2^{(\nu)}(t)$ for $\nu=0,1,2,3,4$ are graphically displayed in figure 2.1.

The convergence behaviour observed in the above example is fairly typical. It is also found in the case of more complicated problems. We summarize:

- The error is not uniform along the interval of integration. The approximation is usually good at the beginning but deteriorates for increasing values of t.

- Each iteration lengthens the time interval where the computed waveform is close to the exact solution.

- The difference between the computed waveform and the exact solution is not necessarily reduced at every time-point. From one iteration to another the error may even increase, especially for large values of t.

It will be shown further on that this qualitative behaviour is in agreement with the general convergence theory.

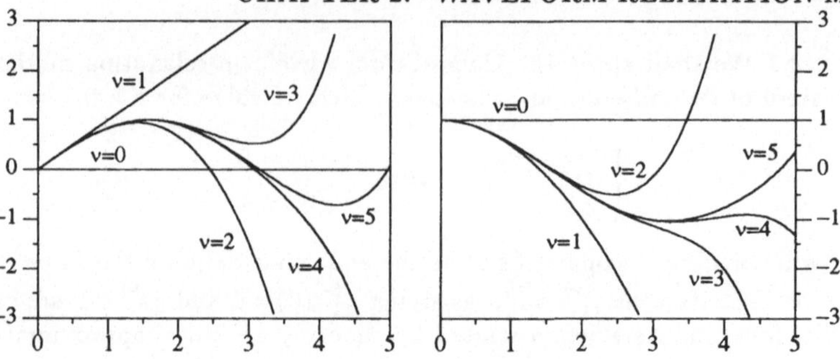

Figure 2.1: Iterates $y_1^{(\nu)}(t)$ (left) and $y_2^{(\nu)}(t)$ (right) for $t \in [0,5]$ (example 2.2.1).

2.3   A classification of waveform methods

The Gauss-Seidel and Jacobi waveform relaxation methods are members of the more general class of so-called *waveform iteration* methods. These methods are characterized by the fact that they iterate on functions, that is to say, they can be defined without reference to time-discretization. In [118], Skeel defines a waveform iteration as a continuous-time iterative method for solving ordinary differential equations in some part of the interval of integration. Analogously, one can define a *discretized waveform iteration* as an iterative method on discrete representations of approximations to the ODE solution along part of the integration interval. In the sequel we shall refer to these methods also by the names of *continuous-time* and *discrete-time* iteration methods.

Let $\mathcal{F}$ denote the waveform iteration operator which maps an "old" iterate into a "new" one, i.e.,

$$y^{(\nu)} = \mathcal{F}\left(y^{(\nu-1)}\right) . \tag{2.6}$$

For our present purposes it suffices to consider operators that are independent of the iteration number ν. In addition, we only consider "one-step" iterations; that is, $y^{(\nu)}$ depends in an explicit way on $y^{(\nu-1)}$ only, and not on the previous iterates. An important class of waveform iteration methods can be characterized by the following iteration formula,

$$\frac{d}{dt}y^{(\nu)} = F(t, y^{(\nu-1)}, y^{(\nu)}), \quad y^{(\nu)}(0) = y_0 . \tag{2.7}$$

The *iteration function* $F(t, u, v)$ is chosen so that $F(t, v, v) = f(t, v)$. This guarantees the consistency of (2.7) with (2.1). Furthermore, the choice of $F(t, u, v)$ is aimed towards achieving the following two (possibly conflicting) goals: the differential equation $\frac{d}{dt}v = F(t, u, v)$ should be easy to solve, and the iteration (2.7) should be rapidly convergent. Some of the more important waveform iteration methods are given below.

• **The waveform Picard method.** The Picard iteration is characterized by the simplest of iteration functions, $F(t, u, v) = f(t, u)$, or, componentwise,

$$F_i(t, u, v) = f_i(t, u_1, u_2, \ldots, u_i, \ldots, u_d) .$$

This well-known iteration is named after E. Picard, who used the iteration in his existence proof for solutions of ordinary differential equations. The scheme is particularly amenable to parallel computing since each differential equation can be integrated simultaneously. In the corresponding discrete-time iteration each equation's right-hand side can be evaluated in parallel across the time-steps.

• **The waveform relaxation methods.** In the relaxation methods $F(t, u, v)$ is chosen so that its Jacobian, $\partial F/\partial v$, is block diagonal or block lower triangular. In that case, the differential equation (2.7) is effectively decoupled into subsystems that can be solved independently or in sequence. The pointwise Jacobi and Gauss-Seidel schemes introduced previously are characterized by,

$$F_i(t, u, v) = f_i(t, u_1, u_2, \ldots, u_{i-1}, v_i, u_{i+1}, \ldots, u_d) \quad \text{(Jacobi)}, \tag{2.8}$$

$$F_i(t, u, v) = f_i(t, v_1, v_2, \ldots, v_{i-1}, v_i, u_{i+1}, \ldots, u_d) \quad \text{(Gauss-Seidel)}. \tag{2.9}$$

Blockwise Jacobi and Gauss-Seidel iterations can be defined in a straightforward way. For instance, a two by two block Jacobi waveform relaxation method is given by function $F(t, u, v)$ specified below. (The number of equations is assumed to be even.)

$$F_{2i-1}(t, u, v) = f_{2i-1}(t, u_1, \ldots, u_{2i-2}, v_{2i-1}, v_{2i}, u_{2i+1}, \ldots, u_d)$$
$$F_{2i}(t, u, v) = f_{2i}(t, u_1, \ldots, u_{2i-2}, v_{2i-1}, v_{2i}, u_{2i+1}, \ldots, u_d).$$

In the pointwise and blockwise relaxation methods groups of unknowns are relaxed in order to make them satisfy groups of equations. This may require the solution of nonlinear systems of ordinary differential equations. These nonlinear equations need not be solved exactly. A good approximation often satisfies to ensure convergence of the waveform iteration. In the *waveform relaxation Newton* methods each equation is first linearized by a Newton-Raphson procedure, and only then solved. The iteration function for the Gauss-Seidel waveform relaxation Newton method is given by,

$$F_i(t, u, v) = f_i(t, v_1, \ldots, v_{i-1}, u_i, \ldots, u_d) + \frac{\partial f_i}{\partial v_i}(t, v_1, \ldots, v_{i-1}, u_i, \ldots, u_d)(v_i - u_i).$$

In the blockwise relaxation methods linearization is applied subsystem per subsystem.

• **The waveform Newton method.** The waveform Newton method implements a global Newton-Raphson linearization of the original system of ordinary differential equations. This leads to the following iteration function,

$$F(t, u, v) = f(t, u) + \frac{\partial f}{\partial v}(t, u)(v - u).$$

It requires analytical differentiation to compute the Jacobian $\partial f/\partial v$, and it does not decouple the system into easily solvable subsystems. Yet, it is rapidly convergent and it allows the parallel evaluation of $f(t, u)$ and $\partial f/\partial v(t, u)$ across time-levels. Note that analytical differentiation can be avoided by numerically approximating the Jacobian. The use of such methods leads to *waveform quasi-Newton methods*, see e.g. [118].

There are several waveform iteration techniques which do not fit the iteration formula (2.7), or even (2.6). E.g., in [118] Skeel studies the *accelerated* and *shifted* waveform Picard methods which require dynamic determination of iteration parameters. In [90] Nevanlinna analyses linearly accelerated waveform methods, in which a new iterate explicitly depends on more than one previous iterate.

2.4 General convergence results

A theoretical analysis of waveform relaxation methods is presented in [147, 148]. There, convergence is proven for certain nonlinear systems which are more general than (2.1) and which are typically encountered in circuit simulation problems. (The left-hand side is pre-multiplied by a matrix which depends on the solution $y(t)$.) We do feel that a closer look at these convergence proofs is both illuminating and required for a good understanding of the waveform relaxation idea. As the mentioned analysis immediately applies to the equations that we consider, we could suffice by referring to the proofs supplied in the above references. However, the generality technically complicates the proofs, and it is not required for our present purposes. Consequently, we shall reformulate some of the ideas and present some simplified or alternative proofs.

2.4.1 The contraction mapping principle

The waveform relaxation method generates an infinite sequence of functions, $\{y^{(\nu)}\}_{\nu=0}^{\infty}$. The convergence of such sequences is generally studied in the context of *Banach* spaces, i.e. complete, normed, linear spaces. The discussion in the current section employs one such space, namely the space of continuous vector-valued functions defined on $[0, T]$, denoted by $C([0, T]; I\!\!R^d)$, or $C[0, T]$ for short. It is a Banach space when equipped with the maximum norm,

$$\| y \|_T \; = \; \max_{t \in [0,T]} \| y(t) \| , \tag{2.10}$$

where $\| \, . \, \|$ denotes any of the usual vector norms in $I\!\!R^d$.

Central to many convergence theorems for sequences in Banach spaces is the *contraction mapping* or *fixed point* principle. The definition of contraction map and the contraction mapping theorem are recalled below. They were taken from [24, p. 20].

Definition 2.4.1 *Let* $(\mathcal{X}, \| \, . \, \|_{\mathcal{X}})$ *be a normed linear space and the associated norm. An operator* $\mathcal{U} : \mathcal{X} \to \mathcal{X}$ *is called a contraction if there exists a* γ*, with* $0 \le \gamma < 1$*, such that*

$$\| \, \mathcal{U}(x) - \mathcal{U}(y) \, \|_{\mathcal{X}} \; \le \; \gamma \, \| \, x - y \, \|_{\mathcal{X}} , \; \textit{for all} \;\; x, y \in \mathcal{X} .$$

Theorem 2.4.1 *Let* $\mathcal{X}$ *be a Banach space and* $\mathcal{U}$ *a contraction. Then there is a unique* $x^\star \in \mathcal{X}$*, such that* $\mathcal{U}(x^\star) = x^\star$*. Moreover, if* $x^{(0)}$ *is any point in* $\mathcal{X}$*, and we define the sequence* $\{x^{(\nu)}\}_{\nu=0}^{\infty}$ *by* $x^{(1)} = \mathcal{U}(x^{(0)})$*,* $x^{(2)} = \mathcal{U}(x^{(1)})$*, ... then* $x^{(\nu)} \to x^\star$ *as* $\nu \to \infty$*.*

2.4.2 Waveform relaxation as a contraction map

In [147, 148], it is shown that the Jacobi and Gauss-Seidel waveform relaxation methods are convergent iterations. The proofs are based on a contraction mapping argument applied in $C[0,T]$, equipped with an exponentially scaled norm,

$$\| y \|_b = \max_{t\in[0,T]} e^{-bt} \| y(t) \| , \qquad (2.11)$$

for some positive value b. In particular, it is demonstrated that the iteration is a contraction for the derivative of the iterates, see e.g. [147, eq. 4.18],

$$\| \frac{d}{dt}y^{(n+1)} - \frac{d}{dt}y^{(k+1)} \|_b \;\leq\; \gamma \| \frac{d}{dt}y^{(n)} - \frac{d}{dt}y^{(k)} \|_b , \quad \text{with } \gamma < 1 . \qquad (2.12)$$

Below we prove a similar result. We demonstrate that the Gauss-Seidel and Jacobi waveform relaxation operators $\mathcal{F}$ are contractions. Our proof differs from the one in the above references in that it proves contractivity for the iterates directly, without passing by (2.12). The proof is very similar to the standard approach for proving existence of solutions to ordinary differential equations, see e.g. [24, p.21 and p.120].

Theorem 2.4.2 *Consider (2.1) where $f(t,y)$ is continuous on $D = [0,T] \times I\!R^d$, and satisfies a Lipschitz condition, that is, $\| f(t,y) - f(t,z) \|\leq L \| y - z \|$, for all $(t,y),(t,z) \in D$ and for some positive constant L. Then, the Jacobi and Gauss-Seidel waveform relaxation methods converge.*

Proof . For notational convenience, we set $\bar{y}$ to denote the result obtained after applying one Gauss-Seidel or Jacobi waveform relaxation step to a continuous function y, i.e., $\bar{y} := \mathcal{F}(y)$. By the continuity assumption, each differential equation in (2.2) or (2.3) can be integrated to yield the following expression for $\bar{y}$,

$$\bar{y}(t) = y_0 + \int_0^t F(s,y(s),\bar{y}(s))\,ds .$$

Introducing a second continuous function z, and setting $\bar{z} := \mathcal{F}(z)$, we derive,

$$\| \bar{y}(t) - \bar{z}(t) \| = \| \int_0^t (F(s,y(s),\bar{y}(s)) - F(s,z(s),\bar{z}(s)))\,ds \|$$

$$\leq \int_0^t \| F(s,y(s),\bar{y}(s)) - F(s,z(s),\bar{z}(s)) \| \, ds . \qquad (2.13)$$

We now use a result, [147, eq. (4.12)], which states that the Lipschitz condition on $f(t,y)$ induces a Lipschitz condition on $F(t,u,v)$. More precisely, there exist positive constants l_1 and l_2 such that for all $u_1, u_2, v_1, v_2 \in I\!R^d, t \in [0,T]$,

$$\| F(t,u_1,v_1) - F(t,u_2,v_2) \|\leq l_1 \| u_1 - u_2 \| + l_2 \| v_1 - v_2 \| .$$

By using this Lipschitz condition for F, (2.13) can be bounded further,

$$\| \bar{y}(t) - \bar{z}(t) \| \leq l_1 \int_0^t \| y(s) - z(s) \| \, ds + l_2 \int_0^t \| \bar{y}(s) - \bar{z}(s) \| \, ds . \qquad (2.14)$$

We multiply left- and right-hand sides by e^{-bt}, and maximize over $[0, T]$,

$$\max_{t\in[0,T]} e^{-bt} \| \bar{y}(t) - \bar{z}(t) \| \leq l_1 \max_{t\in[0,T]} e^{-bt} \int_0^t e^{bs} \max_{\xi\in[0,T]} e^{-b\xi} \| y(\xi) - z(\xi) \| ds$$

$$+ l_2 \max_{t\in[0,T]} e^{-bt} \int_0^t e^{bs} \max_{\xi\in[0,T]} e^{-b\xi} \| \bar{y}(\xi) - \bar{z}(\xi) \| ds .$$

We switch to $\| \, . \, \|_b$-notation and take into account that $\max_{t\in[0,T]} e^{-bt} \int_0^t e^{bs} ds \leq 1/b.$

$$\| \bar{y} - \bar{z} \|_b \leq \frac{l_1}{b} \| y - z \|_b + \frac{l_2}{b} \| \bar{y} - \bar{z} \|_b .$$

For a sufficiently large value of b we get,

$$\| \bar{y} - \bar{z} \|_b \leq \gamma \| y - z \|_b \quad \text{with} \quad \gamma = \frac{l_1/b}{1 - l_2/b} < 1 . \tag{2.15}$$

This proves that $\mathcal{F}$ is a contraction. Convergence follows by Th. 2.4.1. $\square$

Remark 2.4.1 Note that the continuity and Lipschitz assumptions for $f(t, y)$ are sufficient conditions to ensure the existence of a solution of the initial value problem (2.1), see e.g. [71, p. 2]. These conditions are not necessary. The global Lipschitz condition may e.g. be replaced by a local one, see e.g. [24, p.120].

The theorem shows that the error of the approximation gets reduced from iteration to iteration, *when measured in the b-norm*. Indeed, application of equation (2.15) to the iterate $y^{(\nu-1)}$ and the ODE solution y gives,

$$\| y^{(\nu)} - y \|_b \leq \gamma \| y^{(\nu-1)} - y \|_b , \quad \text{for some} \ 0 \leq \gamma < 1 \ \text{and} \ b > 0 . \tag{2.16}$$

Note that this does not require the error to decrease in every iteration *at every point in time*. Indeed, because of the exponential weighting the error may significantly increase, especially for large values of t, without affecting the decrease of the error in the b-norm. Therefore, the theory does not contradict the behaviour observed in example 2.2.1. In addition to (2.16), it can be shown that there is always an interval of time where the error decreases in the maximum norm (2.10), [147, p. 109]. To this end the notion of *strict contractivity* is introduced, [147, p. 105].

Definition 2.4.2 *A differential system is said to have the strict WR contractivity property on $[0, T]$, if the waveform relaxation algorithm applied to the system contracts in the maximum norm on $[0, T]$, i.e.,*

$$\| y^{(\nu+1)} - y^{(\nu)} \|_T \leq \gamma \| y^{(\nu)} - y^{(\nu-1)} \|_T , \quad \gamma < 1 , \quad \nu \geq 1 . \tag{2.17}$$

Theorem 2.4.3 *For any system (2.1) which satisfies the assumptions of the convergence theorem 2.4.2 there exists a $T > 0$ such that the system has the strict WR contractivity property on $[0, T]$.*

Proof . We start from (2.14) with $y = y^{(\nu)}$ and $z = y^{(\nu-1)}$,

$$\| y^{(\nu+1)}(t) - y^{(\nu)}(t) \| \leq l_1 \int_0^t \| y^{(\nu)}(s) - y^{(\nu-1)}(s) \| \, ds + l_2 \int_0^t \| y^{(\nu+1)}(s) - y^{(\nu)}(s) \| \, ds.$$

Taking the maximum over $[0, T]$, we derive,

$$\| y^{(\nu+1)} - y^{(\nu)} \|_T \leq l_1 T \| y^{(\nu)} - y^{(\nu-1)} \|_T + l_2 T \| y^{(\nu+1)} - y^{(\nu)} \|_T .$$

Consequently, for a sufficiently small value of T we get the result,

$$\| y^{(\nu+1)} - y^{(\nu)} \|_T \leq \gamma \| y^{(\nu)} - y^{(\nu-1)} \|_T , \quad \text{with} \quad \gamma = \frac{l_1 T}{1 - l_2 T} < 1 .$$

$\square$

2.4.3 On the order of accuracy

An entirely different approach to analysing convergence is developed in the PhD-thesis of Juang, [66], and in a paper by Gear and Juang, [37]. Their analysis allows to quantify different convergence properties of different iteration methods, and to measure the influence of equation ordering within the Gauss-Seidel procedure. To this end, they define the *order of accuracy* of an approximation, [66, p. 21]. Let $y_i(t)$ be the i-th component of the exact solution $y(t)$ of (2.1), and $z_i(t)$ the i-th component of an approximation.

Definition 2.4.3 *If $z_i(t) - y_i(t) = O(t^{M_i+1})$ over a fixed, finite interval $[0, T]$, then the order of accuracy, $A_c(z_i)$, of $z_i(t)$ is M_i. The order of accuracy of $z(t)$, denoted by $A_c(z)$, is defined as $\min_{1 \leq i \leq d} A_c(z_i)$.*

Loosely speaking, the accuracy order of an approximation is one less than the number of matching terms in the Taylor expansions of the approximation and of the solution to the differential equation. Two basic theorems of Juang are summarized below, [66, Th. 3.2 and Th. 3.6]. They relate the accuracy order of successive waveform iterates.

Theorem 2.4.4 *Let $y^{(\nu)}$ and $y^{(\nu-1)}$ be two successive waveform iterates. Under the conditions mentioned in [66, Th.3.2 and Th. 3.6]), the following holds:*
in the case of Jacobi or Gauss-Seidel waveform relaxation:

$$A_c(y^{(\nu)}) \geq A_c(y^{(\nu-1)}) + 1 , \tag{2.18}$$

in the case of waveform Newton iteration:

$$A_c(y^{(\nu)}) \geq 2(A_c(y^{(\nu-1)}) + 1) . \tag{2.19}$$

As the iteration continues more and more terms of the Taylor expansion of each iterate coincide with terms of the Taylor expansion of the exact solution. This accounts for the phenomena mentioned at the end of section 2.2. In particular, it explains why the accuracy of the approximation is usually higher at the beginning of the integration interval, and why the length of the "converged" interval, where the approximation is close to the solution, grows as more iterations are applied.

Throughout the electrical engineering literature on waveform relaxation methods the importance of the *equation ordering*[2] within the waveform relaxation Gauss-Seidel procedure is stressed. While some update orderings lead to rapidly convergent iterations, others lead to slowly, almost stalling iterative processes. This behaviour is adequately explained by the following theorem of Juang, [66, Th. 4.1].

Theorem 2.4.5 *Under the conditions mentioned in [66, Th.4.1] and in the case of Gauss-Seidel waveform relaxation, the following property is satisfied,*

$$A_c(y_i^{(\nu)}) \geq \min(A_c(y_1^{(\nu)}), \ldots, A_c(y_{i-1}^{(\nu)}), A_c(y_{i+1}^{(\nu-1)}), \ldots, A_c(y_d^{(\nu-1)})) + 1 \ . \qquad (2.20)$$

The accuracy order of each solution component is at least one larger than the smallest accuracy order of any of the "input" variables in the right-hand side of the differential equation associated with the particular component. As mentioned in [66] the "$\geq$" is usually an "$=$" unless there is a fortuitous cancellation. We illustrate the importance of this theorem by an example.

Example 2.4.1 We consider the following system of four differential equations,

$$\frac{d}{dt}y_1 = f_1(y_1, y_4); \quad \frac{d}{dt}y_2 = f_2(y_1, y_2, y_3); \quad \frac{d}{dt}y_3 = f_3(y_1, y_3); \quad \frac{d}{dt}y_4 = f_4(y_2, y_4)$$

We assume the initial conditions, the right-hand side functions and the integration interval to be such that theorem 2.4.5 can be applied and that (2.20) is an equality. The Gauss-Seidel waveform relaxation is started with an initial approximation equal to the initial condition. The accuracy order of each component is then equal to zero.

ν	0	1	2	3	4	5
$A_c(y_1^{(\nu)})$	0	1	3	5	7	9
$A_c(y_2^{(\nu)})$	0	1	3	5	7	9
$A_c(y_3^{(\nu)})$	0	2	4	6	8	10
$A_c(y_4^{(\nu)})$	0	2	4	6	8	10

ν	0	1	2	3	4	5
$A_c(y_1^{(\nu)})$	0	1	4	7	10	13
$A_c(y_2^{(\nu)})$	0	2	5	8	11	14
$A_c(y_3^{(\nu)})$	0	2	5	8	11	14
$A_c(y_4^{(\nu)})$	0	3	6	9	12	15

Table 2.1: Gauss-Seidel waveform relaxation accuracy increase for example 2.4.1 with natural equation ordering (left) and with ordering y_1, y_3, y_2, y_4 (right).

Table 2.1 displays the evolution of the accuracy order of each solution component in the first five iterations. The left part of the table shows the results obtained with natural equation ordering, i.e., the functions are updated in the order y_1, y_2, y_3, y_4. After the iteration has stabilized, the accuracy order of each component is increased by two per iteration. In the right-hand side part of the table the results obtained with

---

[2] Different orderings correspond to different numberings of equations and unknowns.

an alternative update ordering are reported. We first update y_1, then y_3, then y_2, and, finally, $y4$. The accuracy increase per iteration has now become equal to three.

Note that this observed behaviour is closely related to certain properties of the *dependency graph*, which pictures the coupling between the differential equation variables. This important topic is discussed at length in [66].

2.5 Convergence analysis for linear systems

The convergence analysis for nonlinear systems as presented in the previous section is rather restricted. It provides some insight in the qualitative convergence behaviour of waveform relaxation methods. However, it is of little practical use when one is interested in predicting actual convergence rates, e.g., in order to compare the performance of different waveform relaxation variants. The theory for linear systems is much further developed. It is analysed in the papers of Miekkala and Nevanlinna, [84, 85, 88, 89]. We survey some of their results below. But first, we recollect some definitions and properties which can be found in standard functional analysis text-books, e.g. [24, 39, 67].

2.5.1 Functional analysis preliminaries

Some more Banach spaces

The convergence of linear waveform relaxation is studied in more general Banach spaces than the space of continuous functions. In particular, the spaces of complex-valued Lebesgue measurable functions which are p-th power integrable are considered, see e.g. [39, p. 269]. These spaces are denoted by $L_p((0,T); \mathbb{C}^d)$, or $L_p(0,T)$ for short, where T may be equal to ∞. They are Banach spaces with the following norms,

$$\| x \|_p = \sqrt[p]{\int_0^T \| x(t) \|^p \, dt} \, , \quad 1 \le p < \infty \, . \tag{2.21}$$

Here, $\| \, . \, \|$ denotes a norm in $\mathbb{C}^d$ and the integral is the Lebesgue integral. For the sake of completeness, one technicality is to be mentioned. Observe that $\| x \|_p = 0$ does not necessarily imply $x(t) = 0$ everywhere, since $x(t)$ may be different from zero in a set of isolated points (more precisely, in a set of *measure zero*). Consequently, as it stands, (2.21) does not satisfy the requirements of a *norm*. To remedy the situation it is customary not to distinguish between functions that differ only on a set of measure zero. That is, "$x = y$" means $x(t) = y(t)$ *almost everywhere*, and is written as "$x(t) = y(t)$ a.e.". A member of L_p is then an equivalence class of functions equal a.e.

Closely related is the space of *essentially bounded* functions, i.e., the set of Lebesgue measurable functions x such that $\| x(t) \| \le M < \infty$ almost everywhere. This space is denoted by $L_\infty((0,T); \mathbb{C}^d)$, or $L_\infty(0,T)$, and its norm is given by,

$$\| x \|_\infty = \operatorname*{ess\,sup}_{t \in (0,T)} \| x(t) \| := \inf\{M : \| x(t) \| \le M \text{ a.e. on } (0,T)\} \, . \tag{2.22}$$

The discussion on discretized waveform relaxation methods will involve operations on complex-valued, possibly infinite-length sequences $x = \{x_i\}_{i=0}^N$, with $x_i \in \mathbb{C}^d$. The analysis is then considered in the spaces of *p-summable sequences*, $l_p(0..N; \mathbb{C}^d)$. These are Banach spaces with norms given by,

$$\| x \|_p = \begin{cases} \sqrt[p]{\Sigma_i \| x_i \|^p} & 1 \leq p < \infty \\ \sup_i \{\| x_i \|\} & p = \infty \end{cases} \tag{2.23}$$

Spectrum and spectral radius of linear operators

The convergence of relaxation methods for linear systems in finite-dimensional vector spaces is determined by the spectral radius of the iteration operator (a matrix). A similar property holds for iteration methods defined in infinite-dimensional Banach spaces. Before stating the theorem, we shall recall some useful definitions.

Definition 2.5.1 *A linear operator $\mathcal{U}$ in a normed space $\mathcal{X}$ is said to be bounded if there exists a finite constant K such that* $\| \mathcal{U}x \|_{\mathcal{X}} \leq K \| x \|_{\mathcal{X}}$ *, for all $x \in \mathcal{X}$.*

Definition 2.5.2 *The norm of a bounded linear operator $\mathcal{U}$ in $\mathcal{X}$ is defined by,*

$$\| \mathcal{U} \|_{\mathcal{X}} = \sup_{0 \neq x \in \mathcal{X}} \left\{ \| \mathcal{U}x \|_{\mathcal{X}} / \| x \|_{\mathcal{X}} \right\} = \sup_{\|x\|_{\mathcal{X}} = 1} \| \mathcal{U}x \|_{\mathcal{X}} . \tag{2.24}$$

Definition 2.5.3 *Consider a linear operator $\mathcal{U}$ in a normed space $\mathcal{X}$. The spectrum of $\mathcal{U}$, denoted by $\sigma(\mathcal{U})$, consists of those scalars λ for which the operator $\lambda - \mathcal{U}$ does not have a bounded inverse of which the domain is a dense subset of $\mathcal{X}$.*

Definition 2.5.4 *Suppose the spectrum $\sigma(\mathcal{U})$ to be non-empty and bounded. The spectral radius of $\mathcal{U}$ is defined by,*

$$\rho(\mathcal{U}) = \sup_{\lambda \in \sigma(\mathcal{U})} |\lambda| . \tag{2.25}$$

Particular elements of the spectrum are the *eigenvalues* of $\mathcal{U}$, i.e., the scalars λ for which there exist a non-trivial $x \in \mathcal{X}$ such that $\mathcal{U}x = \lambda x$. However, contrary to the case of finite-dimensional vector spaces, the spectrum of $\mathcal{U}$ may also contain other elements, and the spectral radius may be different from the modulus of the largest eigenvalue. The following property further characterizes $\rho(\mathcal{U})$, see e.g. [67, p. 378].

Property 2.5.1 *Consider a bounded linear operator $\mathcal{U}$ in a complex Banach space $\mathcal{X}$. Then,*

$$\rho(\mathcal{U}) = \lim_{n \to \infty} \sqrt[n]{\| \mathcal{U}^n \|_{\mathcal{X}}} . \tag{2.26}$$

We now state the main theorem of this introduction. It deals with the convergence of the *successive approximation scheme*. It is found e.g. in [70, p.149] or [67, p.382].

Theorem 2.5.2 *Let $\mathcal{U}$ be a bounded linear operator in a Banach space $\mathcal{X}$.*

– Suppose $\rho(\mathcal{U}) < 1$. Then, for all $\varphi \in \mathcal{X}$ the successive approximations

$$x^{(\nu)} = \mathcal{U}x^{(\nu-1)} + \varphi , \quad \nu = 1, 2, \ldots \tag{2.27}$$

with arbitrary $x^{(0)} \in \mathcal{X}$ converge to the unique solution of $x - \mathcal{U}x = \varphi$.

    – *Suppose $\rho(\mathcal{U}) > 1$. Then, (2.27) with $x^{(0)} = 0$ cannot converge for all $\varphi \in \mathcal{X}$.*

Note that the theorem does not cover the case $\rho(\mathcal{U}) = 1$. However, in that case the following property applies, see e.g. [67, p.382, remark 1].

Property 2.5.3 *Assume the operator $\mathcal{U}$ is such that all non-zero points of the spectrum are eigenvalues. Then, a necessary and sufficient condition for convergence of the successive approximations (2.27) is that all eigenvalues of $\mathcal{U}$ satisfy $|\lambda| < 1$.*

2.5.2   Waveform relaxation for linear systems

We consider a system of d linear constant-coefficient ordinary differential equations,

$$\frac{d}{dt}x + Ax = f , \quad \text{with} \quad x(0) = x_0 \quad t > 0 , \tag{2.28}$$

where A is a complex $d \times d$ matrix, and x and f are $\mathbb{C}^d$-valued functions of time. A *splitting* is applied to the coefficient matrix, $A = P - Q$, and the system of differential equations is rewritten into an equivalent form,

$$\frac{d}{dt}x + Px = Qx + f , \quad \text{with} \quad x(0) = x_0 \quad t > 0 .$$

In a natural way, this equation leads us to consider the following waveform iteration,

$$\frac{d}{dt}x^{(\nu)} + Px^{(\nu)} = Qx^{(\nu-1)} + f , \quad \text{with} \quad x^{(\nu)}(0) = x_0 \quad t > 0 , \tag{2.29}$$

which is started with an initial approximation $x^{(0)}(t) = x_0$, for $t \geq 0$.

The convergence and the computational complexity of the iteration obviously depend on the nature of the splitting matrices. Some of the splittings (P, Q) that have been proposed in the literature for use with (2.29) are given below. We assume the coefficient matrix A to be decomposed as $-L + D - U$, where D is a diagonal matrix, and L and U are strictly lower respectively strictly upper triangular matrices. Let I denote the identity matrix.

    – Richardson: $P = \frac{1}{\omega}I , \quad Q = \frac{1}{\omega}I - A$

    – Jacobi: $P = D , \quad Q = L + U$

    – Gauss-Seidel: $P = -L + D , \quad Q = U$

    – JOR: $P = \frac{1}{\omega}D , \quad Q = \frac{1-\omega}{\omega}D + L + U$

    – SOR: $P = \frac{1}{\omega}D - L , \quad Q = \frac{1-\omega}{\omega}D + U$

These splittings correspond to the splittings that are typically used when solving linear systems of the form $Ax = b$ by standard relaxation methods. For instance,

$$Px^{(\nu)} = Qx^{(\nu-1)} + b , \quad \text{or} \quad x^{(\nu)} = Gx^{(\nu-1)} + P^{-1}b \quad \text{with} \quad G = P^{-1}Q , \tag{2.30}$$

in which G is called the *iteration matrix*. In [84] this type of iteration is called *static iteration*, whereas (2.29) is called *dynamic* iteration.

It is straightforward to verify that (2.29) with Jacobi or Gauss-Seidel matrix splitting corresponds to the Jacobi or Gauss-Seidel waveform relaxation method as defined in section 2.2. Note also that with use of each of the above-mentioned splittings the iterative procedure consists of solving a sequence of linear constant coefficient ordinary differential equations in a single unknown. These equations are of the form,

$$\frac{d}{dt}x_i^{(\nu)} + p_{i,i}x_i^{(\nu)} = g_i \ , \quad \text{with } x_i^{(\nu)}(0) = x_{i,0} \ , \tag{2.31}$$

where $p_{i,i}$ is the i-th diagonal element of P, and g_i is a sum of known functions.

2.5.3 Continuous-time convergence results

The continuous-time waveform relaxation operator

The solution to the linear ordinary differential equation (2.28) is given by the following well-known formula, see e.g. [24, p. 119],

$$x(t) = e^{-tA}x_0 + \int_0^t e^{(s-t)A} f(s)\, ds \ . \tag{2.32}$$

This formula can be applied to (2.29) in order to derive an explicit relation between successive waveform relaxation iterates. The following successive approximation scheme results,

$$x^{(\nu)} = \mathcal{K}x^{(\nu-1)} + \varphi \ , \tag{2.33}$$

with

$$\begin{cases} \mathcal{K}x\,(t) = \int_0^t e^{(s-t)P}Q\,x(s)\, ds \\[2mm] \varphi(t) = e^{-tP}x_0 + \int_0^t e^{(s-t)P} f(s)\, ds \end{cases} \tag{2.34}$$

Equation (2.33) defines the *continuous-time waveform relaxation operator* $\mathcal{K}$. From (2.34) it can be seen that this operator is a *linear Volterra convolution* operator with a continuous matrix-valued kernel. More precisely,

$$\mathcal{K}x\,(t) = k \star x\,(t) := \int_0^t k(t-s)\,x(s)\, ds \quad \text{with} \quad k(t) = e^{-tP}Q \ . \tag{2.35}$$

Superlinear convergence on finite time-intervals

Because of the contraction mapping argument presented in the previous section one might suspect the convergence of the waveform relaxation method to be *linear*. That is, the error (or its norm) is multiplied in each iteration by a constant bounded by γ, see (2.16), smaller than one. However, it has been shown that the rate of convergence on bounded time intervals is ultimately faster than linear. This *superlinear convergence* result is given in the following theorem, which is proven in [88, p. 333].

Theorem 2.5.4 *Let the constant C be such that $\| k \|_T = C$, where k is the kernel of the waveform relaxation operator. The convergence of (2.29) is characterized by,*

$$\| x - x^{(\nu)} \|_T \;\leq\; \frac{(CT)^\nu}{\nu!} \; \| x - x^{(0)} \|_T \; . \tag{2.36}$$

This result corresponds to a zero value for the spectral radius of $\mathcal{K}$, see [84, p. 461].

Theorem 2.5.5 *Consider $\mathcal{K}$ as an operator in the space of continuous functions on $[0,T]$, equipped with the maximum norm. Then, $\rho(\mathcal{K}) = 0$.*

Note that the spectral radius corresponds to an *asymptotic* value of the convergence factor. This can be understood from property 2.5.1. For large values of C, and for long time intervals $[0,T]$, the multiplier in the right-hand side of formula (2.36) will become very large, before starting to decrease. In those cases the observed convergence behaviour does more closely correspond to the behaviour predicted by the analysis for infinite time-intervals, which is surveyed below.

Convergence on infinite time-intervals

In [84, Th. 2.2], Miekkala and Nevanlinna relate the spectral radius of the integral operator $\mathcal{K}$ to the spectral radius of the Laplace transform of its kernel. It is easily verified that the latter is a complex matrix which is a function of the Laplace transform variable z, and equal to

$$K(z) = (zI + P)^{-1}Q \; , \tag{2.37}$$

where I denotes the identity matrix. The matrix $K(z)$ is frequently called the *symbol* of the integral operator.

Theorem 2.5.6 *Consider $\mathcal{K}$ as an operator in $L_p(0,\infty)$ with $1 \leq p \leq \infty$, and assume that all eigenvalues of P have positive real parts. Then,*

$$\rho(\mathcal{K}) = \sup_{Re(z)\geq 0} \rho(\,K(z)\,) = \max_{\xi \in \mathbf{R}} \rho(\,K(i\xi)\,) \; , \quad \text{with} \;\; i = \sqrt{-1} \; . \tag{2.38}$$

Remark 2.5.1 The second equality follows from the fact that $\rho(K(z))$ satisfies the *maximum principle* (If $f(z)$ is analytic inside and on a simple closed curve Γ, then $|f(z)|$ attains its maximum on Γ, unless $f(z)$ is a constant).

Remark 2.5.2 Note that $(i\xi I + P)^{-1}Q$ evaluated at $\xi = 0$ corresponds to the iteration matrix G of the static iteration (2.30). Consequently, if we assume the conditions of the theorem to be satisfied, $\rho(G) \leq \rho(\mathcal{K})$. The spectral radii coincide, $\rho(G) = \rho(\mathcal{K})$, when the maximum in (2.38) is found at the origin.

In order to determine the spectral radius of $\mathcal{K}$ the spectral radius of the matrix $K(z)$ is to be calculated for every value of z along the imaginary axis. This is generally a very difficult task and defies any further theoretical analysis. However, there are some cases where $\rho(\mathcal{K})$ can be calculated explicitly. One such case arises when A is a consistently ordered matrix with a positive constant diagonal. Since this case is important in view of the discussion in the next chapter, we shall mention some corresponding results.

We set $\mathcal{K}_{JAC}$, $\mathcal{K}_{GS}$ and $\mathcal{K}_{SOR}$ to denote the Jacobi, Gauss-Seidel and SOR waveform relaxation operators. We assume A to be decomposed in the usual way ($A = -L + D - U$). The following corollary, [84, Cor. 4.1], deals with the Jacobi and Gauss-Seidel iterations. It illustrates a case where the maximum in (2.38) is found at the origin.

Corollary 2.5.7 *Let A be a consistently ordered matrix with positive constant diagonal. Then,*

$$\rho(\mathcal{K}_{GS}) = \rho((D - L)^{-1}U) = \rho^2(D^{-1}(L + U)) = \rho^2(\mathcal{K}_{JAC}) \ . \qquad (2.39)$$

The analysis of the SOR waveform relaxation method centers around the following formula which relates the eigenvalues λ_i of the SOR symbol $K(z)$ to the eigenvalues μ_i of the Jacobi static iteration matrix, $D^{-1}(L + U)$, with $D = \delta I$, see [85, eq. 2.23],

$$(1 + z\omega/\delta)\lambda_i + \omega - 1 = \sqrt{\lambda_i\omega\mu_i} \ . \qquad (2.40)$$

For any complex number z, when given the spectral radius μ of $D^{-1}(L + U)$ this formula allows to calculate the spectral radius of $K(z)$. In fact, to determine $\rho(K(z))$ the knowledge of the triple (μ, δ, ω) suffices; precise knowledge of the elements of A is not required. Alternatively, given μ the formula allows to determine what values of z lead to a particular value of $\rho(K(z))$. This allows one to compute *contour* lines of the function $\rho(K(z))$. We have plotted such lines in figure 2.2. The two pictures visualize the function for $\delta=10$, $\mu=0.95$, and for two values of the SOR parameter, $\omega = 1$, and $\omega = 1.3$. Observe that in the case of $\omega=1$ (Gauss-Seidel) the maximum of $\rho(K(z))$ taken over the imaginary axis is found at the origin. This agrees with corollary 2.5.7. In the case of overrelaxation with sufficiently large ω, the maximum is taken in a point away from the origin. The location of this point may be calculated analytically and yields an explicit expression for the spectral radius of the SOR waveform relaxation operator. We refer to [84, Th. 4.2] for the precise formula.

For future reference we mention one further result, which relates the norm of $\mathcal{K}$ to the norm of its symbol, see e.g. [91, p. 491].

Theorem 2.5.8 *Denote by $\| \cdot \|_2$ the l_2-norm and by $\| \cdot \|$ the standard Euclidean vector norm. Assume that all eigenvalues of P have positive real parts. Then,*

$$\| \mathcal{K} \|_2 = \sup_{Re(z) \geq 0} \| K(z) \| = \max_{\xi \in \mathbf{R}} \| K(i\xi) \| \ . \qquad (2.41)$$

2.5.4   Discrete-time convergence results

Linear multistep methods

For the reader's convenience we recall the general *linear multistep formula* for calculating the solution to the ordinary differential equation (2.1), see e.g. [71, p. 11],

$$\frac{1}{\tau} \sum_{j=0}^{k} \alpha_j y_{n+j} = \sum_{j=0}^{k} \beta_j f_{n+j} \ . \qquad (2.42)$$

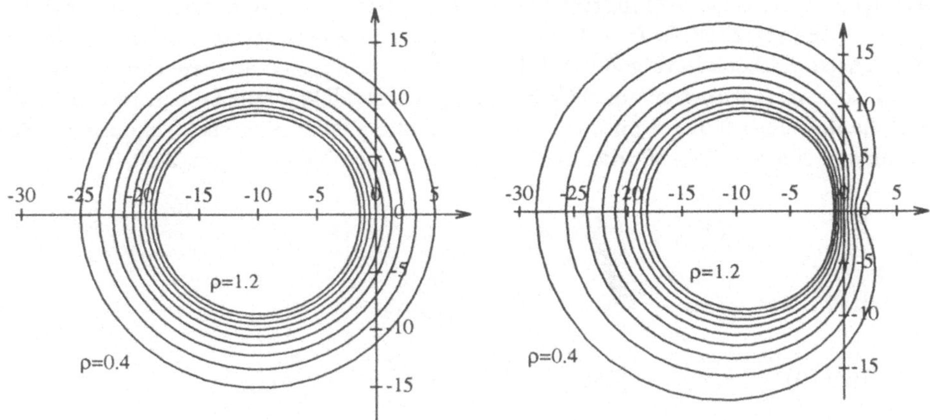

Figure 2.2: Visualization of the function $\rho(K(z))$ by contour lines $\rho = 0.4, 0.5, \ldots, 1.2$. The matrix A is a consistently ordered matrix with positive constant diagonal, and SOR splitting is applied. The pictures correspond to the parameter sets $(\mu, \delta, \omega) = (0.95, 10, 1)$ (left) and $(0.95, 10, 1.3)$ (right).

Here, τ denotes a constant step-size; α_j and β_j are real constants; y_j approximates the ODE solution at the time-level $t_j = j\tau$, and $f_j = f(t_j, y_j)$. We shall further use the "subscript τ" notation to denote sequences of values associated with successive time-levels, e.g. $y_\tau = \{y_i\}_{i=0}^N$, where N denotes the number of time-steps (possibly infinite).

We also introduce the characteristic polynomials of the linear multistep method,

$$a(\xi) = \sum_{j=0}^k \alpha_j \xi^j \quad \text{and} \quad b(\xi) = \sum_{j=0}^k \beta_j \xi^j . \tag{2.43}$$

We adhere to the usual assumptions: $a(\xi), b(\xi)$ have no common roots (irreducibility); $a(1) = 0$ and $a'(1) = b(1)$ (consistency); all roots of $a(\xi)$ are inside the closed unit disk and every root with modulus one is simple (zero-stability). Finally, we recall the definition of the stability region of a multistep method, e.g. [47, p. 257] or [85, p. 576].

Definition 2.5.5 *The stability region S consists of those $\mu \in \bar{C}$ for which the polynomial $a(\xi) - \mu b(\xi)$ (around $\mu = \infty$: $\mu^{-1}a(\xi) - b(\xi)$) satisfies the root condition: all roots satisfy $|\xi_j| \leq 1$ and those of modulus 1 are simple.*

Note that the boundary of the stability region, denoted by ∂S, is a subset of the so-called *root locus curve*, denoted by Γ, i.e.,

$$\partial S \subset \Gamma = \{z = a(\xi)/b(\xi) : |\xi| = 1\} , \tag{2.44}$$

see [47, p. 259]. For most methods used in actual practice, the inclusion "$\subset$" becomes an equality, see e.g. [47, p. 311].

Example 2.5.1 By way of illustration, we have plotted the boundaries of the stability regions of the so-called *backward differentiation formulae* in figure 2.3, see e.g. [47, p. 264]. These methods have good stability characteristics, and are especially appropriate for solving stiff systems of differential equations. They will be considered in the later chapters for solving semi-discretized parabolic partial differential equations. The formulae of the methods of order one to four are given below, [71, p. 242],

$$- \text{BDF}(1): \quad \frac{1}{\tau}(y_{n+1} - y_n) = f_{n+1}$$

$$- \text{BDF}(2): \quad \frac{1}{\tau}(y_{n+2} - \frac{4}{3}y_{n+1} + \frac{1}{3}y_n) = \frac{2}{3}f_{n+2}$$

$$- \text{BDF}(3): \quad \frac{1}{\tau}(y_{n+3} - \frac{18}{11}y_{n+2} + \frac{9}{11}y_{n+1} - \frac{2}{11}y_n) = \frac{6}{11}f_{n+3}$$

$$- \text{BDF}(4): \quad \frac{1}{\tau}(y_{n+4} - \frac{48}{25}y_{n+3} + \frac{36}{25}y_{n+2} - \frac{16}{25}y_{n+1} + \frac{3}{25}y_n) = \frac{12}{25}f_{n+4}$$

Another method which will be used frequently further on is the *trapezoidal rule*, also known (mainly in the context of solving *partial* differential equations) by the name of *Crank-Nicolson method* (CN). The method is a second order accurate time-discretization. Its stability region boundary equals the imaginary axis.

$$- \text{CN}: \quad \frac{1}{\tau}(y_{n+1} - y_n) = \frac{1}{2}f_{n+1} + \frac{1}{2}f_n \ .$$

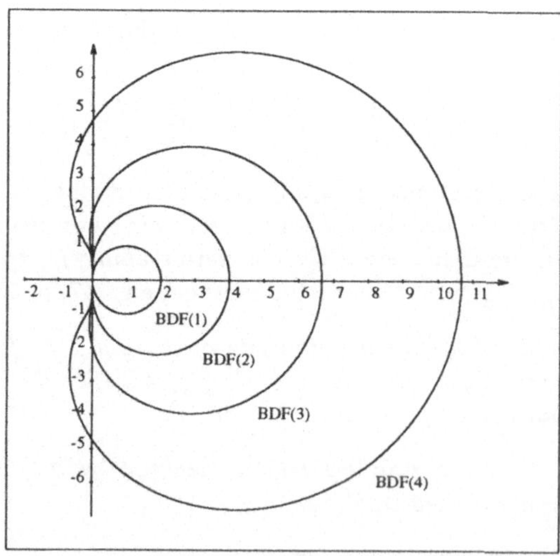

Figure 2.3: Boundary of the stability region for the BDF methods of order 1 up to 4.

The discrete-time waveform relaxation operator

In the analysis presented in [85, 89] it is assumed that a constant time-step is used for all of the differential equations. (Consequently, the theory does not cover the case where the multi-rate property of the waveform relaxation method is exploited.) Application of the linear multistep formula to the basic waveform iteration (2.29) leads to,

$$\frac{1}{\tau}\sum_{j=0}^{k}\alpha_j x_{n+j}^{(\nu)} + \sum_{j=0}^{k}\beta_j P x_{n+j}^{(\nu)} = \sum_{j=0}^{k}\beta_j Q x_{n+j}^{(\nu-1)} + \sum_{j=0}^{k}\beta_j f_{n+j} \ , \ n \geq 0 \ . \tag{2.45}$$

For simplicity's sake, we assume that there are k fixed starting values supplied, i.e., $x_j^{(\nu)} = x_j^{(\nu-1)} = x_j$, for $j < k$ — we do not iterate on the starting values—. Observe that upon convergence, i.e., when $x_\tau^{(\nu)} = x_\tau^{(\nu-1)} = x_\tau$, formula (2.45) becomes,

$$\frac{1}{\tau}\sum_{j=0}^{k}\alpha_j x_{n+j} + \sum_{j=0}^{k}\beta_j P x_{n+j} = \sum_{j=0}^{k}\beta_j Q x_{n+j} + \sum_{j=0}^{k}\beta_j f_{n+j} \ , \ n \geq 0 \ , \tag{2.46}$$

or,

$$\frac{1}{\tau}\sum_{j=0}^{k}\alpha_j x_{n+j} + \sum_{j=0}^{k}\beta_j A x_{n+j} = \sum_{j=0}^{k}\beta_j f_{n+j} \ , \ n \geq 0 \ . \tag{2.47}$$

The solution of the discretized waveform relaxation method is identical to the solution obtained by the corresponding time-stepping method applied to the "non-split" ordinary differential equation (2.28).

Equation (2.45) implicitly relates the *discrete waveforms* $x_\tau^{(\nu-1)}$ and $x_\tau^{(\nu)}$. As in the previous section, this relationship may be written in an explicit way,

$$x_\tau^{(\nu)} = \mathcal{K}_\tau x_\tau^{(\nu-1)} + \varphi_\tau \ . \tag{2.48}$$

In [85, 89] the nature of $\mathcal{K}_\tau$ and of the right-hand side sequence φ_τ is fully described. In particular it turns out that $\mathcal{K}_\tau$ is a *linear discrete convolution* operator. More precisely, there exists a matrix-valued kernel $k_\tau = \{k_i\}_{i=0}^{N}$ with $k_i \in \mathbb{C}^{d\times d}$, such that,

$$(\mathcal{K}_\tau x_\tau)_j = (k_\tau \star x_\tau)_j := \sum_{i=0}^{j} k_{j-i} x_i \ . \tag{2.49}$$

In [89, eq. (4.13) and eq. (4.16)] it is shown that k_τ is related to the symbol of the integral operator $\mathcal{K}$ by the so-called Z-transform (or discrete Laplace transform),

$$\sum_{i=0}^{\infty} k_i z^{-i} = K(\frac{1}{\tau}\frac{a}{b}(z)) \ . \tag{2.50}$$

Convergence on finite time-intervals

The properties of $\mathcal{K}_\tau$ as an operator in the space $l_p(0..N)$ are characterized by the theorem given below, which is proven in [89, Th. 4.1]. It basically states that the spectral radius of $\mathcal{K}_\tau$ is equal to the spectral radius of the symbol $K(z)$ evaluated in a point located on the real axis. The precise location of this point depends on the linear multistep method.

Theorem 2.5.9 *Assume that $\alpha_k/\beta_k \notin \sigma(-\tau P)$, and let the number of time-steps, N, be finite. Then, $\mathcal{K}_\tau$ has the spectral radius,*

$$\rho(\mathcal{K}_\tau) = \rho\left(K\left(\frac{1}{\tau}\frac{\alpha_k}{\beta_k}\right)\right) . \tag{2.51}$$

Remark 2.5.3 The spectral radius is independent of the number of time-steps !

Since the space in which $\mathcal{K}_\tau$ operates is finite-dimensional, the above result may also be derived by standard linear algebra techniques. We outline such a proof below.

Proof . Let x_i denote the value of the discrete-time solution at time-step i. Let $e_i^{(\nu)}$ be the error of the ν-th approximation, i.e., $e_i^{(\nu)} = x_i - x_i^{(\nu)}$. It follows that,

$$\frac{1}{\tau}\sum_{j=0}^{k}\alpha_j e_{n+j}^{(\nu)} + \sum_{j=0}^{k}\beta_j P e_{n+j}^{(\nu)} = \sum_{j=0}^{k}\beta_j Q e_{n+j}^{(\nu-1)} , \quad \text{with} \ \ e_j^{(\nu)} = 0 \ , \ j < k \ ,$$

for $n = 0, \ldots, N - k$. This is rewritten as,

$$\sum_{j=0}^{k} C_j e_{n+j}^{(\nu)} = \sum_{j=0}^{k} D_j e_{n+j}^{(\nu-1)} ,$$

with $C_j = \frac{1}{\tau}\alpha_j I + \beta_j P$ and $D_j = \beta_j Q$. We set $E^{(\nu)} = [e_k^{(\nu)} \ e_{k+1}^{(\nu)} \ \ldots \ e_N^{(\nu)}]^t$, and define the following $(N-k+1) \times (N-k+1)$ block matrices C and D,

$$C = \begin{pmatrix} C_k & & & & & \\ C_{k-1} & C_k & & & & \\ \cdot & C_{k-1} & C_k & & & \\ \cdot & \cdot & \cdot & \cdot & & \\ C_0 & \cdot & \cdot & \cdot & \cdot & \\ & \cdot & \cdot & \cdot & \cdot & \cdot \\ & & C_0 & \cdot & \cdot & C_{k-1} & C_k \end{pmatrix} \quad \text{and } D = \begin{pmatrix} D_k & & & & & \\ D_{k-1} & D_k & & & & \\ \cdot & D_{k-1} & D_k & & & \\ \cdot & \cdot & \cdot & \cdot & & \\ D_0 & \cdot & \cdot & \cdot & \cdot & \\ & \cdot & \cdot & \cdot & \cdot & \cdot \\ & & D_0 & \cdot & \cdot & D_{k-1} & D_k \end{pmatrix}$$

Consequently,

$$C\, E^{(\nu)} = D\, E^{(\nu-1)} , \quad \text{or} \ \ E^{(\nu)} = C^{-1} D\, E^{(\nu-1)} .$$

Convergence of the iteration is therefore determined by the spectral radius of $C^{-1}D$. It can easily be verified that this matrix is a block lower triangular matrix with a constant diagonal the blocks of which are equal to $C_k^{-1}D_k$. By consequence,

$$\rho(C^{-1}D) = \rho(C_k^{-1}D_k) = \rho\left(\left(\frac{1}{\tau}\frac{\alpha_k}{\beta_k}I + P\right)^{-1}Q\right) ,$$

which is identical to (2.51). $\qquad\qquad\qquad\qquad\qquad\qquad\qquad\qquad\qquad\qquad \square$

Convergence on infinite time-intervals

As said before, the spectral radius determines the asymptotic convergence factor. It may be of little use when the convergence factor is to be determined in the early stages of the iteration. Numerical experience shows that this is indeed so, especially when large time-intervals are considered. It turns out that an infinite time-interval analysis is then more appropriate. The theorem below can be found in [85, Th. 3.1].

Theorem 2.5.10 *Suppose* $\sigma(-\tau P) \subset int\ S$, *and consider* $\mathcal{K}_\tau$ *as an operator in* $l_p(I\!N)$ *with* $1 \le p \le \infty$. *Then,*

$$\rho(\mathcal{K}_\tau) = \sup\{\rho(K(z)) \ : \ \tau z \in \bar{\mathbb{C}} \setminus intS\}\ , \tag{2.52}$$

or (by the maximum principle),

$$\rho(\mathcal{K}_\tau) \ = \ \max_{|\xi|=1} \rho\left(K(\frac{1}{\tau}\frac{a}{b}(\xi))\right) \ = \ \max_{\tau z \in \partial S} \rho(K(z))\ . \tag{2.53}$$

In order to determine the spectral radius of $\mathcal{K}_\tau$ by formula (2.53), the spectral radius of the symbol $K(z)$ is to be determined for all values $z = \frac{1}{\tau}\frac{a}{b}(\xi)$, $|\xi| = 1$. These complex values are located on the root locus curve scaled by the constant $\frac{1}{\tau}$. The second equality in (2.53) shows that it suffices to take the maximum over the boundary of the (scaled) stability region of the linear multistep method.

The following corollary, [85, Th. 3.4], can be readily verified by the reader.

Corollary 2.5.11 *If the multistep method is A-stable then* $\rho(\mathcal{K}_\tau) \le \rho(\mathcal{K})$, *where* $\mathcal{K}$ *is the waveform relaxation operator on* $(0, \infty)$.

Finally, for future reference, we recall a formula for the norm of $\mathcal{K}_\tau$, see [91, p. 491].

Theorem 2.5.12 *Suppose* $\sigma(-\tau P) \in int\ S$. *Let* $\| \, . \, \|_2$ *denote the* l_2-norm *and* $\| \, . \, \|$ *the Euclidean matrix norm. Then,*

$$\| \mathcal{K}_\tau \|_2 = \max_{|\xi|=1} \| K(\frac{1}{\tau}\frac{a}{b}(\xi)) \|\ . \tag{2.54}$$

2.6 Waveform relaxation acceleration techniques

The multi-rate integration properties of waveform relaxation are central to its success as an integration method for solving very large systems of ordinary differential equations. Throughout the years many implementations of the method have been made, especially tailored towards specialized circuit simulation problems, see e.g. Lelarasmee [74], Mattison [80], Dumlugol [27], Odent [93], Raman et al. [102], and there are many others. These programs do not naively implement the algorithms as they were presented in the previous sections. Instead, they are extended with various sophisticated acceleration techniques. The more important ones are briefly discussed below.

Ordering of the equations

It has already been mentioned that the ordering of the equations strongly affects the number of iterations required to achieve convergence. The determination of an optimal ordering for arbitrary systems is in general very difficult, or impossible. There are some guidelines though, which very often lead to *good* orderings. In particular one should number the equations in such a way that the Jacobian matrix of the right-hand side $f(t,y)$ is "as lower triangular as possible", and large off-diagonal elements in the upper triangular part should be avoided, [147]. This rule is based on the following observation. If the Jacobian is truly lower triangular, the Gauss-Seidel waveform relaxation method solves the system in one iteration only.

In the VLSI-simulation programs the update ordering is usually aimed at following the signal flow through the electronic circuit. To this end various optimization techniques based on circuit dependency graph and circuit component characteristics are employed. This has led to further waveform relaxation variants. E.g., in the *hierarchical Gauss-Seidel* method certain variables (in tightly coupled loops of the dependency graph) are updated more frequently than others, see e.g. Juang [66].

Partitioning and blockwise iteration

Experience has demonstrated the advantage of merging tightly coupled variables into subsystems of differential equations which are solved as a whole (with standard ODE integration techniques). This leads to more rapidly converging blockwise Gauss-Seidel techniques with blocks of varying size. The blocks need not necessarily be disjoint. An overlapping block waveform relaxation method is analysed by Pohl and Jeltsch in [61].

In the electrical engineering applications the partitioning is naturally associated to the physical system. It is then based on topological circuit properties and quantitative information about circuit components. This "grouping" of unknowns may either be done manually, or automatically. A survey and analysis of various partitioning strategies is presented by Peterson and Mattison in [99], and the references therein.

Windowing

In an implementation it is necessary to store discrete representations of the waveforms. Such a representation may for instance consist of a linked list of (time,value) pairs (one for every function), combined with an interpolation routine to provide intermediate function values. Obviously, the memory requirements of the waveform relaxation methods are very high, especially for large systems and long time-intervals. These requirements can be reduced if the time-interval of interest is divided into several subintervals, so-called *windows*, on which the differential equations are solved in sequence. It is important to choose these windows as long as possible, in order to fully profit from the multi-rate characteristics of the waveform method, and in order to avoid algorithmic *startup* costs, which are inevitably associated with every new window.

Besides affecting considerable savings in memory requirements, *windowing* also affects the computational costs. As was illustrated before, the error of the waveform relaxation iterates is far from uniform. Large errors are found at the end of long time-intervals, and these errors do not necessarily get reduced in magnitude from one iteration to the next. If so, the calculations on these parts of the interval are wasted. The use of a windowing strategy should avoid these superfluous computations. Various *dynamic windowing strategies* have been proposed for automatically selecting appropriate window lengths, e.g. by monitoring the convergence and, based on this information, deciding whether to shorten or lengthen a current window-size, see e.g. [147, 152].

Partial waveform convergence

The computed waveforms are often very accurate near the beginning of a time-window, see e.g. the discussion on the *accuracy order*. In such a *subwindow* additional waveform iterations will not reduce the approximation error any further. Recomputation of these "converged parts" may be avoided by implementing a strategy for locating the starting point of re-integration. Such strategies are usually based on monitoring the input functions in the right-hand side $f(t, y)$. The computation on a particular differential equation is then skipped, as long as the input functions have not changed significantly from their values in previous iterates.

Inaccurate iteration

The waveform relaxation method is an iterative procedure. As in many iterative methods, it is not necessary to compute all iterates to a similar precision in order to ensure convergence. Inaccurate iteration may lead to great computational savings, especially when computing the initial iterates which are far from the converged solution anyway.

In the waveform relaxation method, one could think of calculating the initial iterates with large error tolerance, e.g. by using large step sizes, and gradually reducing the error tolerance, e.g. by refining the step size, as the computation proceeds. Such *non-stationary iterations* have been theoretically studied by Nevanlinna in [89, 91], who suggests a "tolerance game" based on balancing discretization and iteration errors. Related results are also found in the electrical engineering literature, [147, Ch. 6.3].

Parallel implementation

The subsystems that are to be solved in the waveform relaxation method are often decoupled, and therefore, solvable in parallel. This is trivially so in the Jacobi method. Parallelism is also possible with Gauss-Seidel relaxation, at least when adequate ordering schemes are used, e.g. based on the concept of dependency graph *colouring*. One may also consider the use of *block-Jacobi point-Gauss-Seidel* relaxation schemes, in which large subsystems (often called *chains* in the electrical engineering literature) are treated in parallel. The solution to the equations of each subsystem are approximated by a sequential waveform relaxation Gauss-Seidel iteration.

Implementations have been made on a variety of both shared memory and distributed memory parallel computers, see e.g. Peterson and Mattison ([99], Intel hypercube, Symult), Odent ([93], Sequent), Xia ([152], Alliant), Raman et al. ([102], hypercube). In these references various further optimizations have been suggested for increasing the parallelism inherent in the basic waveform relaxation method (like time-segment pipelining, parallel mode evaluation, dynamic decoupling, etc.). We refer to the PhD thesis of Odent, [93], for a very detailed overview.

2.7 Some concluding remarks

The waveform relaxation method has proven to be very successful for solving certain systems of ordinary differential equations that arise in the simulation of very large scale integrated circuits. With a good partitioning and ordering of the equations only very few iterations are required to obtain convergence (typically 5 to 10, [147]). The computation cost is further reduced by taking particular waveform convergence characteristics into account. When compared with standard time-integration software currently used by circuit simulation practitioners, waveform relaxation (often) offers an improvement of one up to two orders of magnitude in speed, [147]. Implementation on a parallel machine is conceptually straightforward —although many technical details complicate the matter—, and reduces the simulation time even further. All of this is obtained at the cost of having to provide a substantial amount of memory.

However, the favourable convergence characteristics strongly depend on the existence of good orderings and partitionings, and our ability of finding them. This has shown to be feasible for circuit simulation problems only by taking the electronic circuit parameters and signal flow properties into account. The method has also been successfully applied for simulating certain chemical distillation processes, see the PhD-thesis of Skjellum, [119]. The differential equations that arise there have similar characteristics to those found in circuits simulation problems, and again, the performance crucially depends on the knowledge of the underlying physical system. The use of waveform relaxation for solving semi-discretized parabolic partial differential equations is the subject of the following chapter. It will be shown that the method is very effective. Yet, again, the very special characteristics of these systems are to be taken into account.

Chapter 3

Waveform Relaxation Methods for Initial Boundary Value Problems

In VLSI-simulation this type of approach, called waveform relaxation method, seems to result in considerable savings in computing time. It would not be surprising if an efficient implementation of the idea appeared also to applications outside circuit equations.

—U. Miekkala and O. Nevanlinna, *in [84, 1987].*

We comment on the use of waveform relaxation techniques for solving parabolic initial boundary value problems. It is illustrated that the Jacobi, Gauss-Seidel and SOR methods do not lead to satisfactory, rapidly convergent algorithms. A linear multigrid acceleration is presented, and illustrated by a numerical example. An analysis of the continuous-time and discrete-time variants is given. The method is extended to nonlinear problems and related to a multigrid method on a space-time grid.

3.1   Introduction and notations

A parabolic initial boundary value problem is generally characterized by a parabolic partial differential equation (or a system of differential equations), a set of boundary conditions, and given initial values,

$$
\begin{cases}
\dfrac{\partial u}{\partial t}(t,x) = \mathcal{L}(u(t,x)) + f(t,x) & t > 0 , \quad x \in \Omega \\[2mm]
\mathcal{B}(u(t,x)) = g(t,x) & t > 0 , \quad x \in \partial\Omega \\[2mm]
u(0,x) = u_0(x) & x \in \Omega
\end{cases}
\tag{3.1}
$$

Here, $\mathcal{L}(.)$ is a linear or nonlinear, uniformly elliptic operator; $\mathcal{B}(.)$ is a linear or nonlinear boundary operator; f and g are known functions of the time and space coordinates, and u_0 is a given function of x. Ω denotes the compact spatial domain. We assume that the solution u exists and that it is unique.

The *numerical method of lines* replaces any spatial derivatives in the above equations by finite differences. After elimination of the boundary conditions, one gets a large set of ordinary differential equations (see e.g. appendix A),

$$\frac{d}{dt}u^h(t) = L^h(u^h(t)) + f^h(t) \ , \quad u^h(0) = u_0^h \ , \quad t > 0 \ . \tag{3.2}$$

We use the superscript-h notation to denote quantities defined on a discrete *mesh* or *grid* Ω^h, which covers the domain Ω. The value h is a measure of the mesh size, e.g., the distance between grid lines or grid points. As before we set d to denote the number of differential equations. Thus, the semi-discrete solution u^h is of the form,

$$u^h(t) = [\ u_1^h(t) \ u_2^h(t) \ \ldots \ u_d^h(t) \ ]^t \ .$$

The system of differential equations (3.2) is known to be stiff for a sufficiently small h. The eigenvalues of the Jacobian of $L^h(.)$ lie close to the real axis and have negative real parts. The computation of the solution normally proceeds by discretizing (3.2) with an appropriate time-integration method which takes the characteristics of the problem into account. In this chapter we shall consider the use of the *backward differentiation formulae* and the *trapezoidal rule* (Crank-Nicolson method). This leads to a fully discrete solution, which we denote by u_τ^h , with

$$u_\tau^h = \{\, u_k^h \,\}_{k=0}^N \ , \quad \text{with} \ \ u_k^h = [\ u_{k,1}^h \ u_{k,2}^h \ \ldots \ u_{k,d}^h \ ]^t \ .$$

N denotes the number of time-steps. In the case of problems defined on two-dimensional rectangular domains we shall often use a more intuitive notation, e.g. $u_{i,j}^h(t)$ and $u_{k,i,j}^h$, to refer to functions and scalars defined at (x_i, y_j) , a coordinate in a discrete mesh.

The disadvantages associated with the above time-stepping approach, in particular w.r.t. parallelism and the exploitation of multi-rate behaviour, have led us to investigate the use of waveform relaxation techniques. The use of "simple" waveform relaxation methods (Jacobi, Gauss-Seidel and SOR) for solving the equations derived by the numerical method of lines is considered in section 3.2. The convergence properties are illustrated with extensive experimental results. Multigrid acceleration of the waveform relaxation method is the subject of section 3.3, where we discuss linear parabolic problems. The convergence properties of both the continuous-time and the discrete-time iteration are studied in section 3.4. Section 3.5 includes a selection of experimental results, which allow to verify the theory. The linear algorithm easily extends to nonlinear parabolic partial differential equations, as we showed in [136, 140]. The nonlinear multigrid method is given in section 3.6. In section 3.7, we argue that a particular version of multigrid waveform relaxation corresponds to standard multigrid on a so-called *space-time grid*. We end the chapter with some concluding remarks.

3.2   Standard waveform relaxation

In this section, we concentrate on a linear model problem: the two-dimensional heat equation defined on the unit square. The treatment of a model problem allows the derivation of analytical expressions which would be difficult to obtain when dealing with more complicated, e.g. variable-coefficient or nonlinear differential equations. Moreover, the case studies in chapter 8 provide ample evidence that the behaviour of the algorithms in more complicated situations is qualitatively very similar.

3.2.1   A model problem: the two-dimensional heat equation

Consider the following model problem,

$$\frac{\partial u}{\partial t} = \frac{\partial^2 u}{\partial x^2} + \frac{\partial^2 u}{\partial y^2} \ , \quad (x,y) \in \Omega = [0,1] \times [0,1] \ , \quad t > 0 \ ,$$

completed with *Dirichlet* boundary conditions and an initial condition chosen in such a way that the analytical solution is given by $u(t,x,y) = 1 + \sin(\pi x/2)\sin(\pi y/2)e^{-\pi^2 t/2}$. The equation is discretized with standard central differences on an equidistant rectangular mesh, with equal mesh size h in x-direction and y-direction,

$$\Omega^h = \{(x_i, y_j) \ : \ x_i = i\,h \ , \ y_j = j\,h \ , \quad \text{with} \ \ 0 \le i,j \le 1/h \ \} \ .$$

This leads to an equation of the following type defined at each interior grid point,

$$\frac{d}{dt}u_{i,j}^h = \frac{1}{h^2}(u_{i-1,j}^h + u_{i,j-1}^h - 4\,u_{i,j}^h + u_{i,j+1}^h + u_{i+1,j}^h) \ , \quad u_{i,j}^h(0) = u_0(x_i, y_j) \ .$$

Note that the right-hand side corresponds to the well-known *five-point star* discretization of the Laplace operator. After elimination of the Dirichlet boundaries, this can be rewritten as a large linear system of $d = (1/h - 1)^2$ equations,

$$\frac{d}{dt}u^h = L^h u^h + f^h \ . \tag{3.3}$$

When the equations are numbered *lexicographically*, the matrix L^h is a well-studied block-tridiagonal matrix. f^h contains the contribution of the boundary functions.

3.2.2   Standard waveform relaxation methods

Following the notation and the ideas expressed in chapter 2, section 2.5.2, a splitting is applied to the coefficient matrix, $-L^h = P^h - Q^h$, and (3.3) is solved iteratively,

$$\frac{d}{dt}u^{h,(\nu)} + P^h u^{h,(\nu)} = Q^h u^{h,(\nu-1)} + f^h \ .$$

As before, we shall use $u^{h,(0)}(t) = u^h(0)$ as the initial iterate. In order not to overload the notation we shall omit the superscript h when dealing with waveform relaxation iterates, e.g., we write $u^{(\nu)}$ instead of $u^{h,(\nu)}$.

The use of the Jacobi, Gauss-Seidel and SOR splittings leads one to solve the following equations. In the case of the Jacobi splitting,

$$\frac{d}{dt}u_{i,j}^{(\nu)} = \frac{1}{h^2}(u_{i-1,j}^{(\nu-1)} + u_{i,j-1}^{(\nu-1)} - 4\,u_{i,j}^{(\nu)} + u_{i,j+1}^{(\nu-1)} + u_{i+1,j}^{(\nu-1)}) \ .$$

Note that the equations at the different grid points can be solved simultaneously.

There are two standard ways of defining a Gauss-Seidel relaxation for partial differential equations. They correspond to different orderings of the equations. In the *lexicographic Gauss-Seidel waveform relaxation* method the iteration is as follows,

$$\frac{d}{dt}u_{i,j}^{(\nu)} = \frac{1}{h^2}(u_{i-1,j}^{(\nu)} + u_{i,j-1}^{(\nu)} - 4\,u_{i,j}^{(\nu)} + u_{i,j+1}^{(\nu-1)} + u_{i+1,j}^{(\nu-1)}) \ .$$

The grid points are visited in a sequential way from the left to the right, and from the bottom to the top. The *red/black Gauss-Seidel waveform relaxation* method on the contrary is a two-stage procedure. First, the equations are updated for which $i + j$ is even (the so-called "red" grid points). This is followed by the update of the equations for which $i + j$ is odd (the "black" grid points),

$$i + j \text{ even}: \quad \frac{d}{dt}u_{i,j}^{(\nu)} = \frac{1}{h^2}(u_{i-1,j}^{(\nu-1)} + u_{i,j-1}^{(\nu-1)} - 4\,u_{i,j}^{(\nu)} + u_{i,j+1}^{(\nu-1)} + u_{i+1,j}^{(\nu-1)})\,,$$

$$i + j \text{ odd}: \quad \frac{d}{dt}u_{i,j}^{(\nu)} = \frac{1}{h^2}(u_{i-1,j}^{(\nu)} + u_{i,j-1}^{(\nu)} - 4\,u_{i,j}^{(\nu)} + u_{i,j+1}^{(\nu)} + u_{i+1,j}^{(\nu)})\,.$$

Observe that the equations in each stage can be solved simultaneously.

The successive overrelaxation method allows the same two variants. We mention the lexicographic SOR waveform relaxation equations, and leave the definition of the red/black SOR waveform method to the reader.

$$\frac{d}{dt}u_{i,j}^{(\nu)} = \frac{1}{h^2}(u_{i-1,j}^{(\nu)} + u_{i,j-1}^{(\nu)} - \frac{4}{\omega}u_{i,j}^{(\nu)} + \frac{4(1-\omega)}{\omega}u_{i,j}^{(\nu-1)} + u_{i,j+1}^{(\nu-1)} + u_{i+1,j}^{(\nu-1)})\,.$$

3.2.3   Some convergence results

The continuous-time iteration

The linear theory of section 2.5 is immediately applicable. We consider the continuous-time iteration first. The superlinear convergence of waveform relaxation on finite time-intervals is based on taking the limit $\nu \to \infty$ in equation (2.36). In this formula a constant C appears. The magnitude of this constant in the case of the above model problem and the use of Jacobi iteration can be calculated as follows (e.g. with the $\|\cdot\|_\infty$ matrix norm). First, we recall the format of the Jacobi splitting matrices (P^h, Q^h) (which satisfy $-L^h = P^h - Q^h$),

$$P^h := \text{diag}(-L^h) = 4/h^2 I \quad \text{and} \quad Q^h := P^h - (-L^h) = 4/h^2 I + L^h\,.$$

$$C := \|\,k\,\|_T = \max_{t\in[0,T]} \|\,e^{-tP^h}Q^h\,\|_\infty = \max_{t\in[0,T]} \|\,e^{-4t/h^2}Q^h\,\|_\infty = \|\,Q^h\,\|_\infty = 4/h^2\,.$$

(The final step requires the knowledge of the precise format of Q^h, which has four non-zero (off-) diagonals of which the elements take the values zero or $1/h^2$.) Thus C is inversely proportional to the square of spatial mesh size. Consequently, when T is not too small, it takes a large number of waveform relaxation steps before the multiplier $(CT)^\nu/\nu!$ in (2.36) starts going to zero. A finite-interval analysis may therefore not be appropriate to describe the convergence behaviour in the initial iterates.

The matrix $-L^h$ satisfies the conditions for applicability of corollary 2.5.7. As such, the spectral radius of the infinite-interval operator is equal to that of the static iteration operator. The latter has a value which is well-known from the theory of relaxation methods for elliptic equations. The result is formulated below, [84, p. 473]. Note that the formulae are valid for the one-, two- and three-dimensional model problem.

Property 3.2.1 *The spectral radii of the (infinite-interval) Jacobi and Gauss-Seidel waveform relaxation operators in the case of the model problem are given by the following formulae, valid for small h,*

$$\rho(\mathcal{K}_{JAC}) \simeq 1 - \pi^2 h^2/2 \quad and \quad \rho(\mathcal{K}_{GS}) \simeq 1 - \pi^2 h^2 \ . \tag{3.4}$$

The SOR waveform relaxation method applied to the model problem is studied by Miekkala and Nevanlinna in [84, Th. 4.2 and p. 473]. In particular, they derive the following property, based on an analysis of equation (2.40).

Property 3.2.2 *The spectral radius of the (infinite-interval) SOR waveform relaxation operator with optimal overrelaxation factor ω_{opt} in the case of the model problem is given by the following formula, valid for small h,*

$$\rho(\mathcal{K}_{SOR,\omega_{opt}}) \simeq 1 - 2\pi^2 h^2 \ , \quad with \quad \omega_{opt} \simeq \frac{4}{3} - \frac{4}{9}\pi^2 h^2 \ . \tag{3.5}$$

Recall that the spectral radius of the static SOR matrix with optimal ω is given by $1 - 2\pi h$ (for small h), which is much smaller than (3.5). Consequently, we may not expect SOR to be as successful for parabolic problems as it was for elliptic ones.

The discrete-time iteration

The discrete-time iteration remains to be considered. When the time-step is sufficiently small the *finite-interval* iteration converges. Indeed, taking the limit in (2.51),

$$\lim_{\tau \to 0} \rho(\mathcal{K}_\tau) = \lim_{\tau \to 0} \rho(\, K(\tfrac{1}{\tau}\tfrac{\alpha_k}{\beta_k})\,) = \lim_{\tau \to 0} \rho(\,(\tfrac{1}{\tau}\tfrac{\alpha_k}{\beta_k}I + P^h)^{-1}Q^h\,) = 0 \ .$$

Analogously, when the time-step is sufficiently large the spectral radius of the finite-interval operator approaches that of the static iteration operator, and (3.4) applies.

$$\lim_{\tau \to \infty} \rho(\mathcal{K}_\tau) = \lim_{\tau \to \infty} \rho(\, K(\tfrac{1}{\tau}\tfrac{\alpha_k}{\beta_k})\,) = \lim_{\tau \to \infty} \rho(\,(\tfrac{1}{\tau}\tfrac{\alpha_k}{\beta_k}I + P^h)^{-1}Q^h\,) = \rho(\, P^{h^{-1}}Q^h\,) \ .$$

Qualitative information about the *infinite-interval* iteration can be derived by considering contour-line figures like fig. 2.2, and plots of the scaled stability region boundaries as in fig. 2.3. By way of illustration we consider the trapezoidal rule and the backward differentiation formulae.

In the case of the trapezoidal rule, the boundary of the stability region is precisely the imaginary axis. Consequently, by a comparison of formulae (2.38) and (2.53): $\rho(\mathcal{K}) = \rho(\mathcal{K}_\tau)$. As such, the formulae in the above properties also apply to the discrete-time iteration. Following corollary 2.5.11, we find that $\rho(\mathcal{K}_\tau) \leq \rho(\mathcal{K})$ for the A-stable BDF(1) and BDF(2) methods, with equality for the Gauss-Seidel and Jacobi methods. Furthermore, when $\tau \to 0$, then $\rho(\mathcal{K}_\tau) \to \rho(\mathcal{K})$ (e.g. for the SOR method) as the complement of the scaled stability region expands to fill up the entire complex right half-plane. The stability region boundary of the higher order backward differentiation methods runs into the complex left half-plane. As such, the convergence properties cannot be related to those of the continuous-time operators in a similarly straightforward way.

3.2.4 Numerical experiments

Successive waveform relaxation iterates

Figure 3.1 depicts waveform relaxation iterates obtained with the red/black Gauss-Seidel method. Of each iterate $u^{(\nu)}$ the component associated with the grid point at the center of the domain is plotted. The upper picture displays the successive iterates for a coarse spatial discretization with $h = 1/6$. The lower picture shows every 20-th iterate obtained with the discretization $h = 1/32$. In both cases the trapezoidal rule is used with $\tau = 1/100$ and 100 time-steps. The fine-grid iteration obviously converges a lot more slowly than the coarse-grid one does.

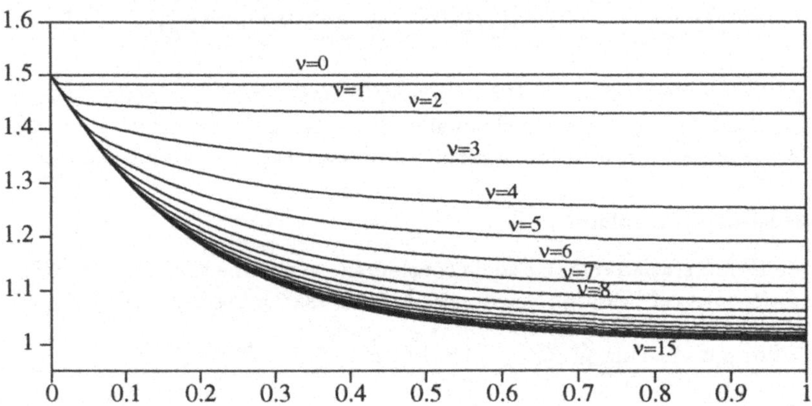

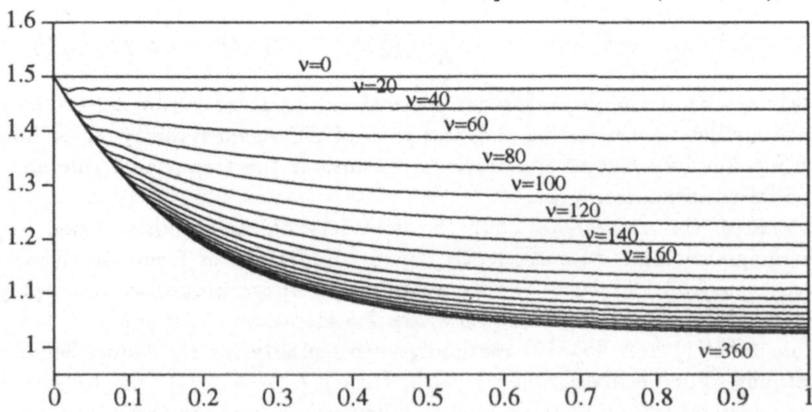

Figure 3.1: Successive Gauss-Seidel waveform relaxation iterates $u^{(\nu)}(t)$, $t \in [0,1]$ for different mesh sizes h. Only the component at the center of the domain is shown.

Measured convergence factors

In table 3.1 we report *averaged convergence factors*, $\bar{\rho}$, for the Jacobi and Gauss-Seidel waveform relaxation methods applied to the discretized heat equation with different mesh sizes h. In each waveform iteration we have determined the l_2-norm of the algebraic error, i.e., $e_T^{(\nu)} = u_T^h - u_T^{(\nu)}$, and calculated the *iteration convergence factor* $\rho^{(\nu)}$ as the following ratio,

$$\rho^{(\nu)} = \parallel e_T^{(\nu)} \parallel_2 / \parallel e_T^{(\nu-1)} \parallel_2 . \tag{3.6}$$

After a sufficiently large number of iterations this factor takes a nearly constant value. An average taken over a large number of iterations in the region of nearly constant behaviour defines the averaged convergence factor, which is the value specified in the table. Observe that the measured values closely correspond to the theoretically determined spectral radii of the infinite-interval waveform relaxation operator.

Table 3.1: Jacobi and Gauss-Seidel waveform relaxation convergence factors.

h	1/5	1/10	1/15	1/20	1/30	1/40	1/50
$\bar{\rho}_{JAC}$	0.803	0.949	0.977	0.986	0.993	0.997	0.998
$1-\pi^2 h^2/2$	0.803	0.951	0.978	0.988	0.995	0.997	0.998
$\bar{\rho}_{GS}$	0.644	0.900	0.955	0.974	0.988	0.993	0.996
$1-\pi^2 h^2$	0.605	0.901	0.956	0.975	0.989	0.994	0.996

Similar convergence factors for the SOR waveform relaxation method are plotted in figure 3.2. They depict the measured value of $\bar{\rho}$ as a function of ω, for three different mesh sizes. The observed minima are in close agreement with formula (3.5). Contrary to the case of using SOR for elliptic problems, waveform SOR does not converge for all $\omega \in (0,2)$. The range of allowable ω is clearly more restricted.

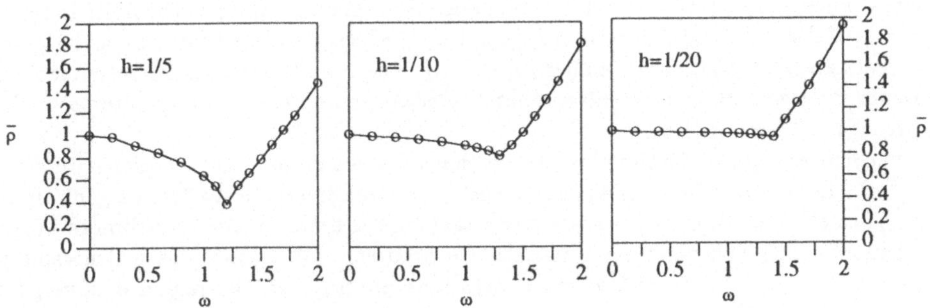

Figure 3.2: SOR waveform relaxation convergence factors as a function ω and h.

Finite-interval versus infinite-interval convergence

In any computer implementation of the waveform relaxation method the number of time-steps is of necessity finite. As such, convergence is determined by the finite-interval convergence analysis. Yet, it has been illustrated above that the numerical results seem to correspond to those predicted by the infinite-interval analysis.

This problem is resolved by considering the *time-level convergence factors*, which correspond to formula (3.6) evaluated for each time-level separately, e.g.,

$$\rho_k^{(\nu)} = \| e_k^{(\nu)} \|_2 \ / \ \| e_k^{(\nu-1)} \|_2 \ . \tag{3.7}$$

In figure 3.3, we have plotted such convergence factors, geometrically averaged over a suitable number of iterations (in casu, over iterations $\nu=21$ to $\nu=28$), for different time-increments. The problem solved is the semi-discretized model problem with $h = 1/16$ and $t \in [0,1]$, and the red/black Gauss-Seidel waveform method is used.

The stable plateau corresponds to the spectral radius value derived in the infinite-time analysis, eq. (3.4). Observe that its height does not depend on τ. The factor measured at the first time-level obviously corresponds to the value specified by the finite-interval analysis. Observe that its magnitude decreases with decreasing τ. In-between the two extremes (convergence factor at the first time-level and the plateau) there is a region with intermediate convergence factors. It can be shown that this region gradually extends, and that the stable plateau recedes backwards as more and more iterations are applied. That is, asymptotically, after a very large number of iterations, the time-level convergence factor at each time-level will correspond to that of the first time-step. This will be illustrated in the next example.

Divergent behaviour

It may happen that the infinite-interval analysis conflicts with the finite-interval analysis. For instance, the former may predict divergence while the latter ensures convergence. Such a case is illustrated in figure 3.4. We consider the SOR waveform relaxation method with parameters $\omega=4/3$ and $h=1/20$, and with BDF(4) time-discretization. It can be shown that the spectral radius of the corresponding continuous-time infinite-interval waveform operator is smaller than one. The spectral radius of the finite interval operator $\mathcal{K}_\tau$ is also below one, as it is equal to $\rho(K(z))$ evaluated for a z on the positive real axis, and, therefore, smaller than $\rho(\mathcal{K})$. The infinite-interval discrete iteration however diverges. It can be shown that the scaled stability boundary intersects the "$\rho(K(z))=1$"-contour-line.

The figure displays the time-level convergence factors for different waveform iterates. The "o"'s in the lower left-hand corner of each picture correspond to "converged" time-levels, i.e., their solution values remain constant (in the finite-precision arithmetic of the implementation). The pictures illustrate the occurrence of oscillations which rapidly explode. However, as more and more iterations are applied, the region of divergent behaviour moves to the right, and, eventually, it is forced out of the finite-length time-window! Afterwards, the iteration converges rapidly. Note that when the same problem is solved with BDF(2) or trapezoidal discretization no such (initially) divergent behaviour occurs.

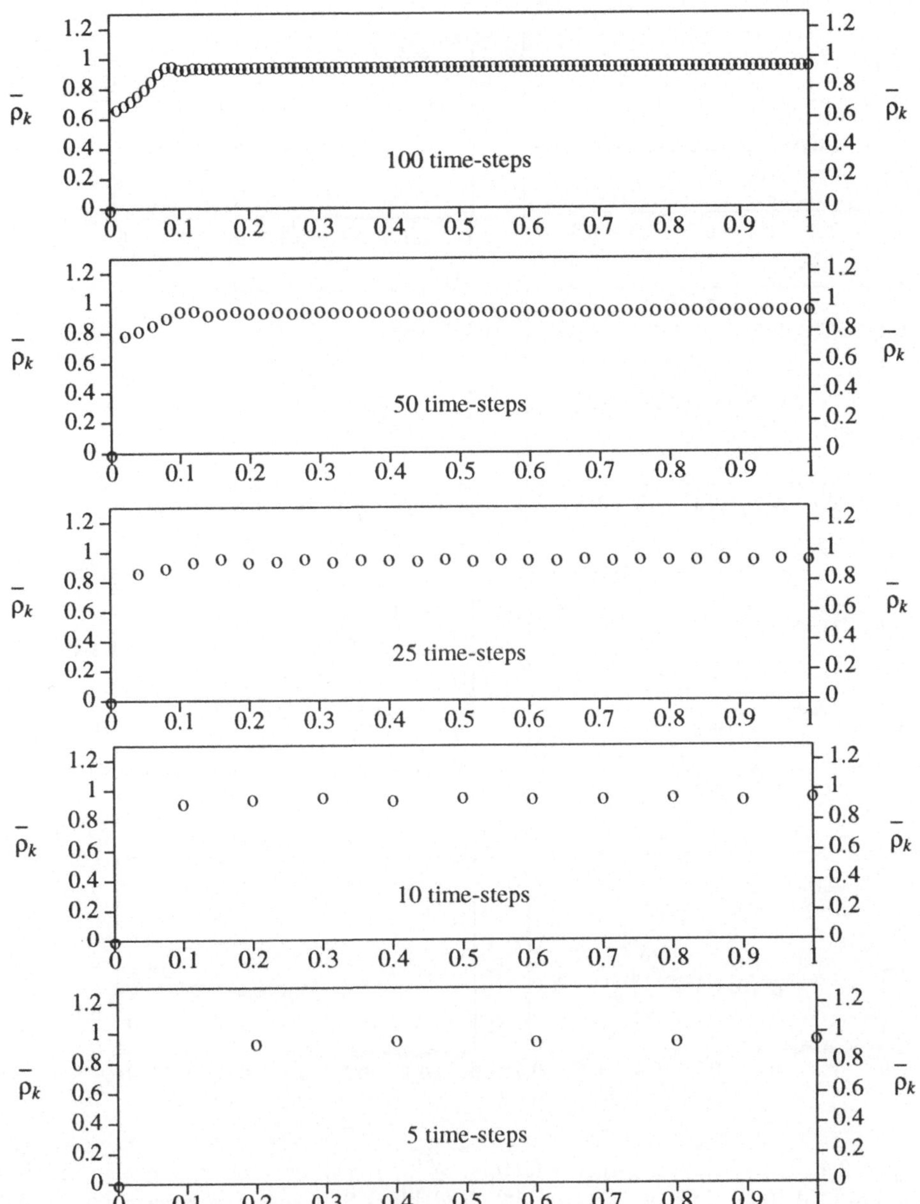

Figure 3.3: Averaged time-level convergence factors plotted for each time-level in the time-window $[0, 1]$, for different values of the time-increment. (red/black Gauss-Seidel waveform relaxation, $h = 1/16$, trapezoidal rule)

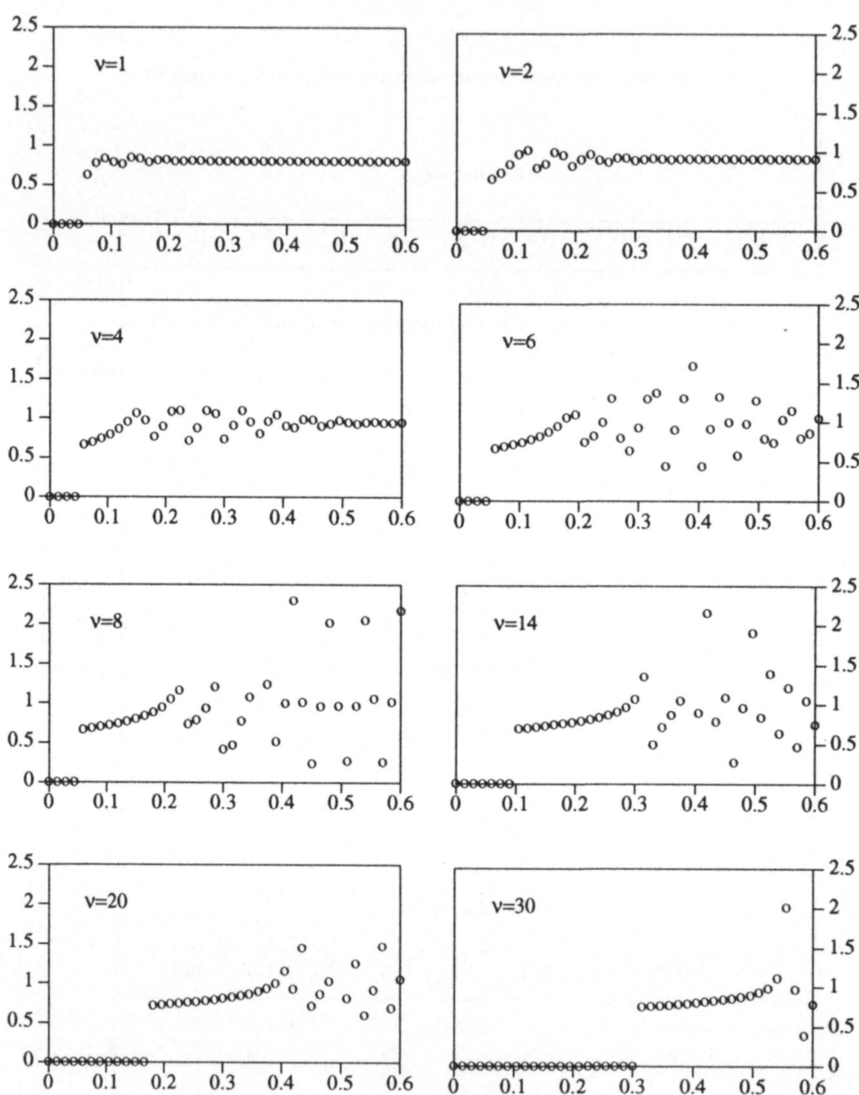

Figure 3.4: Time-level convergence factors, $\rho_k^{(\nu)}$, plotted for each time-level in the time-window $[0, 0.6]$, for different iterates $u_r^{(\nu)}$. (red/black SOR waveform relaxation, $\omega=4/3$, $h=1/16$, $\tau=0.015$, and BDF(4) discretization.)

3.2.5 Further remarks

The Jacobi, Gauss-Seidel and SOR waveform relaxation algorithms are slowly convergent, especially for small spatial mesh sizes. In addition, the trivial selection of the starting approximation leads to large initial errors, except perhaps at the beginning of the time-interval. The reduction of this error requires a very large number of iterations.

The standard acceleration techniques discussed in section 2.6 offer little hope for turning waveform relaxation into a computationally effective procedure. Because of the homogeneous coupling between the unknowns, there is no preferential ordering or partitioning of the differential equations. The use of windowing and partial waveform convergence, in order to restrict the computations to the time-interval where the iteration is doing "useful" work, would lead to the use of small time-windows. As such, this would reintroduce the problems related to parallelism and the exploitation of multi-rate behaviour encountered when using time-stepping methods. Inaccurate iteration and parallel implementation may reduce the cost of each iteration, but they will usually not alter the number of iterations.

3.3 Linear multigrid acceleration

The convergence of the waveform relaxation method can be accelerated following the multigrid idea. The resulting linear *multigrid waveform relaxation* method was first published by Lubich and Ostermann in [78], and independently developed in [129, 140].

We refer to the book by Hackbusch, [44], and the paper by Stüben and Trottenberg, [125], for a detailed analysis of the multigrid method for solving elliptic partial differential equations. A quick overview of the basic idea is presented below.

3.3.1 The multigrid principle

The multigrid method for solving elliptic partial differential equations differs from other iterative techniques in that it uses a *set* of nested discrete meshes (grids). The grid with smallest mesh spacing (the *fine* grid) is the grid on which one wants to obtain the solution, and corresponds to the grid normally considered by single grid methods. The efficiency of the method is based on the interplay of *fine grid smoothing* and *coarse grid correction*. That is, certain error components are attenuated on the fine grid, while other components are attenuated on the coarser grids.

The two-grid method

Consider a linear elliptic partial differential equation, $\mathcal{L}u = f$. Discretization on grid Ω^h leads to a linear system of equations,

$$L^h u^h = f^h \ . \tag{3.8}$$

Let v^h be an arbitrary approximation to the solution u^h, and let e^h denote the corresponding error, i.e., $e^h = v^h - u^h$. This error will usually consist of a combination of smooth error components (low-frequency Fourier modes) and oscillatory components (high-frequency Fourier modes). Note that the error satisfies the so-called *defect equation* given below, where d^h is called the *defect*,

$$L^h e^h = d^h \quad \text{with} \quad d^h = L^h v^h - f^h . \tag{3.9}$$

Application of a standard relaxation method to update approximation v^h usually shows a rapid decrease of the error during the early iterations. Soon after, convergence slows down and the iteration appears to stall. A Fourier analysis shows that the initial phase corresponds to the elimination of the oscillatory error components. Once they have been removed e^h is a smooth function on Ω^h, and the iteration becomes much less effective. Relaxation methods which show such a behaviour are often called *smoothers*. Effective smoothers are, for instance, the Gauss-Seidel relaxation method and the Jacobi overrelaxation method with suitable overrelaxation parameter.

The smooth error components cannot be removed efficiently by applying further relaxation steps. A different technique is called for. If we were able to calculate an approximation to e^h, this could be used to improve v^h. The computation of such an approximation is possible based on using the defect equation. Observe that e^h, being a smooth function on Ω^h, can be represented fairly accurately on a coarser grid Ω^H ($\subset \Omega^h$). This allows the calculation of a *coarse grid approximation* e^H to the fine grid error e^h by solving the following equation on Ω^H,

$$L^H e^H = d^H. \tag{3.10}$$

Here, L^H is a coarse grid discretization of the elliptic operator $\mathcal{L}$, and d^H is a coarse grid approximation of the defect d^h. d^H is calculated by applying a *restriction operator* to d^h. This is an inter-grid transfer operator which transfers information from a fine grid to a coarse grid. It is often denoted by the symbol I_h^H, as e.g. in $d^H = I_h^H d^h$, and it corresponds to an *averaging operator*.

As the number of grid points on Ω^H is usually much smaller than the number of grid points on Ω^h, equation (3.10) can be solved much more rapidly than (3.8). For instance, with the use of *standard coarsening*, i.e., $H = 2h$, the number of grid points is divided by four in going from Ω^h to Ω^H (for two-dimensional problems on regular grids). The computed coarse grid function e^H may then be used to *correct* the current approximation v^h,

$$v^h := v^h - I_H^h e^H .$$

The operator I_H^h is called the *prolongation* operator. It is an *interpolation operator* from the coarse grid to the fine grid. A Fourier analysis shows that the above procedure (defect calculation, coarse grid problem solution, and correction) is very efficient at reducing smooth error components, but hardly changes the oscillatory ones.

The *two-grid cycle* combines the advantages of the smoothing and coarse grid techniques. It start by applying a number of smoothing steps in order to reduce the high-frequency error components, and continues by doing a coarse grid correction in order to eliminate the low-frequency components. Often, this is followed by some additional smoothing in order to further dampen the high frequency errors (possibly reintroduced by the coarse grid correction step).

The multigrid method

The coarse grid problem (3.10) is of similar nature as the original problem, (3.8), defined on the fine grid. It can be solved in a similar way, i.e., by introducing a third grid, coarser than Ω^H, and by executing a number of two-grid cycles. Extension of this idea in a recursive way leads to the *multigrid* algorithm. This algorithm is defined on a nested sequence of grids

$$\Omega^{h_0} \subset \Omega^{h_1} \subset \ldots \subset \Omega^{h_k},$$

where Ω^{h_k} is the fine grid. The coarsest grid Ω^{h_0} usually consists of very few grid points, often only one. Such a grid hierarchy is displayed in figure 3.5.

In the multigrid method a starting approximation is selected for the fine grid solution, u^{h_k}, and a procedure like "procedure mgrid" given in algorithm 3.1 is iteratively called "$mgrid(k, f^{h_k}, u^{h_k})$"[1], until convergence of the approximation. This procedure is completely defined by specifying the grid sequence $\Omega^{h_0}, \ldots, \Omega^{h_k}$, the discretized operators L^{h_i}, the operators $I^{h_{i-1}}_{h_i}$ and $I^{h_i}_{h_{i-1}}$, the nature of the smoothing relaxations, and by assigning values to ν_1, ν_2 and γ_i. The latter defines the nature of the multigrid cycle and the number of times each grid level is visited. A choice of $\gamma_i=1$ leads to the V-cycle. $\gamma_i=2$ gives the W-cycle. (When the Ω^{h_0}-problem is solved exactly, then one can take γ_1 equal to one instead of equal to two, without any change in numerical results.) A third cycle, the F-cycle, does not fit so easily into to above format. Instead, an F-cycle on Ω^{h_i} is defined recursively as follows: its coarse grid part consists of an F-cycle on $\Omega^{h_{i-1}}$ followed by a V-cycle on $\Omega^{h_{i-1}}$. An F-cycle on Ω^{h_1} is just a V-cycle. The three standard cycles are illustrated in figure 3.6. Note that with proper choice of parameters the multigrid iteration is rapidly convergent, with a convergence rate bounded by a small constant which does not depend on the spatial mesh size.

Nested iteration or full multigrid

In order to provide a good approximation for starting the iteration one often implements the so-called *nested iteration* idea. The initial approximation to the fine grid solution is then obtained by interpolation of (an approximation of) the solution obtained on the next coarser grid. A recursive extension of this idea leads to the *full multigrid method*, see algorithm 3.2

Remark that the operator $\bar{I}^{h_i}_{h_{i-1}}$ used in the full multigrid method may be different from the operator $I^{h_i}_{h_{i-1}}$ used in the multigrid procedure. The former is often biquadratic or bicubic, while bilinear interpolation mostly satisfies for the latter. The constant δ_i is a small integer, usually one or two. The full multigrid scheme is illustrated in figure 3.7. There we show schemes that apply one V-cycle or one F-cycle per grid level ($\delta_i = 1$).

Finally, we recall the following important property. Let N be the number of unknowns on the fine grid. With proper choice of parameters the full multigrid method calculates an approximation to the solution of an elliptic problem with algebraic error smaller than the discretization error in $O(N)$ operations.

---

[1]The variables are considered in the computer science sense, i.e., their value may change during the computation.

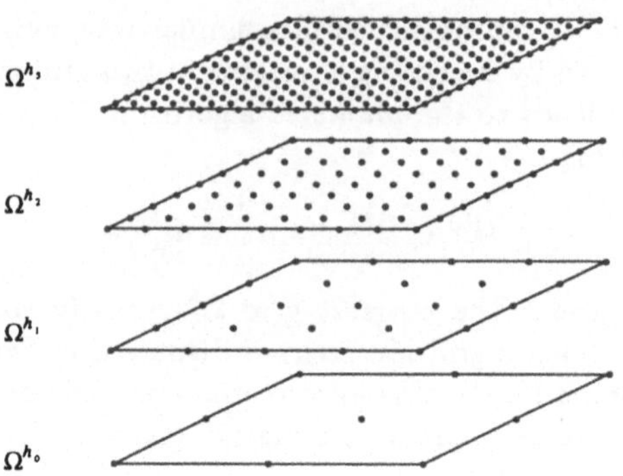

Figure 3.5: A set of nested grids (standard coarsening).

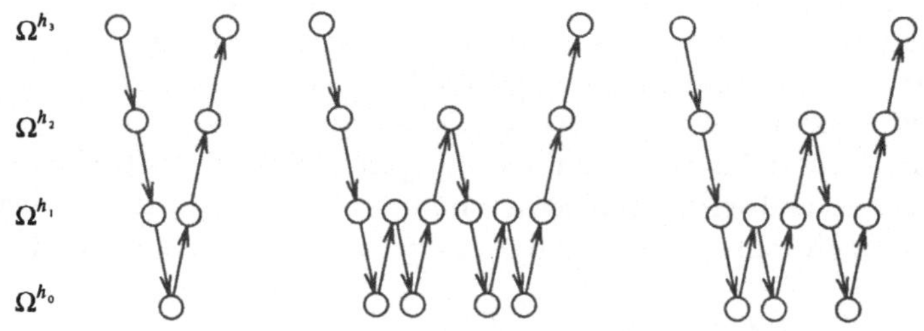

Figure 3.6: Standard multigrid cycles : V, W and F.

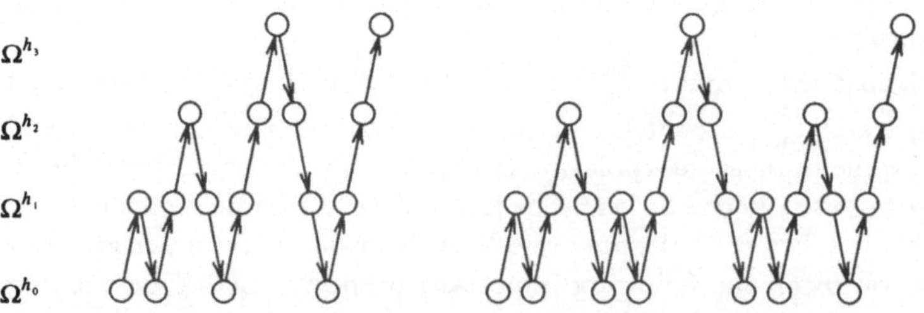

Figure 3.7: Full multigrid with 1 V-cycle (left) or 1 F-cycle (right) per grid level.

Algorithm 3.1: Multigrid method.

PROCEDURE mgrid (i, f^{h_i}, u^{h_i})

if $i=0$ **then** solve $L^{h_0} u^{h_0} = f^{h_0}$.

else
- perform ν_1 smoothing steps.
- compute the defect: $d^{h_i} := L^{h_i} u^{h_i} - f^{h_i}$.
- project the defect on $\Omega^{h_{i-1}}$: $f^{h_{i-1}} := I_{h_i}^{h_{i-1}} d^{h_i}$.
- solve on $\Omega^{h_{i-1}}$: $L^{h_{i-1}} u^{h_{i-1}} = f^{h_{i-1}}$:

  repeat γ_i **times** mgrid $(i\text{-}1, f^{h_{i-1}}, u^{h_{i-1}})$, starting with $u^{h_{i-1}} \equiv 0$.
- interpolate the correction to Ω^{h_i} and correct: $u^{h_i} := u^{h_i} - I_{h_{i-1}}^{h_i} u^{h_{i-1}}$.
- perform ν_2 smoothing steps.

endif

Algorithm 3.2: Full multigrid method.

solve the coarse grid problem: $L^{h_0} u^{h_0} = f^{h_0}$.

for $i = 1$ **to** k
- interpolate the solution on $\Omega^{h_{i-1}}$ to Ω^{h_i}:

  $$u^{h_i} := \bar{I}_{h_{i-1}}^{h_i} u^{h_{i-1}} .$$
- solve the problem on Ω^{h_i}:

  repeat δ_i **times** mgrid (i, f^{h_i}, u^{h_i}) .

endfor

3.3.2   Linear multigrid waveform relaxation

The multigrid principle can be extended to time-dependent problems in essentially the same way as the classical relaxation methods are extended. Each of the elliptic multigrid operators is replaced by a similar operation defined to operate on functions. In this section, we consider the linear equivalent of (3.2),

$$\frac{d}{dt}u^h(t) = L^h u^h(t) + f^h(t) \;, \quad u^h(0) = u_0^h \;, \quad t > 0 \;. \tag{3.11}$$

The multigrid waveform relaxation operators

• **Smoothing.** Pre-smoothing and post-smoothing are performed by applying one or more standard waveform relaxation steps. Particularly straightforward is the use of Gauss-Seidel relaxation (lexicographic or red/black), or the use of a suitably damped Jacobi method (JOR-method). One could also think of using waveform relaxation equivalents of the standard line-relaxation methods.

• **Defect calculation.** The defect of an approximation $\bar{u}^h$ to the solution of the differential equation is defined as,

$$d^h := \frac{d}{dt}\bar{u}^h - L^h \bar{u}^h - f^h \;.$$

This can be elaborated further to get rid of the derivative operator when $\bar{u}^h$ is an iterate of a waveform relaxation process, i.e., when $\bar{u}^h = u^{(\nu)}$ for some $\nu > 0$. Indeed, with use of the splitting matrices P^h and Q^h ($-L^h = P^h - Q^h$),

$$d^h := \frac{d}{dt}u^{(\nu)} - L^h u^{(\nu)} - f^h = \frac{d}{dt}u^{(\nu)} + P^h u^{(\nu)} - f^h - Q^h u^{(\nu)} = Q^h(u^{(\nu-1)} - u^{(\nu)}) \;.$$

As an example we consider the model problem and the use of the red/black Gauss-Seidel smoother. Let v^h be equal to $u^{(\nu-1)} - u^{(\nu)}$. The defect calculation then corresponds to,

$$\begin{cases} i+j \;\; \text{even} \; : \;\; d_{i,j}^h(t) = \frac{1}{h^2}\left(v_{i-1,j}^h(t) + v_{i,j-1}^h(t) + v_{i,j+1}^h(t) + v_{i+1,j}^h(t) \right) \\[2mm] i+j \;\; \text{odd} \;\; : \;\; d_{i,j}^h(t) = 0 \end{cases} \tag{3.12}$$

Note that in complete analogy to (3.9) the error e^h ($= \bar{u}^h - u^h$) satisfies a *defect equation*,

$$\frac{d}{dt}e^h = L^h e^h + d^h \;, \quad e^h(0) = 0 \;, \quad t > 0 \;.$$

• **Restriction.** Assume the coarse grid is derived from the fine grid by standard coarsening ($H = 2h$). Let (I, J) and (i, j) be coarse grid and fine grid indices of the same physical grid point on a two-dimensional domain. The *full-weighting* waveform restriction operator is defined by the following formula, which we write componentwise in *stencil notation* (see e.g. appendix A),

$$u_{I,J}^H(t) = \frac{1}{16} \begin{bmatrix} 1 & 2 & 1 \\ 2 & 4 & 2 \\ 1 & 2 & 1 \end{bmatrix} u_{i,j}^h(t) \;, \quad t \geq 0 \;. \tag{3.13}$$

The *injection* waveform operator is the simplest of restriction operators,

$$u_{I,J}^H(t) = u_{i,j}^h(t) \, , \quad t \geq 0 \, . \tag{3.14}$$

Inbetween is the *half-weighting* restriction, specified below. The formula becomes very simple when used to restrict the defect after a red/black smoothing step since the defect at the black points vanishes. The resulting formula is given between the brackets.

$$u_{I,J}^H(t) = \frac{1}{8} \begin{bmatrix} & 1 & \\ 1 & 4 & 1 \\ & 1 & \end{bmatrix} u_{i,j}^h(t) \quad \left(= \frac{1}{2} \, u_{i,j}^h(t) \right) \, , \quad t \geq 0 \, . \tag{3.15}$$

Prolongation. The prolongation formulae are similar to those used in the multigrid method for solving elliptic problems, yet the arguments are functions instead of scalar values. We do not repeat the bilinear, biquadratic and bicubic interpolation formulae here, but refer to [44, Ch. 3.4].

A two-grid and multi-grid cycle

A *two-grid waveform relaxation cycle* for solving equation (3.11) is stated below. It starts with a fine grid approximation, $u^{(\nu-1)}$, and determines the next iterate, $u^{(\nu)}$, in three steps: pre-smoothing, coarse grid correction, and post-smoothing.

• **Pre-smoothing.** Set $x^{(0)} = u^{(\nu-1)}$. Perform ν_1 standard waveform relaxation steps,

$$\frac{d}{dt}x^{(\nu)} + P^h x^{(\nu)} = Q^h x^{(\nu-1)} + f^h \, , \quad x^{(\nu)}(0) = u_0^h \, , \quad t > 0 \, . \tag{3.16}$$

• **Coarse grid correction.** Compute the defect,

$$d^h = \frac{d}{dt}x^{(\nu_1)} - L^h x^{(\nu_1)} - f^h = Q^h(x^{(\nu_1-1)} - x^{(\nu_1)}) \, . \tag{3.17}$$

Solve the coarse grid equivalent of the defect equation,

$$\frac{d}{dt}v^H = L^H v^H + I_h^H d^h \, , \quad v^H(0) = 0 \, , \quad t > 0 \, . \tag{3.18}$$

Interpolate the correction v^H to Ω^h, and correct the current approximation,

$$\bar{x}^h = x^{(\nu_1)} - I_H^h v^H \, . \tag{3.19}$$

• **Post-smoothing.** Perform ν_2 more smoothing relaxations of type (3.16), starting with $x^{(0)} = \bar{x}^h$, and set $u^{(\nu)} = x^{(\nu_2)}$.

This two-grid cycle can be applied in recursive way to solve the coarse grid problem (3.18). This leads to the *multigrid waveform relaxation* algorithm. This algorithm consists of a repeated application of a procedure like procedure "mgm" displayed in algorithm 3.3 to an initial approximation of the solution on the fine grid.

Algorithm 3.3: Waveform relaxation multigrid.

PROCEDURE mgm (i, f^{h_i}, u^{h_i})

if $i=0$ **then**

- solve $\quad \dfrac{d}{dt} u^{h_0} = L^{h_0} u^{h_0} + f^{h_0}$, e.g., by a standard ODE method.

else

- perform ν_1 smoothing steps.

- compute the defect: $\quad d^{h_i} := \dfrac{d}{dt} u^{h_i} - L^{h_i} u^{h_i} - f^{h_i}$.

- project the defect on $\Omega^{h_{i-1}}$: $\quad f^{h_{i-1}} := I_{h_i}^{h_{i-1}} d^{h_i}$.

- solve on $\Omega^{h_{i-1}}$: $\quad \dfrac{d}{dt} u^{h_{i-1}} = L^{h_{i-1}} u^{h_{i-1}} + f^{h_{i-1}}$, $\quad u^{h_{i-1}}(0) = 0$:

  repeat γ_i times mgm $(i\text{-}1, f^{h_{i-1}}, u^{h_{i-1}})$, starting with $u^{h_{i-1}}(t) \equiv 0$.

- interpolate the correction to Ω^{h_i} and correct: $\quad u^{h_i} := u^{h_i} - I_{h_{i-1}}^{h_i} u^{h_{i-1}}$.

- perform ν_2 smoothing steps.

endif

Full multigrid waveform relaxation

It will be shown later on that the multigrid acceleration solves the problem of the slow convergence of the standard waveform relaxation methods. As with solving elliptic equations, the problem of selecting a satisfactory starting approximation is addressed by considering the nested iteration of full multigrid idea, see Alg. 3.4.

Note a slight difference in the full multigrid algorithms 3.2 and 3.4. The second term in the coarse to fine grid interpolation formula in Alg. 3.4 is needed in order to satisfy the initial condition, $u^{h_i}(0) = u_0^{h_i}$. This "correction" is, of course, not present in the full multigrid algorithm for solving elliptic problems.

A numerical example

We reconsider the model problem, with a spatial discretization $h = 1/32$. In figure 3.8 successive multigrid waveform relaxation iterates are plotted obtained with a multigrid V-cycle, one red/black Gauss-Seidel waveform pre-smoothing step, one similar post-smoothing step, full weighting restriction, bilinear interpolation and a grid sequence obtained by standard coarsening down to a coarse grid with mesh size 1/2. The upper picture shows the result obtained with multigrid cycling, starting from a constant initial profile. The cost of each iterate is about the same as the cost of 8/3 waveform Gauss-Seidel steps. The lower picture shows the starting fine grid approximation obtained with full multigrid, i.e., before the V-cycle on the fine grid is performed. This approximation is computed at the cost of about 8/9 waveform Gauss-Seidel steps. Graphically, the subsequent iterates cannot be distinguished from the initial one.

Algorithm 3.4: Full multigrid waveform relaxation.

solve the coarse grid problem:

$$\frac{d}{dt}u^{h_0} = L^{h_0}u^{h_0} + f^{h_0} \ , \quad u^{h_0}(0) = u_0^{h_0} \ .$$

for $i = 1$ **to** k

    - interpolate the solution on $\Omega^{h_{i-1}}$ to Ω^{h_i}:

$$u^{h_i}(t) := \bar{I}^{h_i}_{h_{i-1}} u^{h_{i-1}}(t) + (u_0^{h_i} - \bar{I}^{h_i}_{h_{i-1}} u_0^{h_{i-1}}) \ .$$

    - solve the problem on Ω^{h_i}:

        repeat δ_i **times** $\mathrm{mgm}(i,f^{h_i},u^{h_i})$

endfor

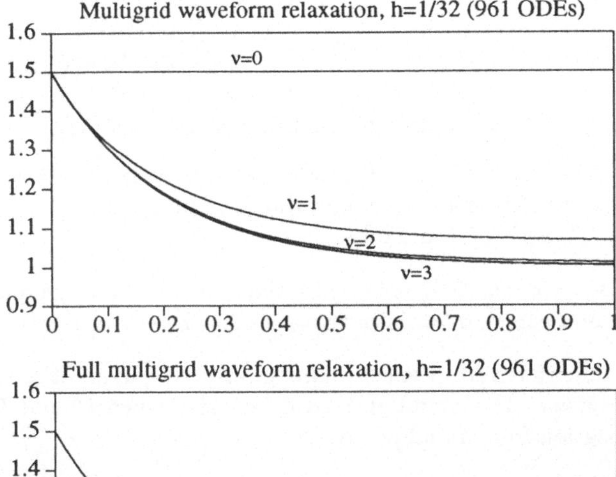

Figure 3.8: Successive multigrid and full multigrid waveform relaxation iterates $u^{(\nu)}(t)$, $t \in [0, 1]$ for $h = 1/32$. Only the component at the center of the domain is shown.

3.4   Convergence analysis

The analysis of linear multigrid waveform relaxation is qualitatively very similar to the analysis of section 2.5. The only change in most formulae is the replacement of the matrix $K(z)$ by a different matrix $M(z)$.

3.4.1   Continuous-time convergence analysis

The continuous-time two-grid waveform relaxation operator

Let $e^{(\nu)}$ be the error of the ν-th waveform relaxation iterate, i.e., $e^{(\nu)} = u^{(\nu)} - u^h$. Denote by $\breve{e}^{(\nu)}(z)$ its Laplace-transform, where z is the complex Laplace-transform variable. Laplace-transforming the equations of the two-grid cycle[2], (3.16) to (3.19), one can derive the relation:

$$\breve{e}^{(\nu)}(z) = M(z)\,\breve{e}^{(\nu-1)}(z)\,, \tag{3.20}$$

with
$$\begin{cases} M(z) = & S^{\nu_2}(z)\,(I - I_H^h(zI - L^H)^{-1}I_h^H(zI - L^h))\,S^{\nu_1}(z) \\ S(z) = & (zI + P^h)^{-1}Q^h \end{cases} \tag{3.21}$$

By considering the inverse Laplace-transform it follows that there is a linear convolution operator $\mathcal{M}$, such that $e^{(\nu)} = \mathcal{M}e^{(\nu-1)}$ ([78, eq. (2.6)]), with

$$\mathcal{M}x\,(t) = m \star x\,(t) := \int_0^t m(t-s)\,x(s)\,ds\,, \quad t > 0\,. \tag{3.22}$$

Its kernel m is defined via its Laplace-transform, which equals $M(z)$.

Remark 3.4.1 When (P^h, Q^h) is a Gauss-Seidel splitting, then $M(z)$ corresponds to the two-grid operator for solving the following (complex) elliptic partial differential equation: $zv^h - L^h v^h = g^h$ ([78, p. 220]).

Remark 3.4.2 As in [78, eq. (2.9)] we require that the entries of the matrix $M(z)$ are rational functions of z vanishing at infinity, with poles having a negative real part.

Note that the remark ensures the continuity of $m(t)$, $t \geq 0$, and the boundedness of $\mathcal{M}$ in any $L_p(\mathbb{R}^+)$-space. The condition is easily satisfied when L^h and L^H are derived by spatial discretization from an elliptic operator, and $\nu_1 + \nu_2 \geq 1$.

Finite-interval analysis

The next theorem is the multigrid equivalent of theorem 2.5.5. Its validity follows from a general functional analysis result, which states that the spectrum of a linear Volterra convolution operator with continuous kernel equals the singleton $\{0\}$, see e.g. [70, p. 33]. For the reader's convenience we outline an elementary proof.

Theorem 3.4.1 *Consider $\mathcal{M}$ as an operator in the space of continuous functions equipped with the maximum norm. Then $\rho(\mathcal{M}) = 0$.*

---

[2]Throughout this section we assume a linear *constant-coefficient* parabolic problem, i.e, L^h and L^H are time-independent matrices.

Proof . Since $\mathcal{M}$ is a bounded operator, we may apply property 2.5.1 to calculate its spectral radius. Let x be in $C([0,T]; \mathbb{C}^d)$. Note that $\mathcal{M}x(t)$ is continuous. Since the vector-norm of a continuous vector-valued function is continuous, it follows that $\| x(t) \|$ and $\| \mathcal{M}x(t) \|$ are continuous.

Consider the following bounds,

$$\| \mathcal{M}x(t) \| = \left\| \int_0^t m(t-s)\, x(s)\, ds \right\| \leq \| m \|_T \int_0^t 1\, ds \ \| x \|_T = \| m \|_T \, t \ \| x \|_T .$$

$$\| \mathcal{M}^2 x(t) \| = \left\| \int_0^t m(t-s)\, \mathcal{M}x(s)\, ds \right\| \leq \| m \|_T^2 \int_0^t s\, ds \, \| x \|_T = \| m \|_T^2 \frac{t^2}{2!} \| x \|_T$$

By induction one can prove that

$$\| \mathcal{M}^n x(t) \| \leq \| m \|_T^n \frac{t^n}{n!} \ \| x \|_T .$$

Maximizing over $[0,T]$ in left- and right-hand side we get,

$$\| \mathcal{M}^n x \|_T \leq \| m \|_T^n \, T^n/n! \ \| x \|_T .$$

Thus, by definition of the operatornorm, Def. 2.5.2,

$$\| \mathcal{M}^n \|_T \ \leq \ \| m \|_T^n \ T^n/n! .$$

Taking the limit $n \to \infty$ in the equation of property 2.5.1 proves the claim. $\quad\Box$

Infinite-interval analysis

The multigrid waveform analogues of theorems 2.5.6 and 2.5.8 are given below. The first equality of each theorem is proven in [78, p. 220]. The second equality follows from the fact that $\rho(M(z))$ and $\| M(z) \|$ are analytic in a set containing the complex right half-plane, and by application of the maximum principle.

Theorem 3.4.2 *Assume the condition in remark 3.4.2 satisfied, and consider $\mathcal{M}$ as an operator on $L_p(\mathbb{R}^+; \mathbb{C}^d)$ with $1 \leq p \leq \infty$. Then,*

$$\rho(\mathcal{M}) = \sup_{Re(z)\geq 0} \rho(M(z)) = \max_{\xi \in \mathbf{R}} \rho(M(i\xi)) , \quad with\ i = \sqrt{-1} . \tag{3.23}$$

Theorem 3.4.3 *Assume the condition in remark 3.4.2 satisfied, and consider $\mathcal{M}$ as operator on $L_2(\mathbb{R}^+; \mathbb{C}^d)$. Then,*

$$\| \mathcal{M} \|_2 = \sup_{Re(z)\geq 0} \| M(z) \| = \max_{\xi \in \mathbf{R}} \| M(i\xi) \| . \tag{3.24}$$

Corollary 3.4.4 *The spectral radius and the norm of the continuous-time two-grid waveform relaxation operator are bounded by the spectral radius resp. norm of the corresponding elliptic two-grid operator,*

$$\rho(\mathcal{M}) \geq \rho(M(0)) \quad and \quad \| \mathcal{M} \|_2 \geq \| M(0) \| . \tag{3.25}$$

Table 3.2: Model problem bounds for the spectral radius and the norm of $\mathcal{M}$.

ν	1	2	3	4	5
$\rho(\mathcal{M})$	0.25	0.162	0.129	0.111	0.098
$\Vert \mathcal{M} \Vert_2$	–	0.437	0.378	0.345	0.322

The above theorems show that the spectral radius and the norm of $\mathcal{M}$ are determined by maximizing the spectral radius and the norm of $M(z)$ over the imaginary axis. This is a calculation that is amenable to a model problem analysis, as was demonstrated by Lubich and Ostermann in [78, p. 224]. There, the two-grid waveform relaxation cycle for the *one-dimensional* heat equation is analysed, with central differencing on a mesh with spacing h, standard coarsening, red/black smoothing, full-weighting and linear interpolation. The authors derive a bound for $\rho(M(z))$ $(Re(z) \geq 0)$,

$$\rho(M(z)) \leq \frac{1}{2} \left| \frac{s}{(1+s)^{2\nu}} \right| \quad \text{with} \quad s = zh^2/2 \quad \text{and} \quad \nu = \nu_1 + \nu_2 \geq 1 . \tag{3.26}$$

They find that the maximum in (3.23) is attained in a point $i\xi$ with $\xi \sim 1/h^2$.

The model problem analysis leads to two properties, [78, p. 223, p. 225], which bound the spectral radius and norm of the two-grid operator. Table 3.2 displays the values of these bounds for different ν. It is conjectured in [78] that these bounds also hold in the case of the two-dimensional heat equation.

Property 3.4.5 *The spectral radius of the infinite-interval two-grid iteration operator for the one-dimensional heat equation with red/black Gauss-Seidel smoothing satisfies,*

$$\rho(\mathcal{M}) \leq \frac{1}{2}\sqrt{\eta(2\nu - 1)} \quad \text{with} \quad \eta(\nu) = \frac{\nu^\nu}{(\nu + 1)^{\nu+1}} , \tag{3.27}$$

with $\nu = \nu_1 + \nu_2 \geq 1$. This is the best possible bound independent of h.

Property 3.4.6 *The norm of the infinite-interval two-grid iteration operator for the one-dimensional heat equation with red/black Gauss-Seidel smoothing satisfies,*

$$\Vert \mathcal{M} \Vert_2 \leq \frac{1}{2}\sqrt{2\sqrt{\eta(2\nu - 2)}} \quad \text{for} \quad \nu = \nu_1 + \nu_2 \geq 1 , \quad \text{if} \quad \nu_1, \nu_2 \geq 1 . \tag{3.28}$$

3.4.2   Discrete-time convergence analysis

The discrete-time two-grid waveform relaxation operator

Let each of the equations (3.16) to (3.19) be discretized with *the same* k-step linear multistep method using the same constant step-size, τ. Assume k starting values supplied. Let $e_\tau^{(\nu)}$ denote the error of the ν-th discrete two-grid waveform iterate. A

Z-transform analysis then proves the existence of a *discrete convolution* operator $\mathcal{M}_\tau$, such that $e_\tau^{(\nu)} = \mathcal{M}_\tau e_\tau^{(\nu-1)}$, with,

$$(\mathcal{M}_\tau x_\tau)_n = (m_\tau \star x_\tau)_n := \sum_{i=0}^{n} m_{n-i} x_i \,, \quad n \geq 0 \,.$$

The kernel m_τ is defined via its Z-transform, [78, eq. (6.8d)] ,

$$\sum_{i=0}^{\infty} m_i z^{-i} = M\left(\frac{1}{\tau}\frac{a}{b}(z)\right) \,. \tag{3.29}$$

Finite-interval analysis

The following theorem can be proven in a similar way as theorem 2.5.9 by using linear algebra arguments. Instead, we choose to follow the lines of a proof (for theorem 2.5.9) given by Nevanlinna in [89, p. 539]. As in section 2.5.4, α_k and β_k denote the coefficients of the highest order term in the characteristic polynomials $a(\xi)$ and $b(\xi)$.

Theorem 3.4.7 *Assume none of the poles of $M(z)$ is equal to $\frac{1}{\tau}\frac{\alpha_k}{\beta_k}$, and let the number of time-steps, N, be finite. Then,*

$$\rho(\mathcal{M}_\tau) = \rho\left(M\left(\tfrac{1}{\tau}\tfrac{\alpha_k}{\beta_k}\right)\right) \,. \tag{3.30}$$

Proof . Since $\mathcal{M}_\tau$ is a linear operator in a finite-dimensional space, its spectrum consists of eigenvalues only. We prove that its spectrum is equal to the spectrum of the matrix $M(\tfrac{1}{\tau}\tfrac{\alpha_k}{\beta_k})$. To this end, we first prove that $\sigma(\mathcal{M}_\tau) \subset \sigma(M(\tfrac{1}{\tau}\tfrac{\alpha_k}{\beta_k}))$.

Let (λ, x_τ) satisfy $\mathcal{M}_\tau x_\tau = \lambda x_\tau$, with $x_\tau = \{x_i\}_{i=0}^{N}$ not identically zero,

$$(\mathcal{M}_\tau x_\tau)_n := \sum_{i=0}^{n} m_{n-i} x_i = \lambda x_n \,, \quad n = 0, \ldots, N \,.$$

Let x_n be the first non-zero element of the sequence x_τ. Then,

$$\sum_{i=0}^{n} m_{n-i} x_i = m_0 x_n = \lambda x_n \,.$$

Thus λ is an eigenvalue of m_0, which we calculate by taking the limit in (3.29),

$$m_0 = \lim_{z \to \infty} \sum_{i=0}^{\infty} m_i z^{-i} = \lim_{z \to \infty} M\left(\tfrac{1}{\tau}\tfrac{a}{b}(z)\right) = M\left(\tfrac{1}{\tau}\tfrac{\alpha_k}{\beta_k}\right) \,.$$

Conversely, we prove that any eigenvalue of $M(\tfrac{1}{\tau}\tfrac{\alpha_k}{\beta_k})$ is also eigenvalue of $\mathcal{M}_\tau$. Let (λ, x) satisfy $M(\tfrac{1}{\tau}\tfrac{\alpha_k}{\beta_k})x = \lambda x$ with $x \neq 0$, and consider the $'N+1'$-element vector $x_\tau = [0\ 0\ \ldots\ 0\ x]^t$. Then, it follows that $\mathcal{M}_\tau x_\tau = \lambda x_\tau$, since,

$$(\mathcal{M}_\tau x_\tau)_n = \sum_{i=0}^{n} m_{n-i} x_i = \begin{cases} 0 & n = 0, \ldots, N-1 \\ \lambda x & n = N \end{cases}$$

Therefore, $\sigma(\mathcal{M}_\tau) = \sigma(M(\tfrac{1}{\tau}\tfrac{\alpha_k}{\beta_k}))$, and the claim (3.30) follows. $\qquad\square$

Remark 3.4.3 The spectral radius is independent of the number of time-steps.

Infinite-interval analysis

The first part of the following theorem is proven in [78, p. 229]. The second part, eq. (3.32), follows by the maximum principle.

Theorem 3.4.8 *Suppose all of the poles of $M(z)$ are in the interior of the scaled stability region $\frac{1}{\tau}S$, and consider $\mathcal{M}_\tau$ as an operator in $l_p(I\!N)$ with $1 \leq p \leq \infty$. Then,*

$$\rho(\mathcal{M}_\tau) = \sup \{\rho(M(z)) \ : \ \tau z \in \bar{\mathbb{C}} \setminus intS\} , \tag{3.31}$$

or, $\qquad \rho(\mathcal{M}_\tau) \ = \ \max_{|\xi|=1} \rho(M(\tfrac{1}{\tau}\tfrac{a}{b}(\xi))) \ = \ \max_{\tau z \in \partial S} \rho(M(z)) .$ $\tag{3.32}$

Corollary 3.4.9 *The spectral radius of the discrete-time two-grid waveform relaxation operator is bounded by the spectral radius of the corresponding elliptic two-grid operator,*

$$\rho(\mathcal{M}_\tau) \geq \rho(M(0)) \tag{3.33}$$

Bounds on the spectral radius of the infinite-interval operator may be derived by taking properties of the stability region into account. An important case is considered in the next theorem, [78, p.229].

Theorem 3.4.10 *If the linear multistep method is $A(\alpha)$-stable, then,*

$$\rho(\mathcal{M}_\tau) \leq \sup_{z \in \Sigma_{\pi-\alpha}} \rho(M(z)) \ = \ \max_{z \in \partial\Sigma_{\pi-\alpha}} \rho(M(z)) , \tag{3.34}$$

with the sector $\Sigma_{\pi-\alpha} = \{z : |arg\, z| \leq \pi - \alpha\} \cup \{0\}$. This is an optimal bound which holds without restriction on the ratio τ/h^2.

Corollary 3.4.11 *If the multistep method is A-stable then $\rho(\mathcal{M}_\tau) \leq \rho(\mathcal{M})$, where $\mathcal{M}$ is the waveform operator on $(0,\infty)$.*

Note that similar relations hold for the operator norm.

The corollary can be used for completing the model problem analysis in the case of the trapezoidal rule and the BDF(1) and BDF(2) time-discretizations. Indeed, it follows that the bounds (3.27) and (3.28) also hold for the discrete-time iteration. In the case of a discretization with an $A(\alpha)$-stable method, the right-hand side of formula (3.34) is to be determined. This requires finding the maximum of $\rho(M(z))$ over two half-lines in the complex plane. From the inequality in theorem 3.4.10 it follows that the bound is an increasing function for decreasing α. Consequently we may expect convergence to be slower with increasing order of the BDF-method.

3.5 Experimental results

To start with, we reconsider the model problem, i.e., the two-dimensional heat equation defined on the unit square with initial values and Dirichlet boundary conditions chosen in such a way that the solution is equal to,

$$u(t,x,y) = 1 + \sin(\pi x/2)\sin(\pi y/2)e^{-\pi^2 t/2} \,.$$

We shall apply the multigrid waveform relaxation method to this problem for different values of the spatial mesh size h and with different time-integration formulae using a constant, global time-step τ. The parameters are as follows: red/black smoothing, full-weighting restriction, bilinear interpolation, standard coarsening down to a coarse grid with mesh size $h=0.5$ (i.e., the coarse grid problem consists of a single ODE).

Figure 3.9 displays the evolution of the maximum norm ($\| \cdot \|_\infty$) of the algebraic error as a function of the iteration index. The lines correspond to different multigrid cycles. For instance, "W(2,1)" denotes the use of W-cycles with 2 pre-smoothing steps and 1 post-smoothing step. The very constant decrease allows the calculation of precise convergence factor values, see table 3.3. Similar factors for other linear multistep methods are given in table 3.4. Note that $\bar{\rho}$ increases, indicating slower convergence, when the order of the BDF method increases. This was to be expected from theorem 3.4.10, and the knowledge that the BDF methods are $A(\alpha)$-stable with $\alpha=90°$ (BDF(1), BDF(2)), $\alpha=88°$ (BDF(3)), $\alpha=73°$ (BDF(4)), and $\alpha=51°$ (BDF(5)).

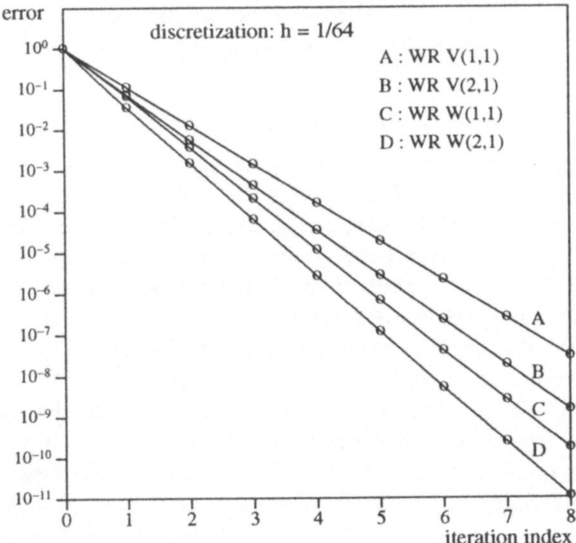

Figure 3.9: Maximum algebraic error of successive multigrid waveform relaxation iterates ($h=1/64$, $\tau=1/100$, 100 time-steps, trapezoidal time-discretization).

Table 3.3: Averaged convergence factor ($h = 1/64$, $\tau = 1/100$, trapezoidal rule).

	V(1,1)	V(2,1)	W(1,1)	W(2,1)
$\bar{\rho}$	0.115	0.079	0.060	0.043

Table 3.4: Multigrid waveform V(1,1)-cycle averaged convergence factors for different linear multistep methods ($h = 1/32$, $\tau = 1/200$, 200 time-steps).

	C.-N.	BDF(1)	BDF(2)	BDF(3)	BDF(4)	BDF(5)
$\bar{\rho}$	0.11	0.10	0.10	0.11	0.23	0.83

Averaged time-level convergence factors, $\bar{\rho}_k$, are depicted in figure 3.10 (V(1,1)-cycle, $h = 1/32$, ρ_k averaged over 5 iterations). They are qualitatively similar to the ones observed in figure 3.3. Quantitatively, they are, of course, much smaller. The iteration at the first time-level converges the fastest; there is a plateau, extending to infinity, with constant convergence factors, and there is a region of intermediate values inbetween. When the time-window is sufficiently large, the measured convergence factor $\bar{\rho}$ corresponds to the value that is dictated by the infinite-interval analysis.

The experiment with 500 time-steps clearly shows that the numerical characteristics of the multigrid waveform relaxation method do not deteriorate as the length of the integration window or the number of time-steps increases.

Below we consider the model problem with an oscillatory function $f(t, x, y)$ added to the right-hand side of the partial differential equation. This function is chosen in such a way that the analytical PDE solution equals,

$$u(t, x, y) = x(1 - x)y(1 - y)\sin(75t) .$$

For this problem the convergence behaviour of the algorithm is not as "straight-lined" as the behaviour shown in figure 3.9. This is illustrated in table 3.5. The iteration convergence factor, (3.6), starts off low, and gradually increases before stabilizing.

From the discrete-time finite-interval analysis in section 3.4.2 we know that eventually $\rho^{(\nu)}$ should decrease again towards its finite-interval asymptotic value, (3.30). The latter is equal to the time-level convergence factor of the first time-level. However, this behaviour is usually not observed in the experiments. The number of iterations before convergence in finite-precision arithmetic (in casu, error $\approx 10^{-16}$) is usually too low to attain this asymptotic convergence rate, especially when the time-window is large. The decrease is visible though when the interval of integration is very small. Most time-levels are then in the region of intermediate time-level convergence factors.

Tables 3.6 and 3.7 report averaged convergence factors obtained with different values of h and τ, a constant number of time-steps equal to 100, and the use of the trapezoidal and BDF(2) time-discretization. It can be seen that for a constant h, a constant number of time-steps and for decreasing τ, the convergence factor starts off relatively low; it then increases and attains a maximum (plateau), before decreasing again. The value of τ for which this maximum is first attained decreases with decreasing h.

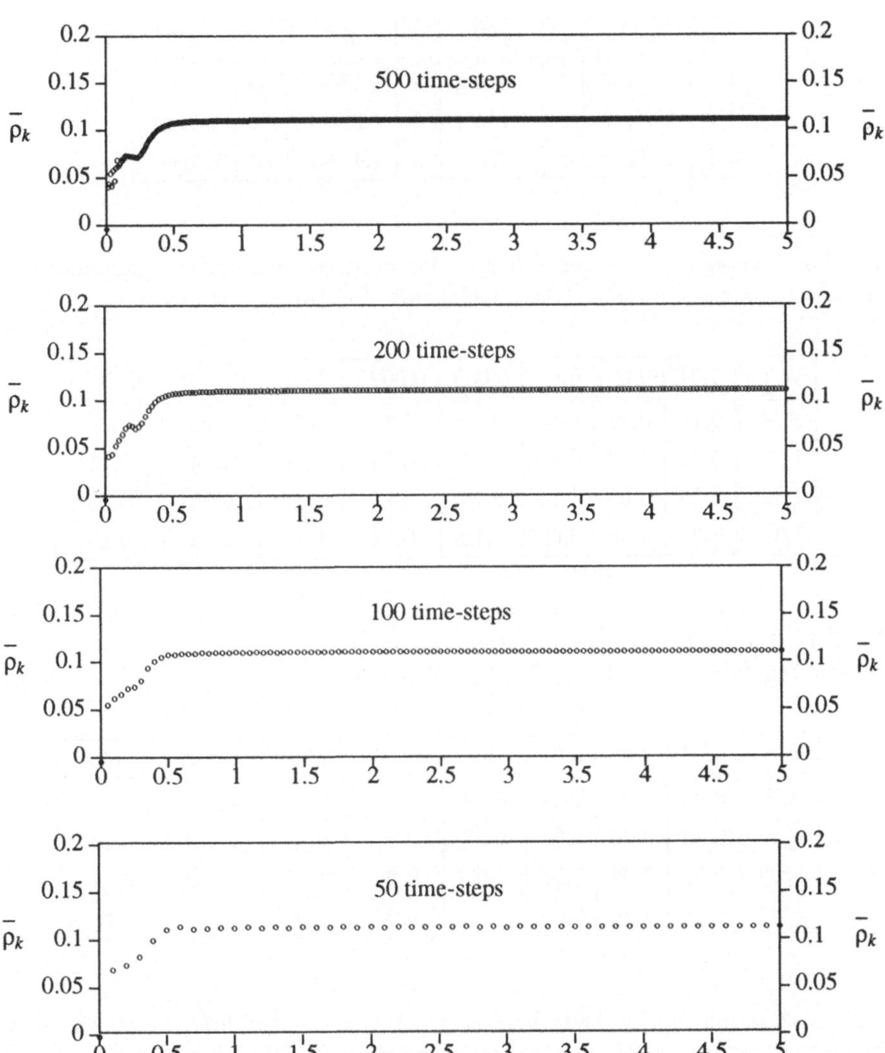

Figure 3.10: Averaged time-level convergence factors plotted for each time-level in the time-window $[0, 5]$, for different values of the time-increment. (multigrid V(1,1)-cycles, $h = 1/32$, trapezoidal rule)

Table 3.5: Iteration convergence factors $\rho^{(\nu)}$, (3.6), of successive multigrid waveform W(1,1)-cycle iterates (trapezoidal rule, $h = 1/16$, $\tau = 1/100$, 100 time-steps).

ν	$\rho^{(\nu)}$	ν	$\rho^{(\nu)}$	ν	$\rho^{(\nu)}$	ν	$\rho^{(\nu)}$
1	3.72e-2	4	8.34e-2	7	1.38e-1	10	1.42e-1
2	4.44e-2	5	1.14e-1	8	1.39e-1	11	1.43e-1
3	5.90e-2	6	1.32e-1	9	1.41e-1	12	1.41e-1

Table 3.6: Averaged convergence factors for multigrid waveform relaxation W(1,1)-cycle as a function of h and τ (trapezoidal rule, 100 time-steps).

h, τ	0.04	0.02	0.01	0.005	0.0025	0.001	0.0005	0.00025
1/8	0.11	0.11	0.10	0.10	0.10	0.07	0.05	0.04
1/16	0.12	0.13	0.14	0.14	0.14	0.14	0.14	0.11
1/32	0.08	0.08	0.14	0.15	0.15	0.15	0.15	0.15
1/64	0.06	0.06	0.09	0.12	0.14	0.15	0.15	0.15

Table 3.7: Averaged convergence factors for multigrid waveform relaxation W(1,1)-cycle as a function of h and τ (BDF(2), 100 time-steps).

h, τ	0.04	0.02	0.01	0.005	0.0025	0.001	0.0005	0.00025
1/8	0.06	0.06	0.09	0.10	0.10	0.07	0.04	0.02
1/16	0.06	0.06	0.06	0.07	0.11	0.13	0.14	0.11
1/32	0.06	0.06	0.06	0.06	0.06	0.10	0.12	0.14
1/64	0.06	0.06	0.06	0.06	0.06	0.06	0.06	0.10

An explanation of this behaviour is given below. But first, recall the relation between the Laplace-transforms of successive errors, (3.20). Asymptotically, any "frequency" component of the initial error (i.e., any $\breve{e}^{(0)}(z)$) converges with the corresponding convergence factor $\rho(M(z))$. The maximum of this factor is found for a "high-frequency" z on the imaginary axis. Consider a constant value of h.

• When τ is sufficiently large, we can apply the property $\lim_{\tau \to \infty} \rho(\mathcal{M}_\tau) = \rho(M(0))$, see e.g. also section 3.2.3. That is, the convergence is essentially similar to that of the elliptic multigrid iteration. Intuitively, we could argue that the high frequency components where the maximum of $\rho(\mathcal{M}(z))$ is found, cannot be represented adequately on the discrete mesh. The convergence factor more closely corresponds to the convergence factor of the low-frequency components. It is as if the maximum in (3.23) is found at the origin instead of at a point $i\xi$ further away on the imaginary axis.

- For smaller τ-values higher frequency components can be represented; ρ increases, and finally reaches a maximum which is not further changed with decreasing τ (at least, if time-windows are assumed that are much longer than the region of intermediate time-level convergence factors). The measured value of the convergence factor corresponds to the value predicted by the continuous-time infinite-interval analysis.

- When τ decreases further while the number of time-levels remains constant (as is the case for the results in the tables), ρ will start to decrease. As the time-window gets shorter and shorter, every time-level will eventually be located in the region of intermediate time-level convergence factors.

In the model problem analysis it was shown that the maximum in (3.23) is found for a value of $z = i\xi \sim i/h^2$. That is to say, the smaller the spatial mesh size, the higher the frequency of the error component which dictates the asymptotic convergence rate. Consequently, with decreasing h it will take a smaller τ to represent the slowest converging error component.

Observe that for a given h and τ the waveform method with BDF(2) discretization is more rapidly converging than the one with the trapezoidal rule. This can be understood from the size and the shape of the stability regions together with theorem 3.4.8.

The model anisotropic problem

We consider the model *anisotropic* problem, with given initial value and with constant Dirichlet boundary conditions,

$$\frac{\partial u}{\partial t} = \epsilon\,\frac{\partial^2 u}{\partial x^2} + \frac{\partial^2 u}{\partial y^2}\ , \quad \text{with}\ \ u(0, x, y) = 1 + sin(\pi x/2)\sin(\pi y/2)\ . \tag{3.35}$$

It is known that standard multigrid with pointwise red/black Gauss-Seidel relaxation does not perform satisfactorily for the model elliptic anisotropic problem when $\epsilon \ll 1$ or $\epsilon \gg 1$. Since the spectral radius of the waveform multigrid operator is bounded by the spectral radius of the elliptic multigrid operator, i.e., $\rho(\mathcal{M}) \geq \rho(M(0))$, we may expect a similar performance degradation. This is confirmed by the numerical results given in table 3.8.

Table 3.8: Multigrid waveform V(1,1)-cycle convergence factors for the model anisotropic problem ($h = 1/32$, $\tau = 0.01$, trapezoidal rule, 100 time-steps).

ϵ	10^{-4}	10^{-3}	10^{-2}	0.1	0.5	1	2	10	10^2	10^3	10^4
$\bar{\rho}$	0.978	0.974	0.934	0.660	0.213	0.116	0.213	0.661	0.935	0.974	0.978

A model convection-diffusion problem

Finally, consider the following model convection-diffusion problem,

$$\frac{\partial u}{\partial t} = \epsilon \left(\frac{\partial^2 u}{\partial x^2} + \frac{\partial^2 u}{\partial y^2} \right) - \frac{\partial u}{\partial x} - \frac{\partial u}{\partial y} \, , \qquad (3.36)$$

defined on the unit square $\Omega = [0,1] \times [0,1]$, with Dirichlet boundary conditions,

$$u(x,1,t) = u(0,y,t) = u(1,y,t) = 0 \quad \text{and} \quad u(x,0,t) = 5\sin(\pi x)\sin(t) \, .$$

The problem is discretized by using central differences for the diffusion terms and stable upwind differences for the convective part, i.e.,

$$\left(\frac{\partial u}{\partial x}\right)_{i,j} \simeq \frac{u_{i,j}(t) - u_{i-1,j}(t)}{h} \quad \text{and} \quad \left(\frac{\partial u}{\partial y}\right)_{i,j} \simeq \frac{u_{i,j}(t) - u_{i,j-1}(t)}{h}$$

It is well-known that the standard multigrid method with pointwise red/black smoothing performs badly as a solver for the corresponding elliptic problem if $\epsilon \ll 1$, see, e.g., [44, p. 220]. The situation is remedied in the elliptic case by using a pointwise *lexicographic* Gauss-Seidel smoother in which the update ordering of the grid points depends on the coefficients of the first order derivatives $\frac{\partial u}{\partial x}$ and $\frac{\partial u}{\partial y}$. In this particular case a good smoother would update the grid points from left to right and bottom to top. Numerical results displayed in table 3.9 show that the convergence behaviour of the waveform algorithm is similar. Note that the lexicographic smoother becomes a direct solver in the limiting case of ϵ equal to zero.

Table 3.9:   Multigrid waveform V(1,1)-cycle convergence factors for the model convection-diffusion problem with pointwise red/black and lexicographic smoothing ($h = 1/32$, $\tau = 0.025$, trapezoidal rule, 40 time-steps,).

ϵ	10^1	10^0	10^{-1}	10^{-2}	10^{-3}	10^{-4}	10^{-5}
$\bar{\rho}_{RB}$	$9.4\ 10^{-2}$	$1.4\ 10^{-1}$	$2.0\ 10^{-1}$	$2.4\ 10^{-1}$	$4.2\ 10^{-1}$	$5.0\ 10^{-1}$	$5.2\ 10^{-1}$
$\bar{\rho}_{LEX}$	$1.4\ 10^{-1}$	$1.6\ 10^{-1}$	$2.0\ 10^{-1}$	$1.0\ 10^{-1}$	$8.1\ 10^{-2}$	$1.1\ 10^{-4}$	$1.3\ 10^{-6}$

3.6   Nonlinear multigrid waveform relaxation

A multigrid waveform relaxation algorithm for solving nonlinear parabolic problems can be derived from the well-known multigrid *full approximation scheme (FAS)*, which is a standard multigrid method for solving nonlinear elliptic partial differential equations introduced by Brandt, [9]. The basic equations of the FAS-method for solving elliptic problems are briefly recalled below. The waveform extension was first described in our paper [140], and more elaborated numerical results were reported in [135, 136].

The FAS method for elliptic problems

Consider a nonlinear elliptic problem $\mathcal{L}(u) = f$. Discretization leads to a nonlinear system of equations,

$$L^h(u^h) = f^h \; . \tag{3.37}$$

Let v^h be an approximation to the solution, and let e^h be the corresponding error, $e^h = v^h - u^h$. This error satisfies a nonlinear *defect equation*,

$$L^h(v^h) - L^h(v^h - e^h) = d^h \; , \quad \text{with} \quad d^h = L^h(v^h) - f^h \; . \tag{3.38}$$

The FAS two-grid cycle is composed of similar steps as the two-grid cycle for solving linear problems. First, a few relaxation steps are executed with a standard nonlinear relaxation method (e.g. nonlinear Gauss-Seidel). Their effect is to dampen the high frequency components of the error. Secondly, a coarse grid approximation to e^h is calculated by solving a coarse grid equivalent of the defect equation. In [9], it is suggested to use the following equation,

$$L^H(v^H) - L^H(v^H - e^H) = d^H \; , \quad \text{with} \quad v^H = \bar{I}_h^H v^h \quad \text{and} \quad d^H = I_h^H d^h \; .$$

The operators $\bar{I}_h^H$ and I_h^H are two possibly different restriction operators. The coarse grid equation may be rewritten into the usual format of equation (3.37),

$$L^H(u^H) = f^H \; , \quad \text{with} \quad u^H = v^H - e^H \quad \text{and} \quad f^H = L^H(v^H) - d^H \; . \tag{3.39}$$

Having solved this equation for u^H, one can correct approximation v^h as follows,

$$v^h := v^h - I_H^h(v^H - u^H) \; . \tag{3.40}$$

This coarse grid step, which dampens the smooth error components, is usually followed by some more post-smoothing nonlinear relaxation steps.

The convergence behaviour of the FAS two-grid cycle, and of the corresponding FAS multigrid and FAS full multigrid methods is very similar to that of the linear multigrid method. Convergence is usually very fast, and the convergence factor is bounded by a value which is independent of the spatial mesh size.

The waveform relaxation FAS method

The waveform FAS method uses the same building blocks, e.g., restriction and interpolation operators, as the linear multigrid waveform relaxation method. The *two-grid FAS waveform relaxation cycle* for solving the nonlinear ODE system (3.2) derived by the numerical method of lines is stated below. It can easily be derived from the corresponding FAS method for solving elliptic problems. The algorithm starts with a fine-grid approximation, v^h, which is updated to provide the next iterate in the standard three steps: pre-smoothing, coarse-grid correction, and post-smoothing.

• **Pre-smoothing.** Perform ν_1 nonlinear Gauss-Seidel waveform relaxation steps (lexicographic or red/black). Note that the ordinary differential equations need not be solved exactly. The use of a waveform relaxation Newton method (section 2.3) with a single Newton linearization will usually be sufficient.

- **Coarse grid correction.** Project the current approximation, v^h, onto the coarse grid Ω^H by using a restriction operator $\bar{I}_h^H$,

$$v^H = \bar{I}_h^H v^h \ . \tag{3.41}$$

Calculate the right-hand side f^H of the coarse grid problem,

$$f^H = \frac{d}{dt} v^H - L^H(v^H) \ - \ I_h^H \left(\frac{d}{dt} v^h - L^h(v^h) - f^h \right) \ . \tag{3.42}$$

The restriction operator I_h^H may be different from operator $\bar{I}_h^H$. However, in order to avoid the derivative calculation in (3.42) the use of the same operator may be advantageous. Indeed, in that case the two derivatives cancel out.

Solve the following coarse grid initial value problem defined on Ω^H,

$$\frac{d}{dt} u^H = L^H(u^H) + f^H \ , \quad u^H(0) = \bar{I}_h^H v^h(0) \ , \quad t > 0 \ . \tag{3.43}$$

Calculate the correction. Interpolate the correction to the fine grid by using a waveform prolongation formula, and correct the current fine grid approximation,

$$v^h := v^h - I_H^h(v^H - u^H) \ . \tag{3.44}$$

- **Post-smoothing.** Perform ν_2 more smoothing relaxations, e.g., again by nonlinear Gauss-Seidel waveform relaxation.

This FAS two-grid cycle may be applied recursively to solve the coarse grid initial value problem. This eventually leads to the *multigrid full approximation scheme waveform relaxation method*. The algorithm is a straightforward adaptation of algorithm 3.3 and therefore not repeated here. Note that the algorithm gives identical results as the linear multigrid waveform relaxation method when applied to a linear parabolic problem.

Numerical examples are given in chapter 8. The convergence factors reported there are very similar in magnitude to the ones given in the previous section.

3.7   A multigrid method on a space-time grid

Multigrid waveform relaxation is a continuous-time iterative method. It can be defined without mention of a time-discretization technique. Actual time-levels and discretization formulae are to be selected only when one starts implementing the method. In the discrete-time analysis of section 3.4.2 we considered the case of using global and constant time-steps. Each function is discretized on the same set of time-levels, be it a function on the fine grid, or on the coarse grid. Although the method is certainly not limited to such an approach, we shall argument in the present section that such a discretization arises in a very natural way.

Consider the following *anisotropic* elliptic partial differential equation defined on a two-dimensional domain in (x, y)-space,

$$-\frac{\partial^2 u}{\partial x^2} - \epsilon \frac{\partial^2 u}{\partial y^2} + \frac{\partial u}{\partial y} = f \ , \quad \text{with} \ \ \epsilon \ll 1 \ . \tag{3.45}$$

In the limiting case, i.e., for $\epsilon = 0$, this equation becomes the one-dimensional parabolic heat equation. Let the problem be discretized on a regular spatial grid Ω^{h_x,h_y} with mesh size h_x in the x-direction and h_y in the y-direction. Multigrid operators for solving this anisotropic elliptic problem are selected as follows.

A factor ϵ which is either very small $\epsilon \ll 1$ or very large $\epsilon \gg 1$ in an operator like $-\frac{\partial^2 u}{\partial x^2} - \epsilon\frac{\partial^2 u}{\partial y^2}$ disturbs the smoothing characteristics of the pointwise relaxation smoothers. For instance, when ϵ is very small highly oscillatory error components in the y-direction are hardly changed by a pointwise smoothing method; only errors in the x-direction are smoothed. The standard remedy is to use either *x-line smoothing*, or *semi-coarsening*, [44, p. 202, p. 226]. The former approach is a blockwise relaxation method, in which the equations corresponding to a grid line extending in the x-direction are grouped during relaxation. In the latter approach a standard pointwise smoother can be used, yet the mesh is only coarsened in the x-direction. The mesh spacing in the y-direction is kept unchanged. The coarse grid is then given by Ω^{2h_x,h_y}. Semi-coarsening is illustrated in figure 3.11. When semi-coarsening is used the restriction and prolongation operators should be altered correspondingly. In particular, one can use operators of which the stencils extend in the x-direction only. These are standard inter-grid operators for one-dimensional elliptic problems.

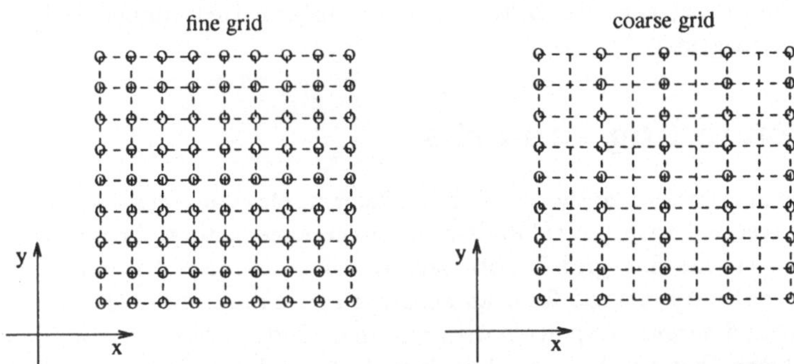

Figure 3.11: Semi-coarsening: fine grid Ω^{h_x,h_y} and coarse grid Ω^{2h_x,h_y}.

The first order derivative term $\frac{\partial u}{\partial y}$ is handled by *backward differencing*, i.e., only information from smaller y-values is used. Central differencing is avoided for reasons of stability. For adequate damping of errors a smoothing scheme is followed which proceeds from low y-values to high y-values. Possible smoothers are, e.g., x-line smoothing from bottom to top, or lexicographic pointwise Gauss-Seidel smoothing from the left to the right and from bottom to top. Also y-line smoothing may be used.

For the above anisotropic problem a suitable combination of a coarsening strategy and a smoother is to be found. It should take the effect of both the factor ϵ and the term $\frac{\partial u}{\partial y}$ into account. Two natural selections are as follows:

- standard coarsening combined with x-line smoothing from bottom to top,

- semi-coarsening combined with a smoothing process from bottom to top.

In the limiting case of $\epsilon = 0$, the anisotropic problem becomes the model parabolic problem. The variable y acts as the time-coordinate t, and h_y becomes the time-increment τ. The discrete mesh is then called a *space-time grid*. The x-line smoother in the first variant becomes a direct solver. In fact, the first variant is then a standard *time-stepping scheme* corresponding to the backward difference approximation used for the first order derivative term. Consider the second variant with y-line smoothing. In this case the multigrid procedure is identical to the *discretized multigrid waveform relaxation method*! The calculation of the discrete solution to the ordinary differential equation at a grid point corresponds to the calculation of a "line" of y-values. In addition, the waveform restriction and prolongation operators, which operate in the spatial direction only, are identical to the restriction and prolongation operators used with semi-coarsened meshes. Semi-coarsening combined with a smoothing process which proceeds time-level per time-level from low y-values to high y-values corresponds to the so-called *parabolic multigrid method* with *sequential smoothing* developed by Hackbusch in [43] and theoretically analysed for a model problem by Burmeister in [13]. In many cases their sequential smoothing procedure is identical to the y-line relaxation smoother. Then, the parabolic multigrid method becomes identical to the discretized multigrid waveform relaxation method, at least when the latter uses constant and global time-steps. The theory surveyed in the current chapter is then immediately applicable to parabolic multigrid.

3.8   Concluding remarks

The multigrid acceleration of the standard waveform relaxation methods has proven to be rapidly convergent with typical multigrid convergence factors. An accurate starting approximation on the fine grid is obtained by using the nested iteration idea. The resulting full multigrid method finds an accurate approximation to the solution of the parabolic partial differential equation with minimal effort. In the case studies of chapter 8 it will be shown that one V-cycle at each spatial grid level is often satisfactory.

Note that we have not yet discussed the arithmetic complexity of the method, nor any parallel implementation issues. In addition, in order to assess the real computational effectiveness of the method we have to compare the multigrid waveform algorithms with standard time-stepping methods. This discussion is postponed until chapter 7, with many examples following in chapter 8. First, we shall consider the extension of the waveform relaxation idea to time-periodic problems (chapter 4), and present a detailed discussion of parallel computing issues (chapters 5 and 6).

Chapter 4

Waveform Relaxation for Solving Time-Periodic Problems

Nothing is as practical as a good theory.
— G. Dahlquist

If we were to wait for convergence proofs and error estimates for the new methods,
most of the computers now in use in technology and industry would come grinding to a halt.

—R. Richtmyer

We extend the applicability of waveform relaxation and discuss its use for solving time-periodic non-autonomous ordinary and partial differential equations. The convergence characteristics of both the continuous-time and the discrete-time time-periodic iteration are analysed. It is shown that the convergence of the method is intimately related to the convergence of the corresponding initial value waveform relaxation method. The multigrid acceleration is discussed. Finally, an algorithm based on a modified shooting method is given for solving autonomous periodic problems.

4.1   Introduction

In this chapter, we consider parabolic partial differential equations,

$$
\begin{cases}
\dfrac{\partial u}{\partial t}(t,x) = \mathcal{L}(u(t,x)) + f(t,x) & (t,x) \in [0,T] \times \Omega \\
\mathcal{B}(u(t,x)) = g(t,x) & (t,x) \in [0,T] \times \partial\Omega
\end{cases}
\tag{4.1}
$$

where the standard initial value condition is replaced by a *T-periodicity* condition,

$$
u(0,x) = u(T,x), \quad x \in \Omega .
\tag{4.2}
$$

The constant T is called the *period* of the solution. We shall usually assume that T is the smallest strictly positive number for which (4.2) is satisfied (unless u is function constant in time in which case (4.2) is satisfied for any T). As before, $\mathcal{L}(.)$ denotes a possibly nonlinear and time-dependent uniformly elliptic operator; $\mathcal{B}(.)$ is a boundary operator, and Ω is the compact spatial domain. It is often convenient and, actually, common practice to extend the domain of the operators $\mathcal{L}(.)$ and $\mathcal{B}(.)$, and of the functions u, f and g, by T-periodicity from $[0,T]$ to the whole of $\mathbb{R}$, for instance, $f(t) = f(t + T)$, $t \in \mathbb{R}$. The T-periodicity condition (4.2) is then replaced by an equivalent condition,

$$u(t,x) = u(t+T,x), \quad (t,x) \in \mathbb{R} \times \Omega . \tag{4.3}$$

The above problem is called a *time-periodic* or *T-periodic* parabolic partial differential equation. If $\mathcal{L}(.)$, $\mathcal{B}(.)$, f and g are independent of the time-variable t, the problem is called *autonomous*. In that case the value of T is unknown and has to be calculated in addition to the periodic solution u. Most of this chapter will deal with *non-autonomous* problems, for which T is known a-priori. Furthermore, we shall usually assume that a solution exists and that it is locally unique. Explicit conditions that guarantee existence and uniqueness are studied elsewhere, e.g. in [107, 124].

Time-periodic differential equations govern the evolution of many systems arising in the physical, biological and social sciences. Consider the oscillations of electrons, heartbeats, vibrations of aircraft wings, the dynamics of economic crises, pulsation of stars, and even certain models of the universe [107, p.86]. The diverse nature and importance of these problems explain the great number of papers and works devoted to this subject. We present a number of examples. The (harmonic) steady-state solution of the one-dimensional heat equation with time-periodic boundary conditions is calculated in [126] by Tee and in [95, 96] by Osborne. In [112, 113], Schippers studies the rotating flow due to an infinite disk performing torsional oscillations at a given angular velocity. This problem is described by the Navier-Stokes equations. By means of the von Karman similarity transformation these equations are reduced to a time-periodic system of parabolic PDEs. In [124], Steuerwalt considers the temperature distribution in a long wire which is stretched across the face of a periodically pulsating meson beam. The same problem is also studied by Hackbusch in [41].

Various techniques for solving time-periodic problems have appeared throughout the literature. We briefly recall these methods in section 4.2 as they will later be compared to a new time-periodic waveform relaxation algorithm. The latter was introduced in the papers [130, 137] for linear problems, and in [136] for nonlinear problems. We briefly review the algorithm in section 4.3. A rigorous convergence analysis of the continuous-time algorithm is presented in section 4.4. The analysis is based on [139]. We show that the convergence of the algorithm is intimately related to the convergence of the corresponding initial value problem, as studied by Miekkala and Nevanlinna in [84]. The discrete-time analogue is considered in section 4.5. As with initial boundary value problems, a multigrid acceleration turns out to be possible. In section 4.6, we take up the analysis of Lubich and Ostermann, [78], and extend their results to the time-periodic case. Finally, we consider the case of *autonomous* problems. In these problems the periodic solution as well as the unknown period T have to be determined. In section 4.7 we report on the use of a modified shooting method which was developed in cooperation with D. Roose, and published in [106].

4.2 Standard time-periodic PDE solvers

Dynamic simulation

A straightforward solution method is based on a dynamic simulation of the studied system. Starting form an arbitrary initial condition — or, better, starting from a physically motivated initial condition such as the fluid state of rest, steady-state temperature distribution, etc. — the system is integrated over a sufficiently large time-interval until the transient effects are eliminated. This approach is taken in [112, 124]. Mathematically, it can be interpreted as Picard's method for computing a fixed point of an operator.

Let $u(t, x; u_0(x))$ denote the solution at time t of (4.1) with the initial condition $u(0, x) = u_0(x)$. Denote by S the operator which maps any $u_0(x)$ to $u(T, x; u_0(x))$, i.e.,

$$S(u_0(x)) \ := \ u(T, x; u_0(x)) \ .$$

Note that S corresponds to the so-called *Poincaré map*. It can easily be seen that the solution of the periodic problem restricted to time-level $t = 0$ corresponds to a fixed point of S, i.e., the solution satisfies $u(0, x) = S(u(0, x))$. If S is a contraction in a suitable Banach space $\mathcal{X}$, the existence of a unique fixed point is guaranteed. The latter may be calculated by successive approximation, i.e., by constructing the Picard sequence,

$$\text{for } u^{(0)}(0, x) \in \mathcal{X}: \quad \text{calculate } u^{(\nu)}(0, x) = S(u^{(\nu-1)}(0, x)), \quad \nu = 1, 2, \ldots \quad (4.4)$$

It is well-known that the computation of a good approximation of a fixed point by successive approximation may be very expensive. In particular, integration over a very large number of periods may be required in the case of slowly decaying transients.

Global discretization

After selection of a *space-time grid*, finite differences or finite elements may be used to discretize the time-periodic problem. This leads to a system of equations, which may be solved by iterative or direct sparse solvers. The dimension of the system is usually very large. E.g., in the case of a problem on a rectangular two-dimensional domain with, say, 100 grid lines in each coordinate direction and a similar number along the time-axis, the number of equations is of the order 10^6. An iterative method for solving this system is used in reference [126]. There, Tee analyses the structure of the coefficient matrix which is obtained when the one-dimensional heat equation with time-periodic boundary conditions is discretized using central finite differences in space and explicit or implicit Euler in time. It is shown that a p-cyclic matrix results, to which a p-cyclic SOR theory applies. The same problem is studied by Osborne in [95, 96], where a specialized direct solver is developed. The use of a direct solver that takes the almost banded structure of the coefficient matrix into account is, for instance, also reported by Holodniok et al. in [57].

Reformulation as a two-point boundary value problem and shooting

After semi-discretization a system of two-point boundary value ordinary differential equations with non-separated boundary conditions results,

$$\frac{d}{dt}u^h(t) = L^h(u^h(t)) + f^h(t) , \quad u^h(0) = u^h(T) . \tag{4.5}$$

The vast literature on two-point boundary value problems is applicable to the above problem. For an overview of this literature see e.g. the recent text-book of Asher, Mattheij and Russell, [3], and the references therein. A particularly popular approach for solving problems like (4.5) is the *shooting* method and also the *multiple shooting* method, which has better stability characteristics. We shall explain the shooting method in some detail in section 4.7, where it is used for solving autonomous periodic problems. The method is basically a Newton-Raphson procedure for calculating the solution, $u^h(0)$, of the *residual equation*,

$$r(u^h(0)) := u^h(0) - u^h(T) = 0 .$$

Evaluation of the residual requires the solution of an initial value problem corresponding to (4.5). In addition, the calculation of a numerical approximation to the Jacobian matrix needed in the Newton steps requires as many initial value ODE-integrations as there are equations. In the case of a semi-discretized parabolic PDE, the ODE system has a very large number of equations, and, consequently, prohibitively many time-integrations of the corresponding initial value problem are needed.

Multigrid method of the second kind

A very fast algorithm for solving time-periodic PDEs was presented by Hackbusch, [41], and by Hemker and Schippers, [52, 113]. They reformulate the periodic problem into an integral equation and apply the *multigrid method of the second kind*. This is no doubt the fastest standard periodic solver; hence we shall discuss it in some more detail. It will further be used to evaluate the performance of our waveform relaxation algorithm. For an in depth analysis we refer to the above papers, and to [44, Ch. 16].

For simplicity, we describe the algorithm for linear problems. Consider (4.1) with the initial condition, $u(0, x) = u_0(x)$. Its solution, restricted to the time-level T, can be written as the outcome of an affine mapping applied to u_0,

$$u_T(x) = \mathcal{T}\, u_0(x) + k(x) .$$

$\mathcal{T}$ is a linear integral operator, so that $\mathcal{T}u_0(x)$ equals the solution at time T of the initial boundary value problem with homogeneous right-hand sides, ($f = 0$ and $g = 0$). The function $k(x)$ equals the solution at time T of the initial boundary value problem with zero initial condition, ($u_0(x) = 0$). With this notation, the periodicity condition (4.2) can be rewritten into the form $u_0(x) = \mathcal{T}u_0(x) + k(x)$. The latter defines a linear Fredholm integral equation of the second kind,

$$y(x) = \mathcal{T}\, y(x) + k(x) , \tag{4.6}$$

where $y(x)$ is an unknown function defined on the *spatial* domain Ω. The determination of y satisfying (4.6) is equivalent to the original problem of finding a u which satisfies the time-periodic equation. Indeed, if y satisfies (4.6), then the solution u of the initial boundary value problem with $u_0 = y$, is the solution of the time-periodic problem.

The integral equation may be solved by the multigrid method of the second kind. In a similar way as in the multigrid method for elliptic equations, (4.6) is discretized on a nested set of grids, Ω^{h_i}, $i = 0, \ldots, k$, which results in a set of discrete equations,

$$Y^{h_i} = T^{h_i} Y^{h_i} + K^{h_i}, \quad \text{on} \quad \Omega^{h_i} . \tag{4.7}$$

The problem on the fine grid Ω^{h_k} is solved by chosing an initial approximation for Y^{h_k} and iteratively calling a procedure like the one outlined in algorithm 4.1. The algorithm has the structure of a multigrid *W-cycle*. It requires the calculation of products like $T^{h_i} Y^{h_i}$ in the smoothing and defect calculation steps. Note, however, that no explicit representation of the discretized integral operator T^{h_i}, a dense matrix, is required. Indeed, application of T^{h_i} to a function Y^{h_i} is equivalent to calculating the solution of one discrete initial boundary value problem defined on Ω^{h_i}. The outcome of $T^{h_i} Y^{h_i}$ may be computed by using standard parabolic solvers, e.g. time-stepping.

In [41], the convergence factor of the algorithm is shown to be of the order $O((h_k)^2)$, where h_k is the fine grid mesh size. The finer the mesh the more rapidly the algorithm converges. (This result requires some mild restrictions on the size of time-increment in order to guarantee a sufficient smoothing behaviour of the time-discretization formula.) As such, one iteration applied to some suitably chosen starting approximation is often sufficient to solve the fine grid problem with small iteration error. It can be shown that the arithmetic complexity of one iteration of the algorithm is similar to that of solving only a few initial boundary value problems on the fine grid.

Algorithm 4.1: Multigrid of the Second Kind

PROCEDURE mgm2nd (i, K^{h_i}, Y^{h_i})

if $i=0$ **then**

    - solve $Y^{h_0} = T^{h_0} Y^{h_0} + K^{h_0}$.

else

    - smoothing: $Y^{h_i} := T^{h_i} Y^{h_i} + K^{h_i}$.

    - compute the defect: $D^{h_i} := Y^{h_i} - T^{h_i} Y^{h_i} - K^{h_i}$.

    - project the defect on $\Omega^{h_{i-1}}$: $K^{h_{i-1}} := I_{h_i}^{h_{i-1}} D^{h_i}$.

    - solve on $\Omega^{h_{i-1}}$: $Y^{h_{i-1}} = T^{h_{i-1}} Y^{h_{i-1}} + K^{h_{i-1}}$.

        repeat 2 times mgm2nd $(i\text{-}1, K^{h_{i-1}}, Y^{h_{i-1}})$, starting with $Y^{h_{i-1}} \equiv 0$.

    - interpolate the correction to Ω^{h_i} and correct : $Y^{h_i} := Y^{h_i} - I_{h_{i-1}}^{h_i} Y^{h_{i-1}}$.

endif

4.3   Time-periodic waveform relaxation

A straightforward extension of standard waveform relaxation leads to a waveform algorithm for solving time-periodic problems. Consider the following system of d first order time-periodic ordinary differential equations,

$$
\begin{cases}
\dfrac{d}{dt}y_1 = f_1(t, y_1, y_2, y_3, \ldots, y_d) \\[2mm]
\dfrac{d}{dt}y_2 = f_2(t, y_1, y_2, y_3, \ldots, y_d) \\[2mm]
\quad \vdots \\[2mm]
\dfrac{d}{dt}y_d = f_d(t, y_1, y_2, y_3, \ldots, y_d)
\end{cases}
\quad \text{with} \quad
\begin{array}{l}
y_1(0) = y_1(T) \\[2mm]
y_2(0) = y_2(T) \\[2mm]
\quad \vdots \\[2mm]
y_d(0) = y_d(T)
\end{array}
\tag{4.8}
$$

As before, the basic waveform idea is to apply a simple iteration on functions. This is performed in such a way that one avoids having to solve the system as a whole. E.g., in the case of the Gauss-Seidel algorithm, the functional iteration reads as follows,

$$
\begin{cases}
\dfrac{d}{dt}y_1^{(\nu)} = f_1(t, y_1^{(\nu)}, y_2^{(\nu-1)}, y_3^{(\nu-1)}, \ldots, y_d^{(\nu-1)}) \\[2mm]
\dfrac{d}{dt}y_2^{(\nu)} = f_2(t, y_1^{(\nu)}, y_2^{(\nu)}, y_3^{(\nu-1)}, \ldots, y_d^{(\nu-1)}) \\[2mm]
\quad \vdots \\[2mm]
\dfrac{d}{dt}y_d^{(\nu)} = f_d(t, y_1^{(\nu)}, y_2^{(\nu)}, y_3^{(\nu)}, \ldots, y_d^{(\nu)})
\end{cases}
\quad \text{with} \quad
\begin{array}{l}
y_1^{(\nu)}(0) = y_1^{(\nu)}(T) \\[2mm]
y_2^{(\nu)}(0) = y_2^{(\nu)}(T) \\[2mm]
\quad \vdots \\[2mm]
y_d^{(\nu)}(0) = y_d^{(\nu)}(T)
\end{array}
\tag{4.9}
$$

This iteration is started with an approximation which is chosen arbitrarily or which is chosen based on experience obtained with related problems. Note that no "natural" initial choice is available here, as it was when solving initial value problems. There the initial approximates were chosen constant and equal to the value of the initial condition.

The complete Gauss-Seidel algorithm is given in Alg. 4.2. The construction of the Jacobi variant is straightforward. An example of a linear time-periodic problem solved with the Gauss-Seidel technique is presented below.

Example 4.3.1 Consider the following 2π-periodic ordinary differential equation:

$$
\frac{d}{dt}\begin{pmatrix} u \\ v \\ w \end{pmatrix}
+ \begin{pmatrix} 2 & 1 & 3 \\ -1 & 3 & 0 \\ 1 & 0 & 3 \end{pmatrix}
\begin{pmatrix} u \\ v \\ w \end{pmatrix}
= \begin{pmatrix} \sin(t) \\ 0 \\ 0 \end{pmatrix}
\quad ; \quad
\begin{pmatrix} u \\ v \\ w \end{pmatrix}(0)
= \begin{pmatrix} u \\ v \\ w \end{pmatrix}(2\pi)
$$

One iteration step of the 2π-periodic Gauss-Seidel waveform relaxation can be formulated as below. (We did not explicitly rewrite the periodicity condition.)

$$
\frac{d}{dt}\begin{pmatrix} u^{(\nu)} \\ v^{(\nu)} \\ w^{(\nu)} \end{pmatrix}
+ \begin{pmatrix} 2 & 0 & 0 \\ -1 & 3 & 0 \\ 1 & 0 & 3 \end{pmatrix}
\begin{pmatrix} u^{(\nu)} \\ v^{(\nu)} \\ w^{(\nu)} \end{pmatrix}
= \begin{pmatrix} 0 & -1 & -3 \\ 0 & 0 & 0 \\ 0 & 0 & 0 \end{pmatrix}
\begin{pmatrix} u^{(\nu-1)} \\ v^{(\nu-1)} \\ w^{(\nu-1)} \end{pmatrix}
+ \begin{pmatrix} \sin(t) \\ 0 \\ 0 \end{pmatrix}
$$

Algorithm 4.2: Time-Periodic Gauss-Seidel Waveform Relaxation

$\nu := 0$; choose $\quad y_i^{(0)}(t)$ for $t \in [0, T]$, $\quad i = 1, \ldots, d$.

repeat

$\qquad \nu := \nu + 1$.

$\qquad$ **for** $i = 1, \ldots, d$

$\qquad\qquad$ solve $\quad \dfrac{d}{dt} y_i^{(\nu)} = f_i(t, y_1^{(\nu)}, \ldots, y_{i-1}^{(\nu)}, y_i^{(\nu)}, y_{i+1}^{(\nu-1)}, \ldots, y_d^{(\nu-1)})$,

$\qquad\qquad$ with $\quad y_i^{(\nu)}(0) = y_i^{(\nu)}(T)$.

$\qquad$ **endfor**

until convergence

The successive iterates are written in the form,

$$\begin{pmatrix} u^{(\nu)} \\ v^{(\nu)} \\ w^{(\nu)} \end{pmatrix} = \begin{pmatrix} u_s^{(\nu)} \sin(t) + u_c^{(\nu)} \cos(t) \\ v_s^{(\nu)} \sin(t) + v_c^{(\nu)} \cos(t) \\ w_s^{(\nu)} \sin(t) + w_c^{(\nu)} \cos(t) \end{pmatrix} \qquad \nu = 0, 1, 2, \ldots$$

The coefficients can be determined analytically. Their values are given below.

ν	$u_s^{(\nu)}$	$u_c^{(\nu)}$	$v_s^{(\nu)}$	$v_c^{(\nu)}$	$w_s^{(\nu)}$	$w_c^{(\nu)}$
0	0.0000	0.0000	0.0000	0.0000	0.0000	0.0000
1	0.4000	-0.2000	0.1000	-0.1000	-0.1000	0.1000
2	0.4400	-0.3200	0.1000	-0.1400	-0.1000	0.1400
3	0.4240	-0.3520	0.0920	-0.1480	-0.0920	0.1480
4	0.4144	-0.3552	0.0888	-0.1480	-0.0888	0.1480
5	0.4118	-0.3539	0.0882	-0.1474	-0.0882	0.1474
6	0.4116	-0.3532	0.0882	-0.1471	-0.0882	0.1471
7	0.4117	-0.3529	0.0882	-0.1471	-0.0882	0.1471
8	0.4117	-0.3529	0.0882	-0.1471	-0.0882	0.1471

Note that the direct calculation of the analytical solution would require solving a linear system of dimension six. This is entirely avoided in the waveform relaxation process in which a sequence of smaller systems of size two by two have to be solved. (In each of the three steps of each iteration two coefficients are updated.)

At any stage of the algorithm the computational task consists of solving a single first-order two-point boundary-value problem of the form,

$$\frac{d}{dt}y = f(t,y) \quad \text{with} \quad y(0) = y(T) \,. \tag{4.10}$$

This equation may be solved for instance by dynamic simulation, by shooting or by discretization. The latter is particularly efficient, here. Since we deal with a single equation, the dimension of the resulting algebraic system is small. In the general nonlinear case, discretization leads to a nonlinear system of equations, which may be solved by a Newton-Raphson procedure. Alternatively, (4.10) may be linearized first around an approximation, $\bar{y}$, of the solution — e.g. the previous iterate —,

$$\frac{d}{dt}y = f(t,\bar{y}) + f_y(\bar{y})(y - \bar{y}) \quad \text{with} \quad y(0) = y(T) \,, \tag{4.11}$$

and solved next, by discretization and the use of a linear system solver. This linearization process may be applied iteratively, until the solution of (4.11) has converged to the solution of (4.10). Or, linearization may be applied only once, which results in a waveform relaxation Newton type iteration. Numerical experiments suggest no performance degradation in the latter case.

In each of the Newton-Raphson approaches it is imperative for convergence to start with a good initial approximation. The latter may be obtained from the solution of a related problem through a continuation procedure, or, in the case of a semi-discretized parabolic partial differential equation, by interpolation of a coarse grid solution.

Finally, we mention the possibility of several further variants, such as blockwise iterations, overlapping-block iterations, successive overrelaxation, etc. The precise algorithmic formulation of the variants is straightforward, and therefore omitted.

4.4 Analysis of the continuous-time iteration

4.4.1 The linear model problem

We consider a linear constant-coefficient time-periodic ordinary differential equation,

$$\frac{d}{dt}x + Ax = f, \quad \text{with} \quad x(0) = x(T) \quad t \in [0,T] \,, \tag{4.12}$$

or, after extension of the domain by T-periodicity,

$$\frac{d}{dt}x + Ax = f, \quad \text{with} \quad x(t) = x(t+T), \quad t \in \mathbb{R} \,. \tag{4.13}$$

A is a complex $d \times d$-matrix. f belongs to a Banach space of functions defined on $[0,T]$, e.g., the space of continuous functions $C([0,T]; \mathbb{C}^d)$, or the space of Lesbesgue measurable functions $L_p([0,T]; \mathbb{C}^d)$. A solution x is defined as an absolutely continuous function which satisfies the T-periodicity condition and which satisfies the differential equation a.e., see e.g. [24, p. 122].

The solution to the differential equation in (4.12) can be written in function of the unknown value of the solution at $t = 0$, as follows,

$$x(t) = e^{-tA}x(0) + \int_0^t e^{(s-t)A} f(s)\, ds \; . \tag{4.14}$$

Using the periodicity condition, $x(0) = x(T)$, we get

$$x(0) = e^{-TA}x(0) + \int_0^T e^{(s-T)A} f(s)\, ds \; ,$$

$$(I - e^{-TA})\, x(0) = \int_0^T e^{(s-T)A} f(s)\, ds \; , \tag{4.15}$$

from which we may formally derive the value of $x(0)$,

$$x(0) = (I - e^{-TA})^{-1} \int_0^T e^{(s-T)A} f(s)\, ds \; .$$

This is used to eliminate $x(0)$ from (4.14) so that we obtain,

$$x(t) = e^{-tA}(I - e^{-TA})^{-1} \int_0^T e^{(s-T)A} f(s)\, ds + \int_0^t e^{(s-t)A} f(s)\, ds \; . \tag{4.16}$$

Note that the same result is found when the derivation is based on the alternative periodicity condition, $x(t) = x(t + T)$.

The questions of existence and uniqueness of problems like (4.12) are studied in great generality in references like [79, Ch. 11]. There, A is assumed to be a general time-periodic operator belonging to a Banach algebra of linear bounded operators. Our lemmata given below may be derived from theorem 112.b in [79]. However, we do not need all of the technical machinery of [79] for our present purposes. Consequently, we give some elementary proofs of the basic properties that we shall need further on. First, we introduce a concept which plays a fundamental role in the analysis.

Definition 4.4.1 ([107]) *A complex matrix A is called non-critical w.r.t. T if the differential equation $\frac{d}{dt}x + Ax = 0$ admits no T-periodic solution except the zero solution. The matrix is called critical w.r.t. T otherwise.*

In the remainder of this chapter we shall often tacitly assume that we deal with T-periodic functions and equations, and therefore omit the "w.r.t. T". The following lemma characterizes non-critical matrices. Let $\sigma(A)$ denote the spectrum of matrix A.

Lemma 4.4.1 *A complex matrix A is non-critical w.r.t. T if and only if*

$$\forall n \in \mathbb{Z} : in\omega \notin \sigma(A), \quad \text{with} \quad \omega = 2\pi/T \quad i = \sqrt{-1} \; .$$

Proof . Consider (4.12) with zero right-hand side, $f \equiv 0$. By (4.15), its solution restricted to $t = 0$ satisfies the homogeneous equation, $(I - e^{-TA})x(0) = 0$. This equation has a unique solution, $x(0) = 0$ if and only if $\det(I - e^{-TA}) \neq 0$. Consequently, $\frac{d}{dt}x + Ax = 0$ has the trivial T-periodic solution only, iff $1 \notin \sigma(e^{-TA})$. The latter is equivalent to the condition $\forall \lambda \in \sigma(A) : 1 \neq e^{-T\lambda}$, from which the condition in the lemma results. $\square$

Lemma 4.4.2 *For any $f \in L_1(0,T)$ [1], the solution to the T-periodic differential equation $\frac{d}{dt}x + Ax = f$ exists and is unique if and only if A is non-critical w.r.t. T.*

Proof . The necessity of the condition is, of course, obvious. If for any $n \in \mathbb{Z} : in\omega \in \sigma(A)$, the homogeneous equation has an infinite number of T-periodic solutions. Therefore, a solution of $\frac{d}{dt}x + Ax = f$ cannot be unique.

The sufficiency part follows from the observation that $\det(I - e^{-TA}) \neq 0$ if and only if A is non-critical. In that case, the solution is given by (4.16), which is well-defined and unique for any f.   □

4.4.2   The time-periodic waveform iteration

In the time-periodic waveform iteration method a splitting (P, Q) is applied to A, i.e., $A = P - Q$. This splitting defines the iteration scheme,

$$\frac{d}{dt}x^{(\nu)} + Px^{(\nu)} = Qx^{(\nu-1)} + f, \quad \text{with} \ \ x^{(\nu)}(0) = x^{(\nu)}(T) , \qquad (4.17)$$

where $x^{(0)}$ is an arbitrary starting iterate. The solution to (4.17) can be constructed in a similar way as we constructed the solution to the problem in the previous section. This immediately leads to the following relation between successive iterates[2],

$$x^{(\nu)} = \bar{\mathcal{K}}x^{(\nu-1)} + \varphi , \qquad (4.18)$$

$$\text{with} \ \ \bar{\mathcal{K}}x\,(t) = e^{-tP}(I - e^{-TP})^{-1}\int_0^T e^{(s-T)P}Qx(s)\,ds + \int_0^t e^{(s-t)P}Qx(s)\,ds \quad (4.19)$$

$$\varphi(t) = e^{-tP}(I - e^{-TP})^{-1}\int_0^T e^{(s-T)P}f(s)\,ds + \int_0^t e^{(s-t)P}f(s)\,ds \quad (4.20)$$

Formula (4.18) formally defines the solution of the time-periodic problem. However, the formula is mainly of theoretical interest and will generally not be used in actual computations. Indeed, as before, the splitting matrices are chosen in such a way that they lead to simplified calculations. Examples are the familiar Jacobi and SOR splittings, which reduce the problem of solving a time-periodic system of d equations into a problem of solving d separate time-periodic equations in one unknown (repeatedly, until convergence). Each of these equations is of the type,

$$\frac{d}{dt}x_i^{(\nu)} + p_{i,i}\,x_i^{(\nu)} = g_i , \quad \text{with} \ \ x_i^{(\nu)}(0) = x_i^{(\nu)}(T) , \qquad (4.21)$$

where $x_i^{(\nu)}$ is the i-th component of $x^{(\nu)}$; $p_{i,i}$ is the i-th diagonal element of P, and g_i is a linear combination of known functions, which are either a given component of f, solution components of $x^{(\nu-1)}$ or solution components of $x^{(\nu)}$ already computed in the current iteration. As with the initial value iteration, the time-periodic SOR iteration is again a sequential process, whereas the JOR is fully parallel.

---

[1] Remark that $C[0,T] \subset L_p(0,T) \subset L_1(0,T)$, $1 < p \leq \infty$.
[2] We use the "—"-notation to denote T-periodic operators ($\bar{\mathcal{K}}$, $\bar{\mathcal{M}}$, ...)

Contrary to the case of solving initial value problems, in the time-periodic case not every splitting defines a sequence of iterates. The theorem below sets the condition for acceptable splittings. For this, the following definition comes in handy.

Definition 4.4.2 *A splitting (P, Q) of a complex matrix is called non-critical w.r.t. T if P is a non-critical matrix w.r.t. T.*

Theorem 4.4.3 *The time-periodic waveform iteration defined by the splitting (P, Q) and the functions $x^{(0)}, f \in L_1(0, T)$ is well-defined, i.e., the iterates $x^{(\nu)}, \nu = 1, 2, \ldots$, exist and are unique if and only if the splitting is non-critical.*

Proof . The proof is an immediate consequence of lemma 4.4.2. Since $x^{(0)}, f \in L_1(0, T)$ and since P is non-critical, existence and uniqueness of $x^{(1)}$ follow. Moreover, since $x^{(1)}$ is continuous on [0,T], it is integrable and belongs to $L_1(0, 1)$. The existence and uniqueness of the other iterates follow by induction. $\quad\square$

4.4.3 The time-periodic integral operator

By construction, $\bar{\mathcal{K}}$ is the linear operator which maps the function v to the function w, where v and w are related by the equation $\frac{d}{dt}w + Pw = Qv$, $w(0) = w(T)$. Some other ways of representing and characterizing $\bar{\mathcal{K}}$ are formulated below. They are helpful for a better understanding of the nature of $\bar{\mathcal{K}}$ and they will come in useful in the remainder of the analysis. They may be proven either directly, starting from the differential equation that defines $\bar{\mathcal{K}}$, or by applying some algebra to formula (4.19). We have opted for the latter. The proofs make use of some elementary matrix properties which are stated and proven below.

Property 4.4.4 *Let A denote a complex matrix and $\alpha, \beta \in \mathbb{C}$; assume that the inverses in the formulae below exist, then*

$$
\begin{aligned}
&a. \quad e^{(\alpha+\beta)A} = e^{\alpha A} e^{\beta A} \\
&b. \quad e^{\beta A}(I - e^{\alpha A})^{-1} = (I - e^{\alpha A})^{-1} e^{\beta A} \\
&c. \quad e^A(I - e^A)^{-1} + I = (I - e^A)^{-1} \\
&d. \quad \int_0^T e^{-tA}\, dt = A^{-1}(I - e^{-TA})
\end{aligned}
$$

Proof .

    a. The result is an immediate consequence of the more general property proven in [145, p. 255]: "$e^{A+B} = e^A e^B$ if $AB = BA$".

    b. $e^{\beta A}(I - e^{\alpha A})^{-1} = ((I - e^{\alpha A})e^{-\beta A})^{-1} = (e^{-\beta A}(I - e^{\alpha A}))^{-1} = (I - e^{\alpha A})^{-1}e^{\beta A}$

    c. l.h.s. $= e^A(I - e^A)^{-1} + (I - e^A)(I - e^A)^{-1} = (e^A + I - e^A)(I - e^A)^{-1} =$ r.h.s.

    d. $\int_0^T e^{-tA}\, dt = A^{-1}(-e^{-tA})|_0^T = A^{-1}(I - e^{-TA})$

$\quad\square$

Theorem 4.4.5 $\bar{\mathcal{K}}$ *is a periodic convolution operator,*

$$\bar{\mathcal{K}}x\,(t) = \bar{k} \star x\,(t) := \frac{1}{T} \int_0^T \bar{k}(t-s)\,x(s)\,ds \qquad (\,t \in I\!\!R\,) \tag{4.22}$$

with a T-periodic matrix-valued kernel, $\bar{k}$, of which the restriction to $[0,T)$ is given by

$$\bar{k}_{\|[0,T)}(t) = T e^{-tP}(I - e^{-TP})^{-1}Q \quad (\,t \in [0,T)\,) \tag{4.23}$$

Proof . Assume first that $0 \le t < T$.

We recall formula (4.19), which defines the operator $\bar{\mathcal{K}}$

$$\bar{\mathcal{K}}x\,(t) = e^{-tP}(I - e^{-TP})^{-1} \int_0^T e^{(s-T)P}Qx(s)\,ds + \int_0^t e^{(s-t)P}Qx(s)\,ds \ .$$

With application of prop. 4.4.4.a and 4.4.4.b the right-hand side can be rewritten,

$$\int_0^T e^{(s-t)P}e^{-TP}(I - e^{-TP})^{-1}\,Qx(s)\,ds + \int_0^t e^{(s-t)P}Qx(s)\,ds \ .$$

By splitting the first integral into two parts, $0 \le s \le t$ and $t < s \le T$, and adding the first term to the second integral we get,

$$\int_0^t e^{(s-t)P}(e^{-TP}(I-e^{-TP})^{-1}+I)\,Qx(s)\,ds + \int_t^T e^{(s-t)P}e^{-TP}(I-e^{-TP})^{-1}\,Qx(s)\,ds.$$

The left-hand term can be rewritten by application of property 4.4.4.c,

$$\int_0^t e^{-(t-s)P}(I - e^{-TP})^{-1}Qx(s)\,ds + \int_t^T e^{-(T+t-s)P}(I - e^{-TP})^{-1}Qx(s)\,ds \ .$$

As $0 \le t-s \le t < T$ (first integral) and $0 \le t \le T+t-s \le T$ (second integral),

$$\bar{\mathcal{K}}x\,(t) = \frac{1}{T} \int_0^t \bar{k}_{\|[0,T)}(t-s)x(s)\,ds + \frac{1}{T} \int_t^T \bar{k}_{\|[0,T)}(T+t-s)x(s)\,ds \ ,$$
$$= \frac{1}{T} \int_0^T \bar{k}(t-s)x(s)\,ds \ .$$

This proves the equivalence of $\bar{\mathcal{K}}x(t)$ and $\bar{k} \star x(t)$ on $[0,T)$. Taking into account that $\bar{\mathcal{K}}x(t)$ and $\bar{k} \star x(t)$ are T-periodic proves their equivalence for $t \in I\!\!R$. $\qquad \square$

Note that $\bar{k}(t)$ is a bounded function which is continuous except at integer multiples of T. Note further that the theorem may also be proven by showing that $\bar{k} \star x\,(0) = \bar{k} \star x\,(T)$ and verifying that the convolution satisfies the defining differential equation of $\bar{\mathcal{K}}$.

Theorem 4.4.6 *The Fourier series coefficients of the kernel $\bar{k}$, $\{\bar{\bar{k}}_n\}_{n=-\infty}^{\infty}$, satisfy*

$$\bar{\bar{k}}_n = (in\omega I + P)^{-1}Q \ , \qquad \omega = 2\pi/T \ , \quad n \in Z\!\!\!Z \ . \tag{4.24}$$

Proof . Following the definition of the Fourier series coefficients, we get

$$\tilde{\bar{k}}_n := \frac{1}{T} \int_0^T \bar{k}(t) \, e^{-in\omega t} \, dt, \qquad \omega = 2\pi/T \, , \quad n \in \mathbb{Z}$$

$$= \int_0^T e^{-tP}(I - e^{-TP})^{-1} Q \, e^{-in\omega t} \, dt$$

$$= \int_0^T e^{-(in\omega I + P)t} \, dt \, (I - e^{-TP})^{-1} Q$$

Since P is non-critical, the matrix $(in\omega I + P)$ is invertible, and we may invoke property 4.4.4.d to arrive at,

$$\tilde{\bar{k}}_n = (in\omega I + P)^{-1}(I - e^{-(in\omega I + P)T})(I - e^{-TP})^{-1} Q$$

$$= (in\omega I + P)^{-1}(I - e^{-PT})(I - e^{-TP})^{-1} Q$$

$$= (in\omega I + P)^{-1} Q \, .$$

$$\square$$

Remark 4.4.1 With use of $K(z)$, defined in (2.37), we get,

$$\tilde{\bar{k}}_n = K(in\omega) \, . \tag{4.25}$$

Theorem 4.4.7 $\bar{\mathcal{K}}$ *is a linear Fredholm integral operator,*

$$\bar{\mathcal{K}}x\,(t) = \int_0^T G(t,s)\,x(s)\,ds \tag{4.26}$$

$$with \qquad G(t,s) = \begin{cases} e^{(s-t)P}(I - e^{-TP})^{-1} Q & 0 \le s \le t \le T \\ e^{(s-t-T)P}(I - e^{-TP})^{-1} Q & 0 \le t < s \le T \end{cases} \tag{4.27}$$

Proof . By definition of $G(t,s)$ it follows that $\int_0^T G(t,s)x(s)\,ds$ can be expanded as,

$$\int_0^t e^{(s-t)P}(I - e^{-TP})^{-1} Qx(s)\,ds + \int_t^T e^{(s-t-T)P}(I - e^{-TP})^{-1} Qx(s)\,ds \, .$$

This is equivalent to an expression found at the end of the proof of Th. 4.4.5. $\square$

Note that $G(t,s)$ is bounded on $[0,T] \times [0,T]$ and continuous everywhere except for the line $s = t$. This is a particular case of a so-called *potential* kernel.

4.4.4   Convergence of the time-periodic waveform relaxation

The error to the solution of (4.17), $e^{(\nu)} = x^{(\nu)} - x$, solves the time-periodic problem,

$$\frac{d}{dt}e^{(\nu)} + Pe^{(\nu)} = Qe^{(\nu-1)} \, , \quad \text{with } e^{(\nu)}(0) = e^{(\nu)}(T) \, .$$

This leads to a linear relation between the errors of successive iterates, $e^{(\nu)} = \bar{\mathcal{K}}e^{(\nu-1)}$. By theorem 2.5.2, the error in the time-periodic waveform iteration will go to zero when the spectral radius of $\bar{\mathcal{K}}$ is smaller than one. Note that the boundedness of $\bar{\mathcal{K}}$ is immediately verified by considering a standard result on the boundedness of convolution operators. The theorem below is given for convolution operators in L_p-spaces, but a similar theorem holds for the space of continuous functions.

Theorem 4.4.8 ([153], p.235) *Let the functions $f(x)$ and $k(x)$ have the period 1; let $k(x) \in L_1(0,1)$ and $f(x) \in L_p(0,1)$ $(1 \leq p \leq \infty)$. Then the periodic convolution $h(x) = \int_0^1 k(x-y)f(y)\,dy$ belongs to $L_p(0,1)$ and $\| h \|_p \leq \| k \|_1 \| f \|_p$.*

Corollary 4.4.9 *The time-periodic operator is bounded in $L_p(0,T)$ and $C[0,T]$.*

The spectral radius of $\bar{\mathcal{K}}$ may now be calculated in a similar way as the spectral radius of the corresponding initial value waveform relaxation operator is calculated in [84]. However, the special nature of $\bar{\mathcal{K}}$ as a convolution operator allows a substantial simplification. The latter is based on the property of *compactness*, see e.g. [39, p. 237].

Definition 4.4.3 *A linear operator $\mathcal{U}$ in a Banach space $\mathcal{X}$ is called compact if for every sequence $\{x_n\}$ in $\mathcal{X}$ with $\| x_n \|_{\mathcal{X}} = 1$, the sequence $\{\mathcal{U}x_n\}$ has a convergent subsequence.*

The compactness of $\bar{\mathcal{K}}$ is established by the following theorem.

Theorem 4.4.10 ([154], p.320) *Consider the linear convolution operator $\mathcal{U}$, defined by $\mathcal{U}f(x) = \int_0^T k(x-y)f(y)\,dy$, $0 < x < T < \infty$. If the kernel is integrable on $(-T,T)$ then $\mathcal{U}$ is compact on any of the spaces $L_p(0,T)$ $(1 \leq p \leq \infty)$ and $C[0,T]$.*

Corollary 4.4.11 *The time-periodic operator is compact in $L_p(0,T)$ and $C[0,T]$.*

The compactness result may also be derived by considering $\bar{\mathcal{K}}$ as a Fredholm integral operator with bounded potential kernel, and by considering theorems that discuss the compactness of such operators, see e.g. [67, p. 328 and p. 397]. Compact operators may in some sense be regarded as infinite-dimensional extensions of the familiar finite-dimensional linear operators. Indeed, some of their characteristics are very similar. An important example is given by the theorem below, which discusses the nature of the spectrum of a linear compact operator.

Theorem 4.4.12 ([154], Th. 3.1) *Let $\mathcal{U}$ be a compact operator acting in an infinite-dimensional space. Then,*

a. *The spectrum of $\mathcal{U}$ consists of zero, and a denumerable set $\lambda_1, \lambda_2, \ldots$, with zero as its only accumulation-point*

b. *every non-zero point λ of the spectrum is an eigenvalue of finite multiplicity.*

We may now state our main result, which relates the spectral radius of $\bar{\mathcal{K}}$, an operator in an infinite-dimensional space, to the spectral radii of the *Fourier series coefficients* of its T-periodic kernel. It is the time-periodic equivalent of the initial value result in theorem 2.5.6, which related the spectral radius of $\mathcal{K}$ to the spectral radius of the *Laplace transform* of its kernel.

Theorem 4.4.13 *Consider $\bar{\mathcal{K}}$ as an operator in $L_p(0,T)$, $1 \leq p \leq \infty$, or $C[0,T]$. Assume (P,Q) is a non-critical splitting. Then,*

$$\rho(\bar{\mathcal{K}}\,) = \max_{n \in \mathbb{Z}} \rho(\,K(in\omega)\,) \quad \text{with} \quad \omega = 2\pi/T \,. \tag{4.28}$$

Proof . Since $\bar{\mathcal{K}}$ is a compact operator, its spectrum, $\sigma(\bar{\mathcal{K}})$, consists of zero and eigenvalues only. The latter can be calculated explicitly as follows.

Let λ be an eigenvalue of $\bar{\mathcal{K}}$ and let x be a non-trivial function which satisfies $\bar{\mathcal{K}}x = \lambda x$. By calculating the Fourier series coefficients of left- and right-hand sides, and taking formula (4.25) into account, we come to the following identity,

$$K(in\omega)\tilde{x}_n = \lambda \tilde{x}_n \ , \quad \forall n \in \mathbb{Z} \ , \tag{4.29}$$

where $\tilde{x}_n$ is the n'th Fourier coefficient of x. (Note that the Fourier coefficients are well-defined for any integrable function.) The left-hand side is derived by applying the Fourier series convolution property (if $f, g \in L_1(0, T)$: $\widetilde{(f \star g)}_n = \tilde{f}_n \tilde{g}_n$). By (4.29), any eigenvalue of $\bar{\mathcal{K}}$ is also an eigenvalue of a matrix $K(in\omega)$, when a non-zero vector $\tilde{x}_n$ exists. The latter must be the case for at least one value of n. Otherwise, x would be equal a.e. to the zero-function.

Conversely, any eigenvalue-eigenvector pair (λ_n, X_n) of the matrix $K(in\omega)$ defines an eigenfunction of $\bar{\mathcal{K}}$ with eigenvalue λ_n. Indeed, consider the function,

$$x_n(t) = X_n e^{in\omega t} \ (\in C[0, T]) \ .$$

This is an eigenfunction since $\bar{\mathcal{K}}x_n(t) = \lambda_n x_n(t)$. This is readily checked by Fourier transforming the left- and right-hand sides.

By consequence, the spectrum of $\bar{\mathcal{K}}$ may be written explicitly as

$$\sigma(\bar{\mathcal{K}}) = \bigcup_{n \in \mathbb{Z}} \sigma(K(in\omega)) \bigcup \{0\} \ , \tag{4.30}$$

from which the conclusion follows. (Note the use of "max" instead of "sup", since $\sigma(\bar{\mathcal{K}})$ is bounded with zero as the only point of accumulation.) $\square$

Remark 4.4.2 $\rho(\bar{\mathcal{K}})$ is the decay factor of the most slowly decaying harmonic in the Fourier expansion of the error. (This follows from $\tilde{e}_n^{(\nu)} = K(in\omega)\tilde{e}_n^{(\nu-1)}$.)

Remark 4.4.3 When (P, Q) defines a Jacobi- or Gauss-Seidel splitting, it follows that $\rho((in\omega I + P)^{-1}Q)$ is equal to the spectral radius of the static iteration for solving a linear system of equations with coefficient matrix "$in\omega I + A$".

The splitting $A = P - Q$ defines a classical (static) iterative method for solving the linear system of equations $Ax = b$, see eq. (2.30). Its convergence properties are related to those of the time-periodic waveform iteration in the following corollary.

Corollary 4.4.14 *Let (P, Q) be a non-critical splitting. It then follows that the spectral radius of the time-periodic operator is bounded from below by the spectral radius of the corresponding static iteration operator.*

Proof . Since P is non-critical, $\rho(\bar{\mathcal{K}})$ exists and P is invertible. Since the spectral radius of the static iteration operator is given by $\rho(P^{-1}Q)$, the conclusion follows.

$\square$

By comparing the result of theorem 4.4.13 to the formula (2.38) we can state an important corollary which links the convergence properties of the standard waveform relaxation method to those of the newly defined time-periodic method.

Corollary 4.4.15 *Let the splitting be such that all of the eigenvalues of P have positive real parts. It then follows that $\rho(\bar{\mathcal{K}}) \leq \rho(\mathcal{K})$, that is, the spectral radius of the waveform relaxation operator for solving the T-periodic ordinary differential equation is bounded by the spectral radius of the infinite-interval waveform relaxation operator for solving the corresponding initial value problem.*

Proof . The condition on the eigenvalues of P is necessary (and sufficient) to guarantee the validity of theorem 2.5.6. It also ensures that the splitting is non-critical, so that $\rho(\bar{\mathcal{K}})$ exists. The validity of corollary then follows immediately by comparison of formula (4.28) with the formula (2.38).   □

A necessary condition for convergence of the time-periodic iteration is the convergence of the corresponding static iteration. A sufficient condition is given by the convergence of the corresponding initial value waveform relaxation. As an immediate consequence several of the convergence related results of the reference [84] may be transposed to the time-periodic case. We shall not repeat these convergence results but refer the reader to the above reference. Note that the spectral radii of $\bar{\mathcal{K}}$ and $\mathcal{K}$ coincide when the maximum in the formula for $\rho(\mathcal{K})$ is found in a value $\xi = n\omega$. The case $n = 0$ is not uncommon. An example of this will be given next. Let A be a consistently ordered matrix with positive constant diagonal $D = \delta I$. We decompose A as $-L + D - U$ where L and U are strictly lower and upper triangular matrices. We set $\bar{\mathcal{K}}_{GS}$ and $\mathcal{K}_{GS}$ to denote the time-periodic and initial value operators corresponding to the Gauss-Seidel splitting of A. Analogously, we define $\bar{\mathcal{K}}_{JAC}$ and $\mathcal{K}_{JAC}$ corresponding to the Jacobi splitting.

Corollary 4.4.16

$$\rho(\bar{\mathcal{K}}_{GS}) = \rho((D-L)^{-1}U) = (\rho(D^{-1}(L+U))^2 = (\rho(\bar{\mathcal{K}}_{JAC}))^2 \qquad (4.31)$$

Proof . In the case of a Jacobi splitting or a Gauss-Seidel splitting, the maximum of $\rho((i\xi I + P)^{-1}Q)$ is found at the origin, [84, Corollary 4.1]. Therefore, $\rho(\bar{\mathcal{K}}_{GS}) = \rho(\mathcal{K}_{GS})$ and $\rho(\bar{\mathcal{K}}_{JAC}) = \rho(\mathcal{K}_{JAC})$. The remainder follows from corollary 2.5.7. □

In the case of an SOR-splitting with sufficiently large overrelaxation parameter the maximum is taken for some value ξ away from zero. In that case $\rho(\bar{\mathcal{K}}_{SOR}) \leq \rho(\mathcal{K}_{SOR})$ where the inequality is strict unless the maximum is attained at an integer multiple of ω.

Finally, we derive a formula for the norm of $\bar{\mathcal{K}}$. As with the spectral radius, also the norm is bounded from below and from above by the norms of the corresponding static iteration and initial value waveform operators.

Theorem 4.4.17 *Denote by $\| \, . \, \|_2$ the L_2-norm, and by $\| \, . \, \|$ the standard Euclidean norm. Consider $\bar{\mathcal{K}}$ as an integral operator in $L_2(0, T)$. Assume P is non-critical. Then,*

$$\| \bar{\mathcal{K}} \|_2 = \max_{n \in \mathbb{Z}} \| K(in\omega) \| \, . \qquad (4.32)$$

Proof . We consider the L_2-space as the proof is based on the Parseval relation. The latter relates the L_2-norm of a function to the l_2-norm of its Fourier coefficients,

$$\| f \|_2^2 := \int_0^T \| f(t) \|^2 \, dt = T \sum_{n=-\infty}^{+\infty} \| \tilde{f}_n \|^2 \; .$$

Application of Parseval's equality to $\bar{\mathcal{K}}x$, with $x \in L_2$, gives,

$$\| \bar{\mathcal{K}}x \|_2^2 = T \sum \| \tilde{\bar{k}}_n \tilde{x}_n \|^2 \; .$$

This can be bounded as follows.

$$\| \bar{\mathcal{K}}x \|_2^2 \le T \sum \| \tilde{\bar{k}}_n \|^2 \| \tilde{x}_n \|^2 \le (\sup \| \tilde{\bar{k}}_n \|)^2 \, T \sum \| \tilde{x}_n \|^2 = (\sup \| \tilde{\bar{k}}_n \|)^2 \, \| x \|_2^2 \; .$$

Consequently, $\| \bar{\mathcal{K}} \|_2 \le \sup \| \tilde{\bar{k}}_n \|$. The "sup" may be replaced by "max", that is, $\| \bar{\mathcal{K}} \|_2 \le \max \| \tilde{\bar{k}}_n \|$, since $\{ \| \tilde{\bar{k}}_n \| \}_{n=-\infty}^{\infty} = \{ \| (in\omega I + P)^{-1} Q \| \}_{n=-\infty}^{\infty}$ is bounded and has zero as the only point of accumulation.

On the other hand, let the maximum be obtained for $\tilde{\bar{k}}_N$, and choose $\tilde{x}_N$ as a vector with $\| \tilde{x}_N \| = 1$ and $\| \tilde{\bar{k}}_N \tilde{x}_N \| = \| \tilde{\bar{k}}_N \|$, then,

$$\| \bar{\mathcal{K}} \|_2 \ge \frac{\| \bar{\mathcal{K}} \, \tilde{x}_N e^{iN\omega t} \|_2}{\| \tilde{x}_N e^{iN\omega t} \|_2} = \frac{\| \tilde{\bar{k}}_N \tilde{x}_N e^{iN\omega t} \|_2}{\| \tilde{x}_N e^{iN\omega t} \|_2} = \frac{\| \tilde{\bar{k}}_N \tilde{x}_N \|}{\| \tilde{x}_N \|} = \| \tilde{\bar{k}}_N \| \; .$$

Therefore, $\quad \| \bar{\mathcal{K}} \|_2 = \| \tilde{\bar{k}}_N \| := \max_{n \in \mathbb{Z}} \| K(in\omega) \| \; .$ $\qquad \square$

4.4.5 The existence of convergent and divergent splittings

We shall complete the discussion on the continuous-time time-periodic waveform iteration with an existence result. One additional definition is formulated first.

Definition 4.4.4 *A non-critical splitting (P, Q) is called convergent (w.r.t. T) if the spectral radius of the T-periodic waveform relaxation operator satisfies $\rho(\bar{\mathcal{K}}) < 1$. It is called non-convergent otherwise.*

The precise meaning of this definition should be thought about by regarding theorem 2.5.2 and property 2.5.3. When the splitting is convergent, the time-periodic iteration will converge for any right-hand side f and any starting approximation $x^{(0)}$. When the splitting is non-convergent ($\rho(\bar{\mathcal{K}}) \ge 1$), there exist f and $x^{(0)}$ for which there will be no convergence. E.g., the iteration may diverge, or there may be a periodic cycle of recurring errors.

The basic existence result is stated next. We refer to our report [138] for a constructive (but rather technical) proof.

Theorem 4.4.18 *Any non-critical complex matrix admits convergent and non-convergent non-critical splittings.*

Remark 4.4.4 When solving intial value problems on $[0, T]$ every splitting is convergent (since $\rho(\mathcal{K}) = 0$). This is obviously not the case for the time-periodic iteration.

4.5 Analysis of the discrete-time iteration

4.5.1 Discretization and solution of the model problem

We fix the time-increment $\tau = T/N$ and formally discretize the differential equation of the model problem (4.12) by a linear multistep method so that we obtain the following difference equation,

$$\frac{1}{\tau} \sum_{j=0}^{k} \alpha_j\, x_{n-k+j} + \sum_{j=0}^{k} \beta_j\, A\, x_{n-k+j} = \sum_{j=0}^{k} \beta_j\, f_{n-k+j} \; . \tag{4.33}$$

(We use a slightly different (but equivalent) notation from the one in (2.42) in order to simplify some further computations.) The introduction of some additional notations will allow us to reformulate this equation into a more suitable form. Let f_τ denote the infinite $\mathbb{C}^d$-valued sequence $\{f_n\}_{n=-\infty}^{\infty}$, with $f_n = f(n\tau)$. Similarly, let x_τ denote the sequence $\{x_n\}_{n=-\infty}^{\infty}$, where x_n is the $\mathbb{C}^d$-vector which approximates $x(n\tau)$. If we write au_τ for the sequence $\{\sum_{j=0}^{k} \alpha_j u_{n-k+j}\}_{n=-\infty}^{\infty}$ and bu_τ for $\{\sum_{j=0}^{k} \beta_j u_{n-k+j}\}_{n=-\infty}^{\infty}$, we may rewrite (4.33) as,

$$\tfrac{1}{\tau} a x_\tau + b A x_\tau = b f_\tau \; . \tag{4.34}$$

The T-periodicity condition of the continuous-time problem is taken into account by requiring N-*periodicity* of the sequences f_τ and x_τ. By this we mean that,

$$\forall n \in \mathbb{Z}: \quad f_{n+N} = f_n \;\; \text{and} \;\; x_{n+N} = x_n \; . \tag{4.35}$$

As such, equation (4.34) becomes an equation in the space of N-periodic $\mathbb{C}^d$-sequences. The latter is a finite-dimensional Banach space, equivalent to $l_p([0..N-1]; \mathbb{C}^d)$. Note that its dimension equals Nd.

The main mathematical tool for analysing existence and uniqueness of a solution to (4.34) will be the concept of a *discrete-time Fourier series* (DFS), see e.g. [100].

Definition 4.5.1 *The DFS transform-pair of an N-periodic complex sequence f_τ is given by*

$$f_n = \sum_{m=0}^{N-1} \hat{f}_m e^{inm(2\pi/N)} \;\; \text{with} \;\; \hat{f}_m = \frac{1}{N} \sum_{n=0}^{N-1} f_n e^{-inm(2\pi/N)} \;\; (i = \sqrt{-1}) \; . \tag{4.36}$$

In the case of a $\mathbb{C}^d$-valued sequence the formulae are applied componentwise. Formula (4.36) basically states that any N-periodic sequence may be uniquely expressed as a linear combination of N "basic" N-periodic sequences, $\varepsilon_{\tau,m}$, i.e.,

$$f_\tau = \sum_{m=0}^{N-1} \hat{f}_m \varepsilon_{\tau,m} \;\; \text{with} \;\; \varepsilon_{\tau,m} = \{e^{inm(2\pi/N)}\}_{n=-\infty}^{\infty} \; .$$

The invertibility of $\tfrac{1}{\tau} a I + b A$ as an operator on the space of N-periodic sequences is specified in the following theorem, which states the *discrete solvability condition*.

Theorem 4.5.1 *For any N-periodic f_τ the solution to the N-periodic difference equation $\frac{1}{\tau}ax_\tau + bAx_\tau = bf_\tau$ exists and is unique if and only if*

$$\frac{1}{\tau}\frac{a}{b}(\xi^m) \notin \sigma(-A), \quad m = 0,\ldots,N-1 \quad \text{with} \quad \xi = e^{i(2\pi/N)} , \tag{4.37}$$

where $a(\xi) = \sum_{j=0}^{k} \alpha_j \xi^j$ and $b(\xi) = \sum_{j=0}^{k} \beta_j \xi^j$.

Proof . Set $(ax_\tau)_n$ to denote the n'th element of ax_τ where x_τ is N-periodic. Since a is an operator which corresponds to a linear combination of N-periodic sequences, it is clear that ax_τ is also N-periodic. We denote its discrete Fourier series coefficients by $\widehat{(ax_\tau)}_m$. They may be calculated as follows,

$$
\begin{aligned}
\widehat{(ax_\tau)}_m &:= \frac{1}{N}\sum_{n=0}^{N-1}(ax_\tau)_n \, e^{-inm(2\pi/N)} \\
&= \frac{1}{N}\sum_{n=0}^{N-1}\Big(\sum_{j=0}^{k}\alpha_j x_{n-k+j}\Big) e^{-inm(2\pi/N)} \\
&= e^{-ikm(2\pi/N)}\sum_{j=0}^{k}\alpha_j e^{ijm(2\pi/N)}\frac{1}{N}\sum_{n=0}^{N-1}x_{n-k+j}e^{-i(n-k+j)m(2\pi/N)} \\
&= \xi^{-km}a(\xi^m)\hat{x}_m \quad \text{with} \quad \xi = e^{i(2\pi/N)}
\end{aligned}
\tag{4.38}
$$

$a(\xi)$ denotes the first characteristic polynomial of the multistep method, (2.43). Analogously, with $b(\xi)$ denoting the second characteristic polynomial, we get,

$$\widehat{(bx_\tau)}_m = \xi^{-km}b(\xi^m)\hat{x}_m . \tag{4.39}$$

Let $\hat{f}_m$ denote the DFS coefficients of f_τ. By calculating the DFS coefficients of the left- and right-hand side of equation (4.34) we derive an identity which is to be satisfied by the Fourier coefficients of any solution x_τ,

$$\left(\tfrac{1}{\tau}a(\xi^m)I + b(\xi^m)A\right)\hat{x}_m = b(\xi^m)\,\hat{f}_m .$$

If $b(\xi^m) = 0$ then necessarily $a(\xi^m) \neq 0$, since $a(\xi)$ and $b(\xi)$ have no common roots. Consequently, $\hat{x}_m = 0$. Taking this into account we may further simplify the condition to,

$$\left(\tfrac{1}{\tau}\tfrac{a}{b}(\xi^m)I + A\right)\hat{x}_m = \hat{f}_m . \tag{4.40}$$

The remainder of the proof is straightforward. When the condition of the theorem is satisfied, the solution equals $\sum_{m=0}^{N-1}\hat{x}_m \varepsilon_{\tau,m}$, with $\hat{x}_m$ uniquely defined by (4.40).

Conversely, when $\frac{1}{\tau}\frac{a}{b}(\xi^m) \in \sigma(-A)$ for some m, equation (4.34) with zero right-hand side has an infinite number of solutions (because (4.40) has an infinite number of solutions). Consequently, a solution of (4.34) cannot be unique. $\quad\square$

Corollary 4.5.2 *If the discrete solvability condition (4.37) is satisfied, the solution of the N-periodic difference equation $\frac{1}{\tau}ax_\tau + bAx_\tau = bf_\tau$ is given by,*

$$x_\tau = \sum_{m=0}^{N-1}\left(\tfrac{1}{\tau}\tfrac{a}{b}(\xi^m)I + A\right)^{-1}\hat{f}_m \varepsilon_{\tau,m} \quad \text{with} \quad \xi = e^{i(2\pi/N)} . \tag{4.41}$$

4.5.2   Reformulation as a linear algebra problem

Equation (4.34) is a linear equation in a finite-dimensional vector-space. It is illustrative to reformulate this equation as a standard linear system of equations in $\mathbb{C}^{Nd}$. To this end we rewrite (4.33) as follows,

$$\sum_{j=0}^{k} C_j x_{n-k+j} = \sum_{j=0}^{k} \beta_j f_{n-k+j} \quad \text{with} \quad C_j = \tfrac{1}{\tau}\alpha_j I + \beta_j A \ .$$

Consider the equations for the N successive unknowns $x_0, x_1, \ldots, x_{N-1}$. When the N-periodicity is taken into account, these equations may be assembled into matrix form and written as,

$$CX = BF \tag{4.42}$$

$$
\begin{aligned}
\text{with} \quad X &= [x_0 \ x_1 \ \ldots \ x_{N-1}]^t \\
F &= [f_0 \ f_1 \ \ldots \ f_{N-1}]^t \\
B &= \text{bcirc}(\beta_k I, 0, \ldots, 0, \beta_0 I, \beta_1 I, \ldots, \beta_{k-1} I) \\
C &= \text{bcirc}(C_k, 0, \ldots, 0, C_0, C_1, \ldots, C_{k-1})
\end{aligned}
$$

where "bcirc" stands for *block-circulant*. The definition of the latter is given below.

Definition 4.5.2 *A block-circulant matrix of type (r,s) is an $rs \times rs$ matrix,*

$$
\text{bcirc}(A_0, \ldots, A_{s-1}) =
\begin{pmatrix}
A_0 & A_1 & A_2 & . & . & . & A_{s-1} \\
A_{s-1} & A_0 & A_1 & . & . & . & A_{s-2} \\
A_{s-2} & A_{s-1} & A_0 & . & . & . & A_{s-3} \\
. & . & . & . & . & . & . \\
. & . & . & . & . & . & . \\
A_1 & A_2 & A_3 & . & . & . & A_0
\end{pmatrix}
\tag{4.43}
$$

where $A_1, A_2, \ldots, A_s$ are square matrices of order r.

Definition 4.5.3 *The polynomial $p_\gamma(z) = A_0 + A_1 z + A_2 z^2 + \cdots + A_{s-1} z^{s-1}$, associated with the s-tuple $\gamma = (A_0, A_1, \ldots, A_{s-1})$, is called the representer of the block-circulant $\text{bcirc}(A_0, \ldots, A_{s-1})$.*

These definitions, together with a wealth of properties and characteristics of circulant and block-circulant matrices can be found in a book by Ph. Davis, [25]. The following theorem, which is the block matrix extension of theorem 3.2.2 in the above reference, will be useful further on.

Theorem 4.5.3 *The spectrum of a block-circulant matrix is given by*

$$\sigma(\text{bcirc}(A_0, \ldots, A_{s-1})) = \bigcup_{m=0,\ldots,s-1} \sigma(p_\gamma(\xi^m)) \quad \text{with} \quad \xi = e^{i(2\pi/s)} \ . \tag{4.44}$$

Proof . Let (λ, X) be an eigenvalue-eigenvector pair of A, with $A = \mathrm{bcirc}(A_0, \ldots, A_{s-1})$ and $X = [x_0 \ldots x_{s-1}]^t$. We write $AX = \lambda X$ (block-)componentwise as,

$$(AX)_j := \sum_{k=0}^{s-1} A_k x_{k+j} = \lambda x_j, \quad j = 0, \ldots, s-1 ,$$

where we have assumed the sequence X to be extended by s-periodicity. By applying the DFS-formulae for x_{k+j} and x_j, we get,

$$\sum_{k=0}^{s-1} A_k \sum_{m=0}^{s-1} \hat{x}_m e^{i(k+j)m(2\pi/s)} = \lambda \sum_{m=0}^{s-1} \hat{x}_m e^{ijm(2\pi/s)}$$

$$\sum_{m=0}^{s-1} \Big(\sum_{k=0}^{s-1} A_k e^{ikm(2\pi/s)} \hat{x}_m \Big) e^{ijm(2\pi/s)} = \sum_{m=0}^{s-1} \lambda \hat{x}_m e^{ijm(2\pi/s)} .$$

By the uniqueness of the DFS-decomposition we get the result,

$$p_\gamma(\xi^m) \hat{x}_m := \sum_{k=0}^{s-1} A_k (\xi^m)^k \hat{x}_m = \lambda \hat{x}_m \quad \text{with} \quad \xi = e^{i(2\pi/s)} .$$

Conversely, with any eigenvalue-eigenvector pair $(\lambda_m, \hat{x}_m)$ of $p_\gamma(\xi^m)$ an eigenvalue-eigenvector pair (λ, X) of A corresponds. Indeed, consider

$$\lambda = \lambda_m \quad \text{and} \quad X = [\hat{x}_m \quad \hat{x}_m \xi^m \quad \hat{x}_m (\xi^m)^2 \quad \cdots \quad \hat{x}_m (\xi^m)^{s-1}] .$$

It follows that $AX = \lambda X$ since for any j,

$$(AX)_j = \sum_{k=0}^{s-1} A_k \hat{x}_m (\xi^m)^{k+j} = \sum_{k=0}^{s-1} A_k (\xi^m)^k \; \hat{x}_m (\xi^m)^j = \lambda_m \hat{x}_m (\xi^m)^j = \lambda X_j .$$

$$\square$$

Building on the above theorem, we can now derive the condition under which a solution of (4.42) exists.

Theorem 4.5.4 *The matrix C is invertible if and only if the discrete solvability condition (4.37) is satisfied.*

Proof . The spectrum of C is given by formula (4.44), with

$$p_\gamma(z) = C_k + C_{k-1} z^{N-1} + C_{k-2} z^{N-2} + \cdots + C_0 z^{N-k} .$$

Consequently, with $\xi = e^{i(2\pi/N)}$ and since $\xi^{Nm} = 1$,

$$p_\gamma(\xi^m) = C_k \xi^{Nm} + C_{k-1} \xi^{(N-1)m} + \cdots + C_0 \xi^{(N-k)m}$$

$$= (\xi^m)^{N-k} (C_k (\xi^m)^k + C_{k-1} (\xi^m)^{k-1} + \cdots + C_0)$$

$$= (\xi^m)^{N-k} \sum_{j=0}^{k} (\tfrac{1}{\tau} \alpha_j (\xi^m)^j I + \beta_j A (\xi^m)^j)$$

$$= (\xi^m)^{N-k} (\tfrac{1}{\tau} a(\xi^m) I + b(\xi^m) A)$$

C is invertible iff $0 \notin \sigma(C)$, or, equivalently, since $\xi^{(N-k)m} \neq 0$,

$$0 \notin \sigma(\tfrac{1}{\tau}a(\xi^m)I + b(\xi^m)A), \quad \text{or} \quad 0 \notin \sigma(\tfrac{1}{\tau}\tfrac{a}{b}(\xi^m)I + A), \quad m = 0,\ldots,s-1 \ .$$

Note that, when $b(\xi^m) = 0$, then $a(\xi^m) \neq 0$ and the condition $0 \notin \sigma(C)$ is satisfied.
$\square$

As expected, the condition derived in the above theorem by linear algebra arguments is identical to the condition for existence of a solution to (4.33) derived in theorem 4.5.1 by DFS properties. Similar analogies can be derived for most of the results that will be discussed in the remainder of this section. It will be possible to either consider the problems and operators as problems and operators in the space of N-dimensional sequences, and to derive results by DFS properties, or to study problems and operators in a standard linear algebra setting and apply standard matrix properties. We opt for the former approach, which more clearly exemplifies the close correspondence to the continuous-time analysis.

4.5.3 The discrete-time time-periodic waveform iteration

Discretization of (4.17) with a fixed time-step τ results in

$$\frac{1}{\tau} \sum_{j=0}^{k} \alpha_j x_{n-k+j}^{(\nu)} + \sum_{j=0}^{k} \beta_j P x_{n-k+j}^{(\nu)} = \sum_{j=0}^{k} \beta_j \{ Q x_{n-k+j}^{(\nu-1)} + f_{n-k+j} \} \ . \tag{4.45}$$

By taking the N-periodicity into account, this can be rewritten as a successive approximation scheme in the space of N-periodic sequences,

$$\tfrac{1}{\tau} a x_\tau^{(\nu)} + b P x_\tau^{(\nu)} = b Q x_\tau^{(\nu-1)} + b f_\tau \ . \tag{4.46}$$

The following theorem, analogous to theorem 4.4.3, deals with existence and uniqueness of the *discrete-time time-periodic waveform relaxation* iterates.

Theorem 4.5.5 *The discrete-time time-periodic waveform iteration defined by the splitting (P,Q) and the N-periodic sequences $x_\tau^{(0)}, f_\tau$ is well-defined, that is, the iterates $x_\tau^{(\nu)}, \nu = 1,2,\ldots$ exist and are unique if and only if P satisfies the discrete solvability condition, $\tfrac{1}{\tau}\tfrac{a}{b}(\xi^m) \notin \sigma(-P), \ m = 0,..,N-1, \ \xi = e^{i(2\pi/N)} \ .$*

Proof . Since $Q x_\tau^{(0)} + f_\tau$ is N-periodic, we may invoke theorem 4.5.1 to conclude that $x_\tau^{(1)}$ exists and that it is unique if and only if P satisfies the solvability condition. An induction argument extends this result to the subsequent iterates. $\square$

If the discrete solvability condition is satisfied, we can introduce a linear operator $\bar{\mathcal{K}}_\tau$ and write the solution of (4.46) as follows,

$$x_\tau^{(\nu)} = \bar{\mathcal{K}}_\tau x_\tau^{(\nu-1)} + \varphi_\tau \tag{4.47}$$

$$\text{with} \qquad \bar{\mathcal{K}}_\tau x_\tau = \sum_{m=0}^{N-1} (\tfrac{1}{\tau}\tfrac{a}{b}(\xi^m)I + P)^{-1} Q \hat{x}_m \varepsilon_{\tau,m} \tag{4.48}$$

$$\varphi_\tau = \sum_{m=0}^{N-1} (\tfrac{1}{\tau}\tfrac{a}{b}(\xi^m)I + P)^{-1} \hat{f}_m \varepsilon_{\tau,m} \tag{4.49}$$

which immediately results from the formula (4.41). In the discussion of the continuous-time iteration, it was shown that the operator $\mathcal{K}$ is a convolution operator. The matching discrete-time result is proven next.

Theorem 4.5.6 $\bar{\mathcal{K}}_\tau$ *is an N-periodic discrete convolution operator,*

$$(\bar{\mathcal{K}}_\tau x_\tau)_n = (\bar{k}_\tau \star x_\tau)_n := \frac{1}{N} \sum_{i=0}^{N-1} \bar{k}_{n-i} x_i \ ,$$

with an N-periodic, matrix-valued kernel $\bar{k}_\tau$, defined by

$$\bar{k}_n = \sum_{m=0}^{N-1} (\tfrac{1}{\tau}\tfrac{a}{b}(\xi^m)I + P)^{-1} Q \ e^{inm(2\pi/N)} \ . \tag{4.50}$$

Proof . By a comparison of (4.50) to (4.36) we can immediately identify the DFS-transform coefficients of $\bar{k}_\tau$,

$$\hat{\bar{k}}_m = (\tfrac{1}{\tau}\tfrac{a}{b}(\xi^m)I + P)^{-1} Q \ . \tag{4.51}$$

By (4.48) and by the DFS convolution property we find that,

$$\forall x_\tau : \quad (\widehat{\bar{\mathcal{K}}_\tau x_\tau})_m = (\tfrac{1}{\tau}\tfrac{a}{b}(\xi^m)I + P)^{-1} Q \hat{x}_m = \hat{\bar{k}}_m \hat{x}_m = (\widehat{\bar{k}_\tau \star x_\tau})_m \ .$$

By the uniqueness of the DFS the correctness of the theorem follows. $\qquad\square$

4.5.4 Convergence of the discrete-time iteration

The convergence of the successive approximation scheme is determined by the spectral radius of $\bar{\mathcal{K}}_\tau$, which is a linear operator in a finite dimensional space. The following theorem relates the spectral radius of $\bar{\mathcal{K}}_\tau$ to the spectral radii of the Fourier coefficients of its discrete convolution kernel. The proof proceeds along the same lines as the proof of the corresponding continuous-time theorem, i.e., theorem 4.4.13. We shall therefore only present the principal ideas.

Theorem 4.5.7 *Let P satisfy the discrete solvability condition. Then,*

$$\rho(\bar{\mathcal{K}}_\tau) = \max_{m=0,\ldots,N-1} \rho(\, K(\tfrac{1}{\tau}\tfrac{a}{b}(\xi^m))\,) \quad \text{with} \ \ \xi = e^{i(2\pi/N)} \ . \tag{4.52}$$

Proof . Since $\mathcal{K}$ is a finite-dimensional linear operator its spectrum consists of eigenvalues only. Let λ be an eigenvalue of $\bar{\mathcal{K}}_\tau$ and let x_τ be a non-trivial N-periodic sequence that satisfies $\bar{\mathcal{K}}_\tau x_\tau = \lambda x_\tau$. By calculating the discrete Fourier series coefficients of left- and right-hand side we get the identity,

$$\hat{\hat{k}}_m \hat{x}_m = \lambda \hat{x}_m \quad \text{with} \quad \hat{\hat{k}}_m = (\tfrac{1}{\tau}\tfrac{a}{b}(\xi^m)I + P)^{-1}Q = K(\tfrac{1}{\tau}\tfrac{a}{b}(\xi^m)) \ .$$

This shows that λ is an eigenvalue of at least one matrix $\hat{\hat{k}}_m$.

On the other hand, any eigenvalue-eigenvector pair (λ_m, X_m) of $\hat{\hat{k}}_m$ defines a non-trivial N-periodic eigen-"sequence" of $\bar{\mathcal{K}}_\tau$ with eigenvalue λ_m, namely

$$x_\tau = X_m \varepsilon_{\tau,m} \ .$$

Therefore,

$$\sigma(\bar{\mathcal{K}}_\tau) = \bigcup_{m=0..N-1} \sigma\big(K(\tfrac{1}{\tau}\tfrac{a}{b}(\xi^m)) \big) \ ,$$

from which (4.52) immediately follows. $\qquad\qquad\square$

As in the continuous-time case we may state two corollaries which relate the convergence properties of the static iteration and of the infinite-interval discrete waveform iteration for solving initial value problems to those of the discrete waveform iteration method for solving time-periodic problems.

Corollary 4.5.8 *Let P satisfy the discrete solvability condition. It then follows that the spectral radius of the discrete time-periodic iteration operator is bounded from below by the spectral radius of the corresponding static iteration operator.*

Proof . The discrete solvability condition ensures the existence of both $\rho(\bar{\mathcal{K}}_\tau)$ and $\rho(P^{-1}Q)$. If $m = 0$ then $(\tfrac{1}{\tau}\tfrac{a}{b}(\xi^m)I + P)^{-1}Q = P^{-1}Q$ because of the consistency condition $a(1) = 0$. Therefore, the conclusion follows. $\qquad\qquad\square$

Corollary 4.5.9 *Let the splitting satisfy: $\sigma(-\tau P) \subset \text{int } S$. It then follows that the spectral radius of the discrete waveform operator for solving the T-periodic ordinary differential equation is bounded by the spectral radius of the infinite-interval discrete waveform operator for solving the corresponding initial value problem.*

Proof . The condition on the splitting is necessary to guarantee the existence of $\rho(\mathcal{K})$, theorem 2.5.10. It also ensures that the splitting is non-critical. The result follows by a comparison of the result of theorem 4.5.7 to the formula (2.53). $\qquad\qquad\square$

The norm of $\bar{\mathcal{K}}_\tau$ may be determined in essentially the same way as it was done for the continuous-time iteration. We shall therefore skip the proof, and only state the result. Since the proof is based on the Parseval relation for discrete Fourier series, only the $l_2([0..N-1]; \mathbb{C}^d)$-space is considered.

Theorem 4.5.10 *Consider $\bar{\mathcal{K}}_\tau$ as an operator in $l_2([0..N-1]; \mathbb{C}^d)$. Then,*

$$\| \bar{\mathcal{K}}_\tau \|_2 = \max_{m=0..N-1} \| K(\tfrac{1}{\tau}\tfrac{a}{b}(\xi^m)) \| \qquad\qquad (4.53)$$

where $\| \, . \, \|_2$ denotes the l_2-norm and $\| \, . \, \|$ the Euclidean matrix norm.

4.5.5 The spectral picture

In order to calculate the spectral radius ρ of any of the waveform relaxation operators defined in this chapter or in the previous chapters, the spectral radius of the matrix $K(z) = (zI + P)^{-1}Q$ is to be calculated for a set of complex values of z, and the maximum has to be determined. More precisely, a general formula holds, of the form

$$\rho = \max_{z \in \Sigma} \rho((zI + P)^{-1}Q) , \qquad (4.54)$$

where the extent of set Σ is recalled in table 4.1.

Table 4.1: Extent of the set Σ for different waveform operators.

	continuous-time	discrete-time
initial value, $[0, T]$	$\{\infty\}$	$\{\frac{1}{\tau}\frac{\alpha_k}{\beta_k}\}$
initial value, $I\!\!R^+$	$\{i\xi : \xi \in I\!\!R\}$	$\{\frac{1}{\tau}\frac{a}{b}(\xi) : \xi = e^{i\theta}, \theta \in [0, 2\pi)\}$
time-periodic, $[0, T]$	$\{in\omega : n \in \mathbb{Z}, \omega = \frac{2\pi}{T}\}$	$\{\frac{1}{\tau}\frac{a}{b}(\xi^m) : \xi = e^{i(2\pi/N)}, m = 0, .., N{-}1\}$

In addition to the sets specified in the table, note that the static iteration leads to the set $\Sigma = \{0\}$. The set of points $\{\frac{a}{b}(\xi)$ with $\xi \in \mathbb{C}, |\xi| = 1\}$ constitutes the root locus of the linear multistep method characterized by the polynomials (a, b). Consequently, the N points, $\frac{1}{\tau}\frac{a}{b}(\xi^m)$, that have arisen in the theoretical analysis of the discrete-time time-periodic waveform iteration lie on the root locus, scaled by a factor $\frac{1}{\tau}$. The precise values of these points may be calculated easily as is illustrated in the following examples.

Example 4.5.1 The characteristic polynomials of the backward Euler method are given by $a(\xi) = \xi - 1$ and $b(\xi) = \xi$. The boundary of the stability region is the set of points z, given by

$$z = \frac{a}{b}(\xi) = 1 - 1/\xi = 1 - e^{-i\theta} = 1 - \cos(\theta) + i\sin(\theta) , \quad \text{with} \quad \theta \in [0, 2\pi) .$$

Consequently, to determine the spectral radius of the discrete-time time-periodic operator, the following N points are to be considered,

$$z = \frac{1}{\tau}\frac{a}{b}(\xi^m) = \frac{1}{\tau}(1 - \cos(m2\pi/N) + i\sin(m2\pi/N)), \quad m = 0..N - 1 .$$

They are equally spaced along a circle of radius $1/\tau$, centered at point $(1/\tau, 0)$.

Example 4.5.2 The characteristic polynomials of the trapezoidal rule or Crank-Nicolson method are $a(\xi) = \xi - 1$ and $b(\xi) = (\xi + 1)/2$. The boundary of its stability region is given by,

$$z = \frac{a}{b}(\xi) = 2(\xi - 1)/(\xi + 1) = 2(e^{i\theta} - 1)/(e^{i\theta} + 1) = 2\tan(\theta/2)i \quad \text{with} \quad \theta \in [0, 2\pi) .$$

The points $\frac{1}{\tau}\frac{a}{b}(\xi^m)$ lie unequally spaced along the imaginary axis, according to the formula,

$$z = \frac{1}{\tau}\frac{a}{b}(\xi^m) = \frac{2}{\tau}\tan(m\pi/N)i, \quad m = 0, \ldots, N - 1 .$$

We illustrate some of these sets in the *spectral picture*, figure 4.1. We depicted the sets for the backward differentiation formulae of order 1 to 4, with $N = 20$ and with $\tau = 1$. The complex values $\frac{1}{\tau}\frac{a}{b}(\xi^m)$ are denoted by a "$\bullet$". The points $\frac{1}{\tau}\frac{\alpha_k}{\beta_k}$ (which are not shown in the picture) are located on the real axis, at values 1 (BDF(1)), $\frac{3}{2}$ (BDF(2)), $\frac{11}{6}$ (BDF(3)), and $\frac{25}{12}$ (BDF(4)).

The use of the spectral picture for a graphical convergence check is illustrated in figure 4.2. We consider the case of a consistently ordered matrix A with constant positive diagonal and the use of the SOR splitting. As was discussed in section 2.5.3, in this case an explicit formula for $\rho((zI + P)^{-1}Q)$ exists. It depends on z and on the tuple (μ, δ, ω), where δ is the diagonal value of A, ω is the overrelaxation factor and μ is the spectral radius of the corresponding Jacobi matrix. In the figure, the function $\rho((zI + P)^{-1}Q)$ is represented by a plot of some of its contour lines ($\rho = 0.4, 0.5, \ldots, 1.0$). This is done for the parameter values $(\mu, \delta, \omega) = (0.951, 200, 4/3)$. (Note that the values of δ and μ correspond to the matrix A obtained by a semi-discretization of the one-dimensional heat equation on $x \in [0, 1]$ with central differences and $h = 0.1$.) Superimposed on the contour lines is the spectral picture, scaled by the factor $\frac{1}{\tau}$, with $\tau = 0.02$. This construction allows the graphical study of the convergence of the waveform iterations and the (approximate) determination of the spectral radii of the waveform operators. Note that the waveform iteration converges when the associated set Σ is contained in the *set of convergence*, that is, the set of points z for which $\rho(K(z)) < 1$.

By looking at the spectral picture the relation between the convergence characteristics of the different operators can easily be verified, in particular the corollaries 4.4.14, 4.4.15, 4.5.8 and 4.5.9. Finally, note the following relations between the different entries of table 4.1. In the limiting case of $T \to \infty$ (with constant τ in the discrete-time case) the time-periodic formulae lead to the initial value formulae. Additionally, note that the continuous-time results "naturally" follow from the discrete-time results. In the time-periodic case this is shown by application of the following property.

Property 4.5.11

For any fixed m: $\quad \lim_{\tau \to 0} \frac{1}{\tau}\frac{a}{b}(\xi^m) = im\omega \quad$ *and* $\quad \lim_{\tau \to 0} \frac{1}{\tau}\frac{a}{b}(\xi^{N-m}) = -im\omega \qquad$ (4.55)

Proof . The result is easily checked by using l'Hospital's rule, the consistency conditions $a(1) = 0$ and $a'(1) = b(1)$ and by taking into account that $\xi = e^{i(2\pi/N)} = e^{i(2\pi/T)(T/N)} = e^{i\omega\tau}$. We calculate the first limit below. The second limit can be calculated similarly.

$$l.h.s. = \lim_{\tau \to 0} \frac{a(e^{im\omega\tau})}{\tau b(e^{im\omega\tau})} = \lim_{\tau \to 0} \frac{a'(e^{im\omega\tau})e^{im\omega\tau}im\omega}{b(e^{im\omega\tau}) + \tau b'(e^{im\omega\tau})e^{im\omega\tau}im\omega} = \frac{a'(1)im\omega}{b(1)} = im\omega$$

$$\square$$

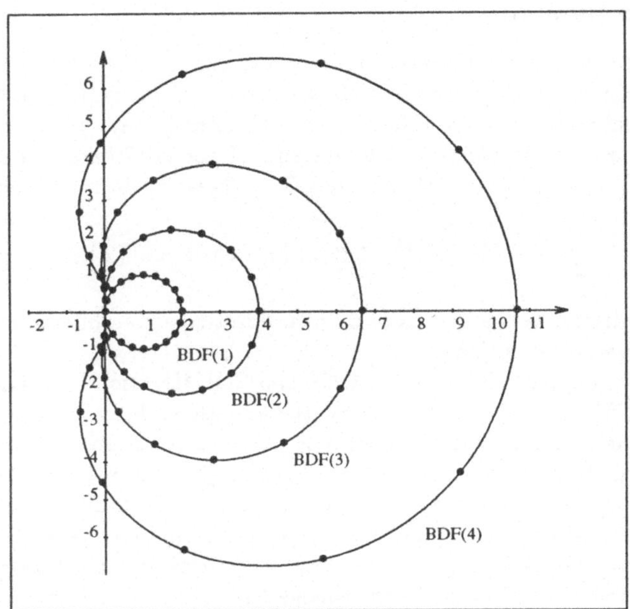

Figure 4.1: Position of the N points $\frac{1}{\tau}\frac{a}{b}(\xi^m)$ for BDF methods ($N = 20, \tau = 1$).

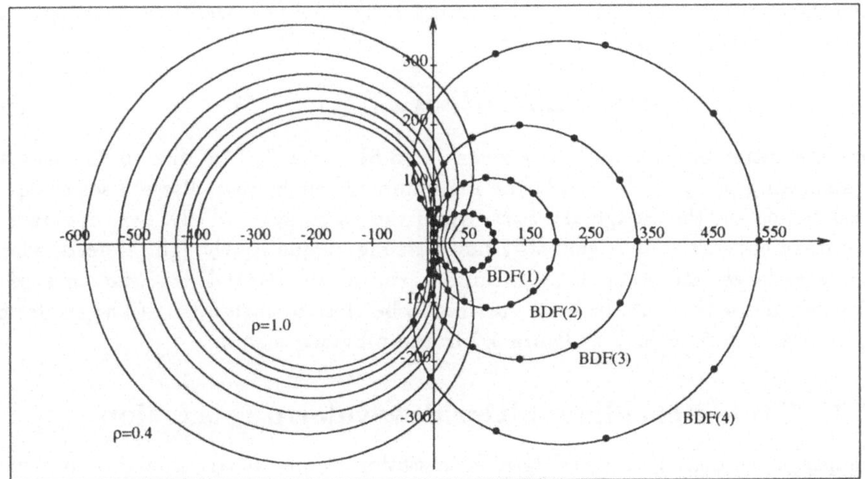

Figure 4.2: Use of the spectral picture for graphical convergence check.

4.6   Multigrid acceleration

4.6.1   Introduction

The time-periodic waveform relaxation method can be used for solving the system of ordinary differential equations which is obtained by semi-discretizing a time-periodic partial differential equation. It will be shown in chapter 8 that the method applies to fairly general nonlinear problems and to systems of partial differential equations. In the current section we shall restrict the analysis to linear problems of the form,

$$\frac{d}{dt}u^h = L^h u^h + f^h, \quad t \in [0,T], \quad u^h(0) = u^h(T) , \tag{4.56}$$

where L^h is the discretization matrix of a linear elliptic partial differential operator; h represents the spatial mesh size.

Classic relaxation methods such as Jacobi and SOR iteration for solving the linear systems derived from elliptic partial differential equations degrade in performance as the mesh size h is reduced. The time-periodic dynamic iteration method suffers from the same problem. Indeed, consider the heat equation $\frac{\partial u}{\partial t} = \Delta u$ in a one-, two-, or three-dimensional domain with Dirichlet boundary conditions. Semi-discretization with the standard five-point star on an equidistant mesh leads to an ODE system coefficient matrix $-L^h$ $(= \text{-}L\text{+}D\text{-}U)$ that satisfies the conditions for application of corollary 4.4.16. Application of (4.31) and the knowledge that $\rho(D^{-1}(L+U)) = cos(\pi h)$, leads to the time-periodic equivalent of (3.4),

$$\rho(\bar{\mathcal{K}}_{JAC}) \simeq 1 - \pi^2 h^2/2 \quad \text{and} \quad \rho(\bar{\mathcal{K}}_{GS}) \simeq 1 - \pi^2 h^2 . \tag{4.57}$$

In the case of SOR waveform relaxation a spectral radius may be found which differs from that of the static iteration. For small h and for the overrelaxation parameter value which minimizes $\rho(\mathcal{K}_{SOR})$ we get,

$$\rho(\bar{\mathcal{K}}_{SOR,\omega_{opt}}) \leq \rho(\mathcal{K}_{SOR,\omega_{opt}}) \simeq 1 - 2\pi^2 h^2 , \tag{4.58}$$

where the latter approximation is given in (3.5). The "$\leq$" is due to the fact that for calculating $\rho(\bar{\mathcal{K}}_{SOR})$ by (4.28) the maximum is sought in a discrete set of equally spaced points on the imaginary axis, while the calculation of $\rho(\mathcal{K}_{SOR})$ involves the whole imaginary axis. It is generally not a strong inequality though. Indeed, when ω is not very large, say order of magnitude 1, the set of points $\Sigma = \{in\omega : n \in \mathbb{Z}\}$ is sufficiently dense in order for both maxima to be close to each other. (The reader may wish to have a second look at figure 4.2 in the previous section.)

4.6.2   Time-periodic multigrid waveform relaxation

In chapter 3 we discussed a multigrid acceleration of the initial value waveform relaxation method. A small modification of this algorithm has led us to a similar multigrid extension for the time-periodic case. This extension was first proposed in our paper [130]. Numerical examples of linear and nonlinear problems, a complexity analysis of a parallel implementation and a theoretical discussion followed in subsequent papers, [135, 136, 139, 137]. The basic steps of the algorithm are again very similar to those of the "elliptic" multigrid method, yet all of the operations are defined w.r.t. T-periodic functions instead of w.r.t. scalar values. Below, we formulate the two-grid iteration of the time-periodic multigrid scheme for linear problems. The two-grid cycle starts from an approximation $u^{(\nu-1)}$ to the solution u of (4.56), and calculates the next iterate $u^{(\nu)}$ in the following series of three steps:

● **Pre-smoothing.** Set $x^{(0)} = u^{(\nu-1)}$ and perform ν_1 time-periodic relaxation steps defined by the splitting $-L^h = P^h - Q^h$,

$$\frac{d}{dt}x^{(\nu)} + P^h x^{(\nu)} = Q^h x^{(\nu-1)} + f^h, \quad x^{(\nu)}(0) = x^{(\nu)}(T), \quad \text{for} \quad \nu = 1, \ldots, \nu_1 \ . \qquad (4.59)$$

The splitting corresponds to e.g. lexicographic or red/black Gauss-Seidel.

● **Coarse-grid correction.** Compute the defect,

$$d^h = \frac{d}{dt}x^{(\nu_1)} - L^h x^{(\nu_1)} - f^h = Q^h(x^{(\nu_1-1)} - x^{(\nu_1)}) \ . \qquad (4.60)$$

Solve the time-periodic coarse-grid problem on Ω^H,

$$\frac{d}{dt}v^H = L^H v^H + I_h^H d^h \ , \quad v^H(0) = v^H(T) \ . \qquad (4.61)$$

Correct the current approximation ,

$$\bar{x}^h = x^{(\nu_1)} - I_H^h v^H \ . \qquad (4.62)$$

● **Post-smoothing.** Perform ν_2 smoothing relaxations starting with $x^{(0)} = \bar{x}^h$,

$$\frac{d}{dt}x^{(\nu)} + P^h x^{(\nu)} = Q^h x^{(\nu-1)} + f^h, \quad x^{(\nu)}(0) = x^{(\nu)}(T), \quad \text{for} \quad \nu = 1, \ldots, \nu_2 \ , \qquad (4.63)$$

and set $u^{(\nu)} = x^{(\nu_2)}$ afterwards.

Since the coarse-grid problem (4.61) is of similar nature as the fine-grid problem, it may be solved analogously. Recursive application of the above idea, extended with a suitable direct method for solving the problem on the coarsest grid leads to the time-periodic multigrid algorithm. Nonlinear time-periodic problems are again tackled with a waveform extension of the full approximation scheme. The necessary changes to the multigrid algorithm for solving nonlinear initial value problems are trivial, and therefore not given here. Examples of its use are given in chapter 8. In the time-periodic *nested iteration* or *full multigrid method*, the initial approximation $u^{(0)}$ is obtained by interpolation of the time-periodic approximation found on a coarser grid. The algorithm is similar to alg. 3.4. However, since there is no initial condition to be satisfied, the special correction in the coarse-grid to fine-grid interpolation may be skipped.

4.6.3   Analysis of the continuous-time iteration

The continuous-time analysis is qualitatively very similar to the analysis in section
3.4. Let $e^{(\nu)}$ denote the error of the ν'th iterate, and let $\tilde{e}_n^{(\nu)}$ be its n'th Fourier-series
coefficient. When the equations of the two-grid cycle are "Fourier-transformed", one
easily derives the following relation between the Fourier coefficients of $e^{(\nu)}$ and $e^{(\nu-1)}$

$$\tilde{e}_n^{(\nu)} = M(in\omega)\,\tilde{e}_n^{(\nu-1)} \tag{4.64}$$

$$\begin{cases} M(z) = S^{\nu_2}(z)\,(I - I_H^h(zI - L^H)^{-1}I_h^H(zI - L^h))\,S^{\nu_1}(z) \\[2mm] S(z) = (zI + P^h)^{-1}Q^h \end{cases} \tag{4.65}$$

Note that the matrix $M(z)$ is identical to the symbol of the convolution operator
that arose in the discussion of the initial value multigrid waveform relaxation, formula
(3.21). We formally write $e^{(\nu)} = \bar{\mathcal{M}}e^{(\nu-1)}$. The nature of $\bar{\mathcal{M}}$ is specified below.

Theorem 4.6.1 *Let $M(z)$ be given by (4.65); let the splitting (P^h, Q^h) and the matrix
L^H be non-critical and assume $\nu_1 + \nu_2 \geq 1$. Then, $\bar{\mathcal{M}}$ is a linear T-periodic convolution
operator, that is,*

$$\bar{\mathcal{M}}x(t) := \bar{m} \star x\,(t) := \frac{1}{T}\int_0^T \bar{m}(t - s)\,x(s)\,ds \;,$$

*of which the matrix-valued kernel, $\bar{m}$, is the periodic function ($\in L_1(0,T)$) whose
Fourier series coefficients are given by,*

$$\tilde{\bar{m}}_n = M(in\omega)\;. \tag{4.66}$$

The proof is based on the convolution property, see page 97. Before we present the
proof we prove a lemma on which the integrability of $\bar{m}$ is based.

Lemma 4.6.2 *Let $R(z)$ be a rational function of which none of the poles belongs to the
set $\{in\omega : n \in \mathbb{Z}, \omega = 2\pi/T, i = \sqrt{-1}\}$, and which vanishes at infinity. It follows that
$\{R(in\omega)\}_{n=-\infty}^{\infty}$ are the Fourier coefficients of a T-periodic function which is bounded
and continuous on $(0,T)$.*

Proof . Since $R(z)$ is rational and vanishing at infinity, it may be written by partial
fraction expansion as a finite sum of rationals of the form,

$$r(z) = \frac{\eta}{(z + \mu)^m}$$

where η is a complex number, μ is complex but different from $in\omega$ for all integers
n, and m is a positive integer. If $m = 1$ then $\{r(in\omega)\}_{n=-\infty}^{\infty}$ are the Fourier
coefficients of the T-periodic function whose restriction to $[0,T)$ equals,

$$\eta T(1 - e^{-\mu T})^{-1}e^{-\mu t},$$

which is bounded and continuous on $(0,T)$. (Note that $e^{-\mu T} \neq 1$.) If $m > 1$, then
$\sum_{n=-\infty}^{\infty}|r(in\omega)| < \infty$. By a standard property, e.g. [17, Th. 15.10], it follows
that $\{r(in\omega)\}_{n=-\infty}^{\infty}$ are the Fourier coefficients of a periodic function which is
continuous on $\mathbb{R}$. By the linearity of the Fourier transform the conclusion follows.

$\square$

Building on the above lemma and on the uniqueness of the Fourier series transformation, we can now prove theorem 4.6.1.

Proof . The inverse of a $d \times d$-matrix $zI - A$ is given by,

$$(zI - A)^{-1} = \frac{1}{\det(zI - A)} \operatorname{adj}(zI - A)$$

Since $\det(zI - A)$ is a polynomial of exact degree d in z and since the entries of $\operatorname{adj}(zI - A)$ are polynomials of degree at most $d-1$, all of the entries of $(zI - A)^{-1}$ are rational functions vanishing at infinity. Moreover, their poles are eigenvalues of A. Applying this result to the matrices $zI - L^H$ and $zI + P^h$ and taking the assumptions of the theorem into account, we may conclude that each of the matrix entries of $M(z)$ satisfies the conditions of the lemma. Consequently, $\{M(in\omega)\}_{n=-\infty}^{\infty}$ are the Fourier coefficients of a function $(\bar{m})$ which is integrable on $(0,T)$. Consequently we may apply the convolution property to $\bar{m} \star x$, and taking also the definition of $\bar{\mathcal{M}}$ (4.64) into account, we get,

$$\forall x \in L_1(0,T): \quad (\widetilde{\bar{\mathcal{M}}x})_n = M(in\omega)\,\tilde{x}_n = (\widetilde{\bar{m} \star x})_n \ .$$

Therefore $\bar{\mathcal{M}}$ is identical to the convolution operator with kernel $\bar{m}$. $\square$

Note that the conditions of the theorem are easily satisfied when L^h and L^H are derived by discretization of a real elliptic operator. The main theorem about the convergence of the two-grid operator $\bar{\mathcal{M}}$ can now be stated.

Theorem 4.6.3 *Assume the conditions of theorem 4.6.1 are satisfied. Consider* $\bar{\mathcal{M}}$ *as an operator in* $L_p(0,T)$, $1 \leq p \leq \infty$, *or* $C[0,T]$. *Then,*

$$\rho(\bar{\mathcal{M}}) = \max_{n \in \mathbb{Z}} \rho(\,M(in\omega)\,) \quad with \quad \omega = 2\pi/T, \quad i = \sqrt{-1}. \tag{4.67}$$

Proof . Since $\bar{\mathcal{M}}$ is a convolution operator with integrable kernel, it is compact by theorem 4.4.10. Its spectrum consists of eigenvalues and zero only. The spectrum may be derived explicitly, in a similar way as in the proof of theorem 4.4.13,

$$\sigma(\bar{\mathcal{M}}) = \bigcup_{n \in \mathbb{Z}} \sigma(M(in\omega)) \bigcup \{0\} \ ,$$

(compare with formula (4.30)) which directly leads to (4.67). $\square$

Consequently, we may conclude that $\rho(\bar{\mathcal{M}})$ is bounded from below by the spectral radius of the elliptic multigrid operator. By comparing formula (4.67) to the corresponding formula (3.23), we may in addition conclude that the spectral radius of the time-periodic operator is bounded from above by the spectral radius of the initial value operator. Note that this bound is independent of the value of the period T. A corollary of property 3.4.5 exemplifies the above relation.

Property 4.6.4 *The two-grid operator for the T-periodic one-dimensional heat equation, with red-black Gauss-Seidel smoothing, satisfies*

$$\rho(\bar{\mathcal{M}}) \leq \frac{1}{2}\sqrt{\eta(2\nu - 1)} \quad with \quad \eta(\nu) = \frac{\nu^{\nu}}{(\nu + 1)^{\nu+1}} \ , \tag{4.68}$$

with $\nu = \nu_1 + \nu_2 \geq 1$. This is the best possible bound which is independent of h and T.

4.6.4   Analysis of the discrete-time iteration

We discretize (4.59) to (4.63) with a linear multistep method with constant time-increment $\tau = T/N$. This defines a discrete-time two-grid cycle very similar to the continuous-time two-grid cycle. It differs from the latter in that the operations are defined on N-periodic sequences instead of on T-periodic functions. Let the multistep method again be specified by its characteristic polynomials $a(\xi), b(\xi)$ and let a, b denote the corresponding linear combination operators defined in section 4.5.1. The successive steps of one two-grid cycle read as follows. They map $u_\tau^{(\nu-1)}$ to $u_\tau^{(\nu)}$.

- **Pre-smoothing.** Set $x_\tau^{(0)} = u_\tau^{(\nu-1)}$ and perform ν_1 discrete-time relaxations,

$$\tfrac{1}{\tau} a x_\tau^{(\nu)} + b P^h x_\tau^{(\nu)} = b Q^h x_\tau^{(\nu-1)} + b f_\tau^h, \quad \nu = 1, \ldots, \nu_1 \ .$$

- **Coarse-grid correction.**

  - *Calculate the defect:* $b d_\tau^h = \tfrac{1}{\tau} a x_\tau^{(\nu_1)} - b L^h x_\tau^{(\nu_1)} - b f_\tau^h = b Q^h (x_\tau^{(\nu_1-1)} - x_\tau^{(\nu_1)}) \ .$

  - *Solve the coarse-grid problem:* $\tfrac{1}{\tau} a v_\tau^H = b L^H v_\tau^H + I_h^H b d_\tau^h \ .$

  - *Correct the current approximation:* $\bar{x}_\tau^h = x_\tau^{(\nu_1)} - I_H^h v_\tau^H \ .$

- **Post-smoothing.** Perform ν_2 post-smoothing relaxations starting with $x_\tau^{(0)} = \bar{x}_\tau^h$ and set $u_\tau^{(\nu)} = x_\tau^{(\nu_2)}$ afterwards.

Let $\hat{e}_m^{(\nu)}$ denote the m-th discrete Fourier series coefficient of the error to the ν-th iterate. DFS-transforming the equations of the discrete-time two-grid cycle, we get,

$$\hat{e}_m^{(\nu)} = M(\tfrac{1}{\tau}\tfrac{a}{b}(\xi^m))\, \hat{e}_m^{(\nu-1)} \ , \quad \xi = e^{i(2\pi/N)} \ , \quad m = 0, \ldots, N-1 \ , \tag{4.69}$$

with M given by (4.65). The derivation of formula (4.69) requires that,

$$\tfrac{1}{\tau}\tfrac{a}{b}(\xi^m) \notin \sigma(-P^h) \cup \sigma(L^H) \ . \tag{4.70}$$

That is, none of the poles of the matrix elements of $M(z)$ equals $\tfrac{1}{\tau}\tfrac{a}{b}(\xi^m)$. Formula (4.69) defines a linear operator $\bar{\mathcal{M}}_\tau$ for which, $e_\tau^{(\nu)} = \bar{\mathcal{M}}_\tau e_\tau^{(\nu-1)}$. The nature of this operator and its spectral radius are characterized in the following two theorems. We skip their proofs, since they are almost identical to the proofs of the corresponding theorems 4.5.6 and 4.5.7. (Replace matrix K by M, and kernel $\bar{k}$ by $\bar{m}$.)

Theorem 4.6.5 *Assume (4.70). $\bar{\mathcal{M}}_\tau$ is an N-periodic discrete convolution operator,*

$$(\bar{\mathcal{M}}_\tau x_\tau)_n = (\bar{m}_\tau \star x_\tau)_n := \frac{1}{N} \sum_{i=0}^{N-1} \bar{m}_{n-i} x_i \ ,$$

with an N-periodic, matrix-valued kernel $\bar{m}_\tau$, defined by,

$$\bar{m}_n = \sum_{m=0}^{N-1} M(\tfrac{1}{\tau}\tfrac{a}{b}(\xi^m))\, e^{inm(2\pi/N)} \quad with \quad \xi = e^{i(2\pi/N)} \ .$$

Theorem 4.6.6 *Assume (4.70). Then,*

$$\rho(\bar{\mathcal{M}}_\tau) = \max_{m=0,\ldots,N-1} \rho\left(M\left(\tfrac{1}{\tau}\tfrac{a}{b}(\xi^m)\right)\right) \quad \text{with} \quad \xi = e^{i(2\pi/N)} . \tag{4.71}$$

The spectral radius of $\bar{\mathcal{M}}_\tau$ is bounded by the spectral radius of the corresponding discrete-time infinite-interval two-grid operator for solving initial boundary value problems. This immediately allows us to transpose some of the results obtained by Lubich and Ostermann in [78] to the time-periodic case. We formulate the theorem corresponding to theorem 3.4.10.

Theorem 4.6.7 *If the linear multistep method is $A(\alpha)$-stable then*

$$\rho(\bar{\mathcal{M}}_\tau) \leq \max_{z\in\partial\Sigma_{\pi-\alpha}} \rho(M(z)) \quad \text{with } \Sigma_{\pi-\alpha} = \{z : |arg z| \leq \pi - \alpha\} \cup \{0\} . \tag{4.72}$$

This bound holds without restriction on h, τ and T.

4.6.5 Numerical example

In order to illustrate the algorithms described in this chapter and to verify some of the results we shall calculate the solution of a parabolic partial differential equation with a time-periodic right-hand side. Other examples of time-periodic problems are treated in chapter 8. Consider the following problem on $\Omega = [0,1] \times [0,1]$ for $t \in [0,1]$,

$$\frac{\partial u}{\partial t} = \frac{\partial^2 u}{\partial x^2} + \frac{\partial^2 u}{\partial y^2} + f, \quad \text{with } u(0,x,y) = u(1,x,y), \tag{4.73}$$

with homogeneous Dirichlet boundary conditions, and with f equal to the "sawtooth"-function. The latter is defined as the periodic function of which the restriction to $[0,T)$ satisfies $f_{|[0,T)}(t,x,y) = t$. Here, the period T is equal to 1. The problem is discretized on a spatial mesh with mesh-size h equal to 1/4, 1/8, 1/16, 1/32 and 1/64. The resulting systems of ordinary differential equations are solved with time-periodic Jacobi and red-black Gauss-Seidel waveform relaxation, and also with the time-periodic multigrid waveform relaxation method. In the latter method we applied V-cycles with red-black Gauss-Seidel pre-smoothing and post-smoothing steps, standard coarsening down to a grid with only one grid point, full weighting restriction and bilinear interpolation. The trapezoidal rule was used for time-discretization, with a constant time-step $\tau = 1/100$. A constant zero profile was chosen as the starting iterate.

In table 4.2 we report the observed averaged convergence factors. The ones for the Gauss-Seidel and Jacobi methods closely correspond to the factors that can be obtained by evaluation of the theoretically derived formula (4.57). The multigrid convergence factors are clearly bounded by a constant less than one, independent of h.

Figure 4.3 pictures the evolution of the l_2-norm of the discrete residual for successive iterates of the multigrid algorithm, with different numbers of smoothing iterations. Note the constant slope of the lines. Finally, in figure 4.4 we have plotted several iterates, $u^{(k)}(t,x,y)$, evaluated at $(x,y) = (1/2,1/2)$, for $t \in [0,1]$ and $h = 1/16$. With the Gauss-Seidel method about 120 iterates are to be calculated, before they can no

Table 4.2: Averaged convergence factors for time-periodic waveform relaxation (with theoretical values for Jacobi and Gauss-Seidel methods).

h	1/4	1/8	1/16	1/32	1/64
Jacobi	0.640	0.916	0.979	0.995	0.999
$1 - \pi^2 h^2/2$	0.692	0.923	0.981	0.995	0.999
Gauss-Seidel	0.498	0.853	0.962	0.990	0.997
$1 - \pi^2 h^2$	0.383	0.846	0.961	0.990	0.998
V(1,1)	0.063	0.105	0.116	0.119	0.120

longer be distinguished from one another in the figure. One iteration of the multigrid method suffices. In the case of the full multigrid procedure – which is not shown in the picture – the plot of the initial approximation on the fine grid coincides with the plots of the subsequent iterates.

4.7 Autonomous time-periodic problems

4.7.1 Introduction

In this final section, we are interested in calculating periodic solutions to systems of *autonomous* parabolic partial differential equations of the form,

$$\frac{\partial u}{\partial t} = \mathcal{L}(u), \quad (t,x) \in I\!\!R \times \Omega \, ,$$

extended with suitable boundary conditions. The above system is called *autonomous* when the elliptic operator $\mathcal{L}(.)$ and the boundary conditions are not explicitly time-dependent. Since there is no external "forcing"-function from which information about the period of the solution can be obtained, that period is generally an additional unknown. Any non-constant solution of an autonomous problem generates an infinite number of "neigbouring" solutions. Indeed, when $u(t,x)$ is a solution so is any function of the form $u(t + \eta, x)$, $\eta \in I\!\!R$. To generate a particular unique solution, a supplementary so-called *phase-condition* is to be added to the system of equations.

Below, we present a solution technique which is based on a modification of the shooting method suggested by D. Roose. The algorithm, combined with the waveform relaxation method, was published in [106]. The performance of the algorithm is superior to that of most other approaches suggested in the literature, e.g. dynamic simulation, global discretization and (standard) shooting. Moreover, the basic idea is appealingly simple, and opens up promising directions for future research. We shall present the method in its current status of development. That is, there is little theoretical foundation, and experience is limited to a few examples.

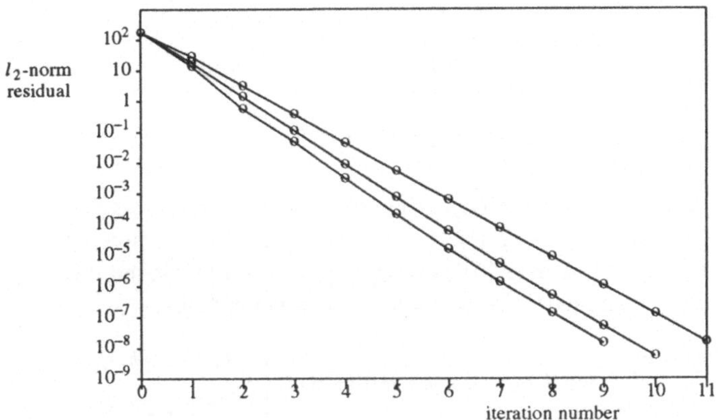

Figure 4.3: Evolution of the discrete residual, for V(1,1)-, V(2,1)-, and V(3,1)-cycles (top, middle, bottom), with $h= 1/32$, $\tau= 1/100$ and trapezoidal rule.

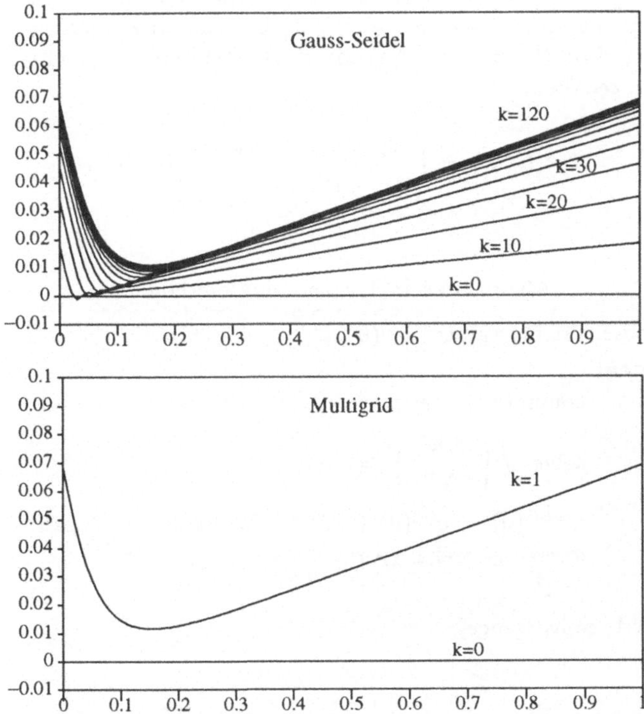

Figure 4.4: Waveform iterates $u^{(k)}(t, 1/2, 1/2)$, $t \in [0, 1]$ (one period), with $h = 1/16$.

4.7.2   Shooting with coarse grid Jacobian approximation

Semi-discretization and elimination of the boundary conditions lead to a large non-linear *autonomous* system of n ordinary differential equations,

$$\frac{d}{dt}u^h = L^h(u^h)\,, \quad t \in \mathbb{R}\,, \tag{4.74}$$

where u^h denotes the vector of unknown functions located at the grid points. Let T denote the (unknown) period of the (unknown) periodic solution u^h. By using the transformation $s = Tt$, and after adding a phase-condition, equation (4.74) is reformulated into the "standard" two-point boundary value form,

$$\left\{ \begin{array}{l} du^h/ds = TL^h(u^h) \\ dT/ds = 0 \end{array} \right. \quad \text{with boundary conditions} \quad \left\{ \begin{array}{l} u^h(1) - u^h(0) = 0 \\ P(u^h, T) = 0 \end{array} \right. \tag{4.75}$$

The latter boundary condition represents the phase-condition and is, e.g., of the form $P(u^h, T) := u_i^h(0) - c = 0$ for some $1 \le i \le n$ and for a suitably chosen constant c. We refer to [98, 117] for a description of other phase-conditions, in particular for conditions that are appropriate in the context of a continuation procedure.

Let $\phi(u^h(0), T)$ denote the solution at $s = 1$ of the differential equation in (4.75) with $(u^h(0), T)$ as the initial condition at $s = 0$. The shooting algorithm, written out in alg. 4.3, is based on the use of a Newton or Newton-like iteration for solving the so-called *residual equation*,

$$r(u^h(0), T) := \left(\begin{array}{c} \phi(u^h(0), T) - u^h(0) \\ P(u^h, T) \end{array} \right) = 0\,. \tag{4.76}$$

Algorithm 4.3: The shooting method.

> choose starting values $u^{(0)}(0), T^{(0)}$; $\nu := 0$.
> **repeat**
> > compute the residual $r(u^{(\nu)}(0), T^{(\nu)})$.
> >
> > solve $\tilde{J} \left(\begin{array}{c} \Delta u^h \\ \Delta T \end{array} \right) = -r$.
> >
> > $u^{(\nu+1)}(0) := u^{(\nu)}(0) + \lambda \Delta u^h$ (λ: damping factor).
> > $T^{(\nu+1)} := T^{(\nu)} + \lambda \Delta T$.
> >
> > $\nu := \nu + 1$.
> **until** convergence

The computation of the residual in each iteration step requires a time-integration of the differential equation from $s = 0$ to $s = 1$, with initial condition $u^h(0) = u^{(\nu)}(0)$ and with $T = T^{(\nu)}$. The matrix $\tilde{J} \in \mathbb{R}^{(n+1)\times(n+1)}$ represents (an approximation of) the

Jacobian matrix of first-order partial derivatives of r. Theoretically, this matrix can be computed by integration of certain variational equations, see e.g. [57, 98]. However, it is often more economical to approximate the matrix by finite differencing. Denote the k-th column of the Jacobian by $J_{.k}$, with

$$J_{.k} = \frac{\partial r}{\partial u_k^h} \ (k = 1, \ldots, n) \ \text{ and } \ J_{.n+1} = \frac{\partial r}{\partial T} \ .$$

By finite differencing these columns are numerically approximated, for $k = 1, \ldots, n$,

$$J_{.k} \approx \frac{1}{\delta}(\ r(u^h(0) + \delta e_k, T) - r(u^h(0), T) \ ) \ . \tag{4.77}$$

e_k is the vector with 1 as its k-th element and zero elements elsewhere. δ is a small real scalar. A similar formula holds for $J_{.n+1}$. The first n components of the $(n+1)$-element vector $J_{.k}$ are given by,

$$\frac{1}{\delta}(\ \phi(u^h(0) + \delta e_k, T) - \phi(u^h(0), T) - \delta e_k \ ) \ . \tag{4.78}$$

The calculation of the first n entries of each column of $\tilde{J}$ corresponds to one time-integration of (4.74). To avoid the cost of frequent matrix set-up, and to reduce the computational cost of the linear system solver in alg. 4.3, $\tilde{J}$ is often kept fixed during a number of shooting iterations. This leads to a chord-Newton iteration.

It should be clear that the standard shooting approach can be prohibitively expensive, especially when fine meshes are used for semi-discretization and when the PDE is two- or three-dimensional. To reduce the cost of Jacobian set-up a *coarse grid Jacobian approximation* was proposed in [106]. We briefly explain the basic idea.

For notational convenience, we assume a one-dimensional parabolic problem on the unit interval, discretized in space with mesh size $h = 1/2^p$, i.e., $n = 2^p - 1$. Equation (4.78) indicates that the first n components of $J_{.k}$ $(k = 1, \ldots, n)$ can be considered to be approximately proportional to the perturbation of the state of the system after integration over time T, caused by a perturbation of the initial condition located in gridpoint x_k. In particular $J_{i,k}$ gives the resulting perturbation in gridpoint x_i. If the mesh size is sufficiently small, a perturbation of the initial condition in point x_k will have nearly the same effect in gridpoint x_i, as the effect in gridpoints x_{i+1} and x_{i-1} caused by a perturbation of the initial condition in grid points x_{k+1} and x_{k-1} respectively. In other words, the values of $J_{i,k}$, $J_{i-1,k-1}$ and $J_{i+1,k+1}$ are closely related.

A coarse grid approximation of the Jacobian matrix can now be obtained by computing $J_{.k}$ for k odd using (4.78) and by approximating $J_{.k}$ for k even by using a *shifted interpolation* between $J_{.k-1}$ and $J_{.k+1}$. With linear interpolation this leads to,

$$J_{i,k} = \tfrac{1}{2} \left(J_{i-1,k-1} + J_{i+1,k+1} \right) \ (i = 2, \ldots, n-1) \ ; \quad J_{1,k} = J_{2,k+1} \ ; \quad J_{n,k} = J_{n-1,k-1} \ .$$

Column $J_{.n+1}$ must be computed by numerical differentiation. One can even further reduce the number of time-integrations required, by computing the columns that are associated with an even coarser grid, e.g. having $2^{p-1} - 1$ equidistant points (in that case, however, extrapolation is required for some grid points near the PDE domain boundary, see [106]).

This technique can be generalized immediately to partial differential equations defined on a two-dimensional domain. In that case the columns associated with gridpoint $x_{k,l}$ can be computed by a shifted bilinear interpolation of the columns associated with the neighbouring gridpoints, i.e. $x_{k-1,l}$, $x_{k+1,l}$, $x_{k,l-1}$ and $x_{k,l+1}$.

4.7.3   The use of waveform relaxation within shooting

The time-integrations in the shooting procedure may be performed by using waveform relaxation, and in this particular case by using the multigrid algorithm. Besides the advantages of *high parallel efficiency* and *efficient vectorization*, which will be discussed in chapter 7, another advantage can be brought about.

Within a shooting approach a sequence of time-integrations is performed for the same differential equations with slightly perturbed initial conditions. In this case, the multigrid waveform relaxation algorithm can be started with the solution profile over the whole time interval, obtained in a previous step. This will significantly reduce the number of iterations needed to achieve a converged solution. This effect is illustrated in the example below. Note that classical time-stepping methods do not allow to exploit the availability of a good approximation of the solution over the whole time interval.

4.7.4   A numerical example: the Brusselator

We consider the *Brusselator model*, a model that has attracted considerable interest in the study of bifurcation phenomena. The two-dimensional Brusselator is described by a system of nonlinear partial differential equations defined over the unit square $\Omega = [0,1] \times [0,1]$:

$$\begin{cases} \dfrac{\partial X}{\partial t} = \dfrac{D_X}{L^2}\left(\dfrac{\partial^2 X}{\partial r^2} + \dfrac{\partial^2 X}{\partial s^2}\right) - (B+1)X + X^2 Y + A \\[3mm] \dfrac{\partial Y}{\partial t} = \dfrac{D_Y}{L^2}\left(\dfrac{\partial^2 Y}{\partial r^2} + \dfrac{\partial^2 Y}{\partial s^2}\right) - X^2 Y + BX \end{cases} \tag{4.79}$$

The functions $X(t,r,s)$ and $Y(t,r,s)$ denote chemical concentrations. The homogeneous concentrations A and B and the diffusion coefficients D_X and D_Y are considered to be fixed control parameters; L denotes the reactor length and is often used as a bifurcation parameter. For the control parameters we use the following values : $D_X = 0.004$, $D_Y = 0.008$, $A = 2.0$ and $B = 5.45$, and we consider constant Dirichlet boundary conditions $X = A$, $Y = B/A$. It can be verified that for all values of L, a homogeneous steady state solution exists, equal to the boundary conditions. It is shown in [48] that a supercritical Hopf bifurcation occurs at $L \approx 0.72$; furthermore, stable periodic spatially symmetric solutions exist for $L > 0.72$, with $T \approx 3.7$ close to the Hopf point.

The modified shooting technique was used to calculate the periodic solution of the Brusselator model for the values of $L = 0.9$ and $L = 1.1$. We used a second order finite difference discretization on an equidistant mesh, with mesh sizes $h = 1/8$, $1/16$, resulting in a system of ODEs of dimension 98 and 450 respectively. For both gridsizes the coarse grid Jacobian approximation was obtained by computing the columns

Table 4.3: Computation of periodic solutions by shooting (two-dimensional Brusselator with $L = 1.1$): convergence history of the damped Chord-Newton iteration with required number of waveform V(1,1)-cycles.

$h = 1/8$ (7 × 7 gridpoints)				$h = 1/16$ (15 × 15 gridpoints)			
iter. step	$\| \text{resid.} \|$	$\| \Delta u^h \|$	number of cycles	iter. step	$\| \text{resid.} \|$	$\| \Delta u^h \|$	number of cycles
0	0.26 E-1		5	0	0.51 E-1		9
	coarse grid approx of J		3[†]		coarse grid approx of J		3[†]
1	0.15 E-01	0.42 E-01	5	1	0.30 E-01	0.12 E 00	8
2	0.11 E-01	0.31 E-01	5	2[‡]	0.37 E-01	0.80 E-01	8
3	0.78 E-02	0.20 E-01	5	2	0.74 E-02	0.40 E-01	8
4	0.53 E-02	0.14 E-01	5	3	0.72 E-02	0.15 E-01	8
5	0.34 E-02	0.91 E-02	4	4[‡]	0.92 E-02	0.17 E-01	7
	...	...		4	0.13 E-02	0.85 E-02	7
10	0.24 E-03	0.69 E-03	4	5[‡]	0.17 E-02	0.31 E-02	6
11	0.12 E-03	0.37 E-03	3	5	0.30 E-03	0.15 E-02	6
12	0.62 E-04	0.19 E-03	3	6[‡]	0.38 E-03	0.70 E-03	6
13	0.29 E-04	0.94 E-04	3	6	0.67 E-04	0.35 E-03	6

†: average number of cycles for the 19 time-integrations needed to compute $\tilde{J}$
‡: increase of $\|$ residual $\|$: result rejected; damping factor decreased.

corresponding to a 3 × 3-grid by numerical differencing. The remaining columns were computed by interpolation (and extrapolation near the PDE domain boundary). A total of 19 time-integrations were needed to calculate $\tilde{J}$, which was kept fixed during the chord-Newton iteration.

The convergence history for the $L = 1.1$ problem is shown in table 4.3. As an initial approximation we used the solution to the $L = 0.9$ problem. As the shooting iteration proceeds, better starting profiles for the waveform relaxation process become available. This leads to a reduction in the number of multigrid cycles needed to achieve the requested accuracy for the time-integration (algebraic error equal to 10^{-7}). For the time-integrations needed to compute $\tilde{J}$, by using formula (4.78) with $\delta = 10^{-4}$, very good starting values are always available; hence only very few cycles are required.

The convergence history for the $L = 0.9$ problem is similar and therefore omitted. Its solution is shown in figure 4.5 as a series of plots of $X(t, r, s)$, evaluated at an equidistant set of points along the time-axis.

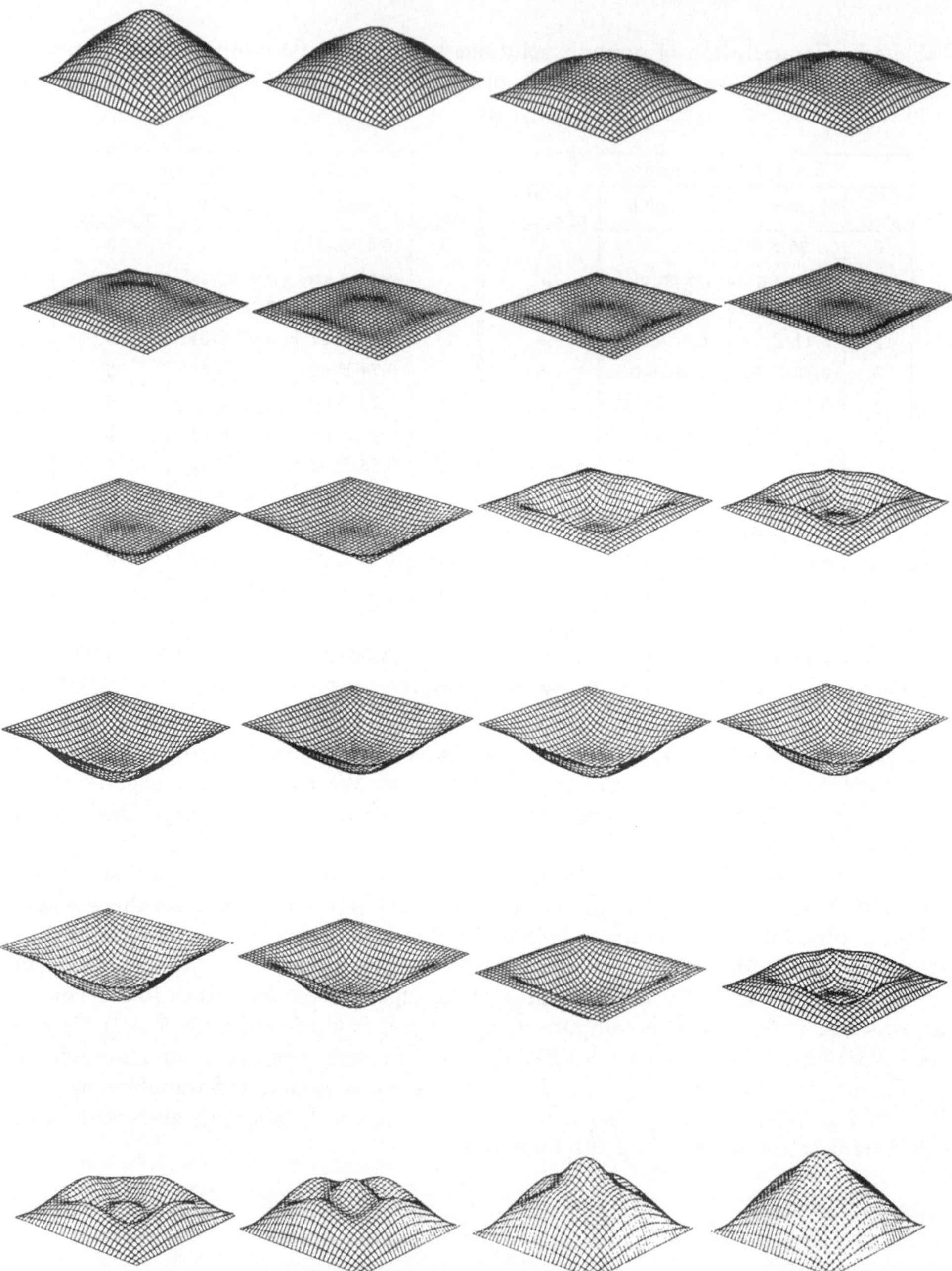

Figure 4.5: Periodic solution of the two-dimensional Brusselator with $L=0.9$ and $h=1/32$, represented by $X(t_i, r, s)$ $(i = 0, \ldots, 23)$ $(\Delta t_i \approx 0.16$, period $T \approx 3.9)$.

Chapter 5

A Short Introduction to Parallel Computers and Parallel Computing

A major concern which is frequently voiced in connection with computing machines,
particularly in view of the extremely high speed which may now be hoped for,
is that they will do themselves out of business rapidly; that is, that they will
out-run the planning and coding which they require and, therefore, run out of work.

—John von Neumann, *"The future of high speed computing",*
address presented at the IBM Seminar on Scientific Computing, November 1949.

This chapter provides a short elementary introduction to parallel computers and parallel computing, and is a prerequisite for a good understanding of the following chapters. We present some architectural multiprocessor characteristics with the emphasis on distributed memory multicomputers. We discuss the hypercube topology in particular, and we illustrate its connectivity by recalling a number of well-known properties. A particular multicomputer with hypercube topology, the Intel iPSC/2, which will be used for the experiments in later chapters, is discussed in some more detail. Finally, a number of important performance parameters, such as speedup and parallel efficiency, are defined.

5.1   Introduction

It is becoming increasingly difficult for the developers of traditional mono-processor supercomputers to meet the computational requirements of complex present-day scientific and engineering problems. One reason is the very high cost involved in the development of faster computing elements. Another reason has to do with physical barriers such as the finite speed of light, which necessitates further miniaturization in order to decrease signal travel times, and engineering limitations w.r.t. interconnection contacts and chip surface cooling, which inhibit such miniaturization. As a result, during the last decade a lot of effort has been put into the development of systems consisting of a number of cooperating processors. Some of these designs use a small number of very powerful (vector)-processors. Others interconnect a large number of off-the-shelf microprocessors, or custom-designed processors with mini-computer performance. Still others provide a massive number of very simple (often 1 bit) processing elements.

The programming of such parallel machines differs significantly from that of monoprocessor or sequential computer systems. Various new programming issues have arisen. They involve techniques for distributing complex tasks onto several processors in such a way that load-balance is maintained. They deal with methods to share and pass information between processors, e.g., by mutual exclusion, synchronization and message passing constructs. New parallel languages and language extensions have been proposed. These issues, together with a description of past and present accomplishments in the field of parallel computing are discussed at length in several books, e.g., by Hockney and Jesshope [56], by Fox et al. [31], and by Quinn [101].

In this chapter we would like to present a brief overview of some topics that are important for a good understanding of the remainder of the text. Our discussion deals with one parallel computer in particular, the Intel iPSC/2 hypercube, which is the machine used in our experiments. A classification of parallel computers is given in section 5.2. This will allow the reader to situate the above machine in the wide range of possible hardware designs. Section 5.3 discusses the hypercube interconnection topology. Some hardware and software characteristics of the Intel iPSC/2 are reported in section 5.4. This section is necessary to appreciate the timing results given in later chapters. Basic parallel performance parameters, such as speedup and efficiency are defined in the final section, section 5.5.

5.2   Classification of parallel computers

The multiprocessor used in our study is a *general purpose, large grain-size, distributed memory, MIMD* multiprocessor with *hypercube* interconnection topology. We shall briefly elaborate on this definition and clarify each of its constituent components.

• **Classification of Flynn.**   In 1966 Flynn proposed a computer classification scheme based on *instruction streams* and *data streams*, that is, based on the way the processors operate and handle the data, [30]. He discerns the following four classes.

SISD (single instruction stream – single data stream). This is a conventional computer, in which a single data stream is processed by a single instruction stream. That is, there is a single instruction unit which sequentially processes the data. This type of computer is also called a serial *von Neumann* computer.

SIMD (single instruction stream – multiple data stream). Here, a single instruction stream processes many data streams. Typical realizations are the array computers, in which a multitude of small processors execute an identical sequence of operations in lock-step mode. Other examples are pipelined vector processors.

MISD (multiple instruction stream – single data stream). We quote from [56, p.28]: "This class seems to be void, because it implies that several instructions are operating on a data item simultaneously". Some authors include pipeline computers in this class, though it is doubtful whether this is appropriate.

MIMD (multiple instruction stream – multiple data stream). Here, different processors work on different data streams. This class includes all multiprocessor systems in which each processor executes its own program on its own data.

Flynn's classification does not fully satisfy the needs of computer architects because it is not fine enough and because the interpretation of the class MISD is not clear. A further difficulty occurs if a computer contains both parallelism and pipelining. A processor of a present-day MIMD machine is often an SIMD vector processor itself.

● **Distributed memory and shared memory machines.** In shared memory architectures processors are connected via a bus or via an interconnection network to a global common memory. They communicate by accessing the shared data in the common address space. A fundamental limitation of these configurations is that they are not scalable to a very high number of processors. This is due to the limited system bus or network capacity and the occurrence of memory access contention and collisions. Typical examples of this class consist of, say, 2 to 20 processors (e.g., Sequent, Alliant).

In distributed memory multiprocessors each processor possesses a local memory, and data is exchanged by passing messages over an interconnection network. This type of architecture does not suffer as much from the scalability problem. When the number of processors increases so will the number of communication links and the total communication bandwidth. MIMD distributed memory machines have been built with more than a thousand processors (e.g., Ncube/2). Commercial SIMD distributed memory machines typically have even larger numbers of processors (e.g., the Connection Machine, with 65 536 processors).

● **Small and large grain-size machines.** Grain-size is sometimes considered a property of a parallel algorithm, e.g., [101, p.61]. It is then defined as the relative amount of work done between synchronizations or communications. *Fine grain-size programs* are programs in which very few instructions are executed inbetween two successive synchronization or communication steps. They are opposed to *coarse grain-size programs* in which substantial program sections are executed before any such interprocessor activity becomes necessary.

Others define grain-size as a property of a parallel machine, e.g., [31, p.22]; more precisely, it is then related to the amount of local memory. *Small grain-size processors* are processors with very little memory, say a few Kbyte. They are mostly of SIMD type. *Large grain-size processors* typically have half a Mbyte or more local memory. Note that fine grain-size programs are usually executed on small grain-size processors, which provide the necessary hardware and software for very fast synchronization and communication. Coarse grain-size programs often require the memory provided only on large grain-size computers. The performance of the communication system is then often of lesser importance.

- **General-purpose versus special-purpose machines.**   These characterizations are self-explanatory. Most modern multiprocessors are general-purpose machines. Some however have been built with one particular application in mind, e.g. the GF-11, an IBM product designed for solving quantum field theory problems at Gflop speeds (1 Gflop $= 10^9$ floating point operations per second). An important class of special purpose parallel computers are the so-called *systolic arrays*. They are SIMD, small grainsize, parallel machines characterized by a fixed topology and by processing elements that are specialized towards one particular application, such as matrix multiplication or image deconvolution.

- **Interconnection topologies.**   With a growing number of processors in a distributed memory machine it is getting more and more difficult to connect every processor to every other processor. When the *full interconnect* or *crossbar* is no longer feasible, one has to resort to restricted interconnection topologies with a smaller number of links. Popular topologies, because of the simplicity of the hardware implementation, are the linear array and the ring, the multi-dimensional array, the tree, and the hypercube.

5.3   The hypercube topology

In the current section we briefly characterize the hypercube topology. We recall its definition and mention without proof some of the basic properties. More detailed discussions can be found in a technical report by Saad and Schultz [108], and in papers by Johnsson [62], Chamberlain [16], and Chan and Schreiber [20].

5.3.1   Definition and properties

Definition 5.3.1 *An n-dimensional cube (hypercube, or boolean cube) is a graph of 2^n nodes labeled from 0 to $2^n - 1$ in such a way that there is an edge between any two nodes if and only if the binary representation of their labels differs in precisely one bit.*

This definition is illustrated in figure 5.1 where we show hypercubes of dimension zero up to four. We have numbered each node by its *label*. We shall often refer to that label as the *node number*, or the *physical node number*, when talking about the hypercube as the topology of a multiprocessor rather than as the topology of a graph. An equivalent definition is given below. It is constructive and defines the topology in a recursive manner.

Definition 5.3.2 *A zero-dimensional hypercube is just one node. A k-dimensional hypercube, with k greater than or equal to one, consists of two (k-1)-dimensional hypercubes with links between the corresponding processors in each half.*

The hypercube topology differs from many other topologies in that the distance between any two nodes is very small compared to the total number of nodes in the network. Equally important is that this is realized with a small number of connections per node. Both characteristics are particularized by the following two properties.

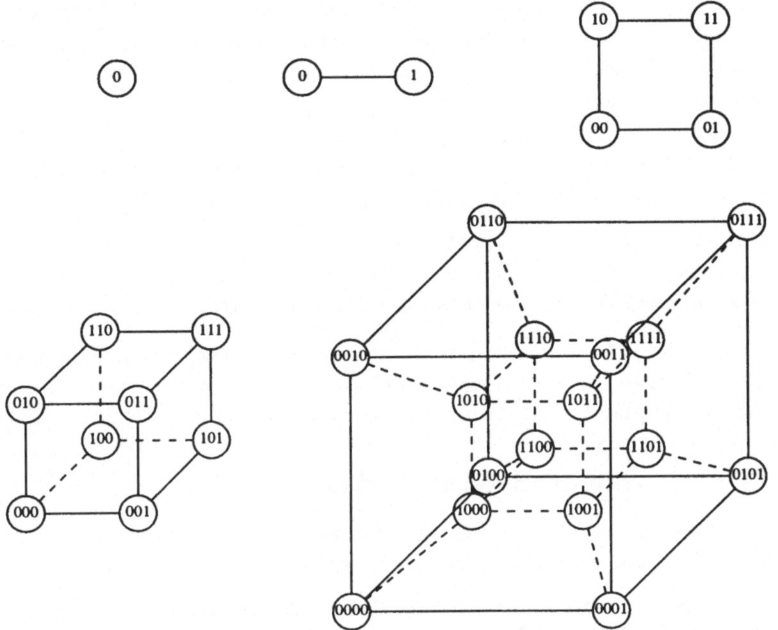

Figure 5.1: The zero-dimensional to four-dimensional hypercube topology

Property 5.3.1 *The degree of an n-dimensional hypercube equals n; each node is directly connected to precisely n other nodes.*

Property 5.3.2 *The diameter of an n-dimensional hypercube is n; to reach a node from any other node one needs to cross at most n internode connections.*

We shall denote the binary representation of the label of a node by the following bit sequence "$(a_{n-1}a_{n-2}\ldots a_1a_0)_2$", where each a_i is either zero or one. For notational convenience we shall usually omit the "$(\,\cdot\,)_2$" when labelling nodes in a figure. With definition 5.3.1, it is straightforward to identify the neighbouring nodes, i.e., the nodes connected by an edge to a given node.

Property 5.3.3 *The n neighbours of the node with label $(a_{n-1}a_{n-2}\ldots a_1a_0)_2$ are nodes $(a_{n-1}\ldots \bar{a}_i\ldots a_0)_2$, for all $i \in \{0, 1, \cdots, n-1\}$, where $\bar{a}_i$ denotes the complement of a_i.*

Example 5.3.1 The neighbours of node $(0110)_2$ in a 4-cube are the nodes $(0111)_2$, $(0100)_2$, $(0010)_2$ and $(1110)_2$. This is easily verified by looking at figure 5.1.

The following property discusses the distance between two given nodes, i.e., the minimum number of edges in a path connecting them.

Property 5.3.4 *The minimum distance between two nodes in an n-cube is given by the number of different bits in the binary representation of their labels.*

Example 5.3.2 The distance between $(0010)_2$ and $(1001)_2$ equals 3. Paths of minimal length between both nodes can easily be derived by successively complementing the bits where the binary sequences differ. One such path is given below,

$$(0010)_2 \longrightarrow (0011)_2 \longrightarrow (0001)_2 \longrightarrow (1001)_2 \ .$$

Finally, it will sometimes be important to be able to identify certain collections of nodes, the so-called *subcubes*, which inherit the hypercube properties.

Property 5.3.5 *Let Σ be the symbol set $\{0, 1, *\}$, where $*$ is a "don't care" symbol. Then, every subcube of a hypercube can be represented by a string of symbols in Σ. Such a string is called the address of the corresponding subcube.*

Example 5.3.3 Two three-dimensional subcubes which make up the four-dimensional cube of figure 5.1 have the addresses "$0 * **$" (outer cube) and "$1 * **$" (inner cube). The right-hand side plane of the outer cube consists of the nodes of subcube "$0 * *1$".

5.3.2   Binary reflected Gray codes

The use of so-called binary reflected Gray codes will be of importance when optimizing the processor allocation in various parallel applications. The definition is stated below.

Definition 5.3.3 *The k-bit binary reflected Gray code (BRG-code) denoted by G_k is recursively defined: $G_1 = \{0, 1\}$; from the i-bit code $G_i = \{g_0, g_1, g_2, \ldots g_{2^i-1}\}$, the $(i+1)$-bit code is generated as follows: $G_{i+1} = \{0g_0, 0g_1, 0g_2, \ldots, 0g_{2^i-1}, 1g_{2^i-1}, 1g_{2^i-2}, \ldots, 1g_0\}$.*

Example 5.3.4

$$G_2 = \{00, 01, 11, 10\}$$
$$G_3 = \{000, 001, 011, 010, 110, 111, 101, 100\}$$

The following two well-known properties are required for further reference. A proof of property 5.3.7 in given in [62, p. 138].

Property 5.3.6 *The binary reflected Gray code is periodic, that is, the first and the last element in the code differ by precisely one bit.*

Property 5.3.7 *If $\{g_0, g_1, g_2, \ldots g_{2^k-1}\}$ is the k-bit binary reflected Gray code, then g_j and $g_{(j+2^i) \bmod 2^k}$ differ in exactly 2 bits for $i \geq 1$.*

5.3.3   Topology embedding onto the hypercube

One of the main reasons for the interest in hypercubes is that many classical topologies can be efficiently embedded onto it. This is important since certain classes of algorithms fit particularly well on certain classes of topologies. For instance, algorithms based on the divide and conquer strategy fit well on the binary tree interconnection network. Matrix algorithms and various algorithms arising from problems in computational physics map naturally onto one-, two-, or three-dimensional arrays. When implementing these algorithms it is important that an embedding of their natural topology onto the hypercube is known.

- **Embedding of a linear array.** The embedding problem consists of finding a one-to-one mapping of linear-array-nodes onto hypercube-nodes in such a way that neighbouring relations are respected. More precisely, neighbouring nodes of the array should be mapped onto neighbouring nodes of the hypercube. The *binary embedding* strategy which maps array node i onto the hypercube node whose label is the binary representation of i, clearly does not respect the neighbouring relation. For instance, node one is then not directly connected to node two. They are at a distance two apart, since their hypercube labels differ in two bits. A strategy which does satisfy our requirement is the so-called *binary reflected Gray code embedding*,

$$\text{array node} : i \longrightarrow \text{hypercube node} : g_i . \tag{5.1}$$

Node i is then mapped onto the hypercube node whose label equals g_i, the i'th element in a BRG code. The embedding of an 8-element array onto a 3-cube is illustrated in figure 5.2. Note that by the BRG periodicity property 5.3.6 the neighbouring relation is also satisfied between the first and last array element (indicated by the presence of the dashed line in the figure).

- **Embedding of a multi-dimensional array.** A two-dimensional array with 2^k rows and 2^l columns can be embedded in a $(k+l)$-dimensional hypercube by assigning k bits to the row index and l bits to the column index. An example is given by the following rule,

$$\text{array node} : (i,j) \longrightarrow \text{hypercube node} : g_i \parallel g_j . \tag{5.2}$$

The node with logical coordinates (i,j) is mapped onto the hypercube node whose node number equals the concatenation ("$\parallel$") of g_i, the i-th number in a k-bit BRG code, and g_j, the j-th number in an l-bit BRG code. This is illustrated for a 2×4 array in figure 5.2.

Application of property 5.3.5 shows that any row or column of the array is mapped onto a subcube of the hypercube. Their addresses contain l "don't cares" (rows) or k "don't cares" (columns). In addition, the periodicity property establishes the existence of end-around connections for any row or column. Finally, the principle explained for embedding of a two dimensional array can be straightforwardly generalized to multi-dimensional arrays.

- **Embedding of trees.** The embedding of an unbalanced 3-ary tree rooted at node 0 is illustrated in figure 5.2. This tree topology is particularly useful. On a multiprocessor with hypercube topology it specifies the necessary interconnections to gather information from all of the nodes onto one node by the so-called *subcube induction* principle. In the above example the information flow would be as follows,

$$* * 1 \longrightarrow * * 0 ; \quad *10 \longrightarrow *00 ; \quad 100 \longrightarrow 000 .$$

The general embedding rule for such trees is, for instance, described in [82, p. s242].

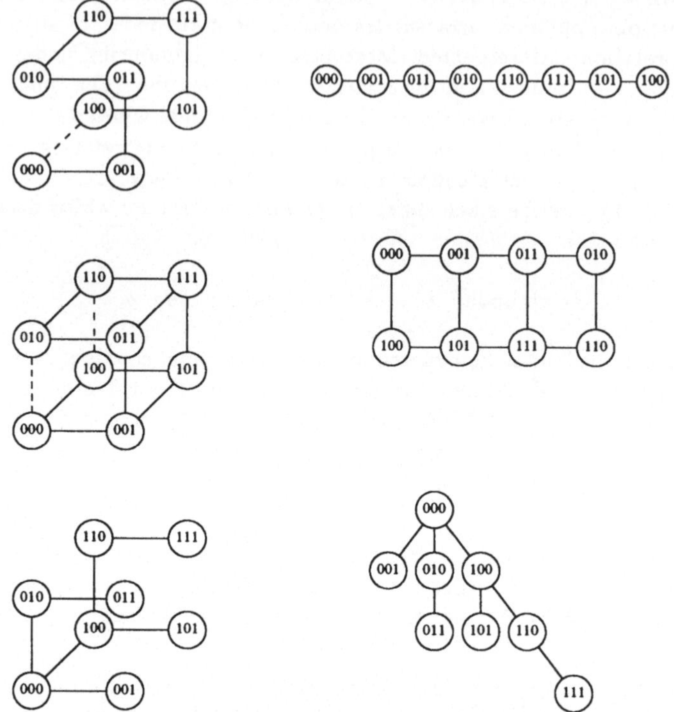

Figure 5.2: Topology embedding onto a three-dimensional hypercube

5.4   The Intel iPSC/2 hypercube multiprocessor

A hypercube multiprocessor is a parallel computing engine in which the processing
elements are thought to lie on the nodes of an n-dimensional cube with physical interconnections spanning along the edges. The hypercube is the interconnection topology
favoured by many hardware manufactures. This is because it balances node connectivity, communication diameter, algorithm embeddability, and programming ease. The
first such machine to be built was the Cosmic Cube, developed in 1983 by G. Fox and
C. Seitz at Caltech [116]. A few years later, machines with hypercube topology became commercially available from Intel, Ncube, Ametek, Floating Point Systems and
Thinking Machines. The machine used in our experiments is the second generation
machine from Intel, the Intel iPSC/2.

5.4.1   System overview

The basic node of the Intel iPSC/2 consists of an Intel 80386 processor extended with
a 80387 co-processor. In addition, each node contains up to 16 Mbyte of memory, a
64 Kbyte cache and a separate communication processor, called the *Direct Connect*

Module or *DCM*. The DCM supervises eight full duplex communication channels with performance of 2.8 Mbyte per second and per channel. Message routing and the set-up of communication paths between nodes is largely done by hardware.

The nodes run a small operating system, the Node eXecutive/2 or NX/2, which provides for process management and message passing. A total of 20 processes can be handled on each processor. Communication between processes on the iPSC/2 is *asynchronous*. There is no rendez-vous between processes at the moment of message passing. When the destination process is not yet expecting a message at the time of its arrival, the message data is temporarily buffered by the operating system. The communication primitives exist in a *blocking* and *non-blocking* version. In the blocking mode, the sending or receiving process is halted until the message has been sent or received. In the non-blocking mode the process is allowed to proceed with further computations while the DCM handles the message request.

The cube is controlled by a so-called *intermediate host computer*, the *System Resource Manager* or *SRM*. This is a small computer which operates standard software development tools, e.g., compiler, loader, debugger, etc. In addition, it provides for node monitoring and diagnostics, host to node communication, and it controls *cube sharing*, i.e., it allows different users to operate different subcubes simultaneously.

The cube is often programmed by providing two programs. A *host program* runs on the SRM or on another computer connected via a network to the SRM. This program takes care of I/O operations such as input from the keyboard and output to disks or graphics devices. A *node* program runs on each of the nodes, and performs the actual computations. Often the node programs are identical. Each node runs the same code. However, at any time nodes may be executing different sections of the code, according to their node numbers and their local data.

5.4.2 Some computation benchmarks

An extensive benchmark of various communication and computation parameters is found in the paper of Bomans and Roose [8]. We recollect some of their results.

• Some double precision floating point operation times are presented in table 5.1. They were obtained with the use of the Greenhills C Compiler. Note that for sequences of floating-point operations the arithmetic unit can retain intermediate results in registers, and that a certain amount of pipelining is exploited in the mathematical co-processor. For these operations a much higher flop rate is attained than in the case of single operations (e.g. 211 Kflops for the sum of three products, compared to 151 Kflops for a single addition and 117 Kflops for a single multiplication). The timing results for the copy operations include indexing and loop overhead, i.e., the cost of incrementing a counter for each pass through the loop in which the copy operation is executed.

• The cost of sending a message between neighbouring processors as a function of the message length is often modeled as follows,

$$t(n) = t_{startup} + n t_{send} , \qquad (5.3)$$

where n is the number of bytes transferred. On the iPSC/2, different values for $t_{startup}$ and t_{send} are found for short and long messages, see table 5.2. Note that $t_{startup}$ is

Table 5.1: Computation parameters for the Intel iPSC/2

$+$	$\times$	$\times + \times + \times$	a[i] = b[i],	a[i][j] = b[i][j]
6.64 μs	8.52 μs	23.65 μs	3.5 μs	5.4 μs

Table 5.2: Communication parameters for the Intel iPSC/2

	$t_{startup}(\mu s)$	$t_{send}(\mu s)$
short messages ($\leq$ 100 bytes)	350	0.2
long messages ($>$ 100 bytes)	660	0.36

rather large. It is therefore extremely important to structure the parallel algorithms in such a way that as few communication steps as possible are used.

• Messages to non-neighbouring nodes (so-called *multi-hop* messages) do not interfere with the computation going on in the intermediate nodes along the communication path. Communication with a far away node is therefore almost as fast as communication with a neighbouring node, provided there is a free communication path to that node. This is e.g. illustrated in [8, fig.4]. The same reference illustrates the potential for communication with calculation overlap when the non-blocking primitives are applied. When the message length is sufficiently large a substantial fraction of the cost associated with the nt_{send} term in (5.3) can be overlapped, see [8, fig.8]. Finally, we want to point at the full duplex nature of the channels, i.e., communication can occur simultaneously in both channel directions. This is advantageous when information is to be exchanged by two neighbouring processes. It is shown in [8, table 9] that the cost of an exchange, i.e., two simultaneous send/receive pairs, is only about 60% of that of two consecutive send/receive pairs. Note however that this is valid only when both processors start their exchange operations at about the same time. Otherwise, due to some architectural details, the overlap will be limited to part of $t_{startup}$, [15, 115], unless special software precautions are taken (e.g. use of the so-called "force-types").

5.5 Parallel performance parameters

Speedup and *efficiency* are two measures of the quality of an algorithm implemented on a parallel computer. The speedup achieved by a parallel algorithm running on p processors is often defined as the ratio of the time taken by that parallel computer executing the fastest known sequential algorithm and the time taken by the same parallel computer executing the parallel algorithm on p processors. The efficiency is equal to the speedup divided by p. With n denoting the problem size, this reads like,

$$\bar{S}(n,p) = \frac{T_{best}(n)}{T(n,p)} \quad \text{and} \quad \bar{E}(n,p) = \frac{\bar{S}(n,p)}{p} = \frac{T_{best}(n)}{p\,T(n,p)} . \tag{5.4}$$

Practical considerations limit the usefulness of these definitions. First of all, it is often very difficult to determine what algorithm is the best sequential one; this may depend on the problem size n, on the particular hardware used, on implementation issues, etc. Moreover, the notion of "best" algorithm may change in time, as better algorithms become available. Also, a good implementation of that algorithm is not always available. All this has led to an alternative and more practical definition of speedup. It is called the *relative speedup*, the *multiprocessor speedup*, or *parallel speedup* opposed to (5.4) which defines the *absolute speedup*.

$$S(n,p) = \frac{T(n,1)}{T(n,p)} \quad \text{and} \quad E(n,p) = \frac{S(n,p)}{p} = \frac{T(n,1)}{p\,T(n,p)} \; . \tag{5.5}$$

Here, $T(n,1)$ is the execution time of the parallel algorithm on a single processor. These will be the definitions used in the remainder of the text. For annotation in figures and tables, we shall further denote speedup and efficiency by S_p and E_p, where p is replaced by the actual number of processors used in the experiment.

Note that (5.5) does not give any information about the quality of an algorithm. It solely measures how well an algorithm has been parallelized. As such, it should always be complemented with data which indicate the *numerical efficiency* of the parallel algorithm. This could for instance be the following ratio of single processor execution times: $T_{seq}(n)/T(n,1)$, where $T_{seq}(n)$ is the execution time of a good sequential algorithm, or of a well-known numerical library routine.

If we assume that a p-processor machine cannot execute more than p times faster than a single processor machine, we obviously have that $S_p \leq p$ and $E_p \leq 100\%$. (An example of *superlinear* speedup, $S_p > p$, will be given and explained in section 6.7.) We enumerate some overheads that may cause a deviation from linear speedup.

• **the sequential fraction.** The speedup achievable on a parallel computer can significantly be limited by the existence of a small fraction of inherently sequential code which cannot be parallelized. This is expressed by *Amdahl's law*, see [2]:

> Let α be the fraction of operations in a computation that must be performed sequentially, where $0 \leq \alpha \leq 1$. The maximum speedup achievable by a parallel computer with p processors is then limited as follows,

$$S_p \leq \frac{1}{\alpha + (1-\alpha)/p} \leq \frac{1}{\alpha} \; . \tag{5.6}$$

For example, when 10% of the code must be executed sequentially, the maximum speedup is limited by 10, independent of the number of processors available.

Amdahl's law has been a central argument of people doubting the usefulness of massively parallel systems. Their criticism is justified as long as one considers solving a particular problem of a fixed size (i.e., with a constant value of α). In actual practice, however, this is rarely the case, as problem sizes tend to scale with the number of processors and with the computing power available. (Large-scale parallel processors are used to solved bigger problems than the ones solved on small-scale parallel processors.)

For many computational problems the sequential fraction α rapidly goes to zero as the problem size increases. Consequently, when problem scaling is in effect, α depends on the number of processors, and (5.6) looses much of its significance. An alternative to Amdahl's law was formulated in the paper by Gustafson, Montry and Benner [40].

Let $\bar{\alpha}$ denote the sequential fraction of the time spent during a computation on a parallel system with p processors. The maximum speedup achievable is then limited as follows,

$$S'_p \leq p\,(1 - \bar{\alpha}) + \bar{\alpha}\,. \tag{5.7}$$

Here, the "speedup" S'_p is usually called the *scaled speedup*. It is equal to the ratio T'_1 over T_p, where T'_1 is the time the parallel program would take to run on a single processor if sufficient resources (memory) were available. In actual practice $\bar{\alpha}$ is often a small number, and very high speedups are attainable on large-scale parallel processors. Several examples are reported in the above-mentioned paper.

- **non-optimal algorithm and algorithmic overhead.** The best sequential algorithm may often be difficult or impossible to parallelize. We shall encounter an example of such an algorithm in chapter 6 (Gaussian elimination for solving tridiagonal linear systems). In that case the parallel algorithm may have a larger operation count than the sequential one. Additionally, in order to avoid communication overhead a programmer may wish to duplicate some calculations on different processors, rather than having one processor doing the calculation and then distributing the result.

- **software overhead.** Parallelization based on the data decomposition approach often results in an increase of software overheads such as the overheads associated with indexing, procedure calls, etc. Also, this approach usually results in shorter loops, thus restricting vector lengths. This reduces the potential gain of using vectorization.

- **load imbalance.** The execution time of a parallel algorithm is determined by the execution time of the processor having the largest amount of work. As soon as the computational workload is not evenly distributed, load imbalance will result, and processor idling will occur. That is, processors will waste valuable computation time while waiting for other processors to finish a particular computation.

- **communication overhead.** Finally, any time spent in communicating results between processors is pure overhead.

Obviously, it is one of the major goals of parallel computing research to develop algorithms with low sequential fractions, which have similar complexity as the best sequential ones, which can efficiently be distributed onto a large number of processors and which require small communication overhead. Some of the many subtle issues involved towards achieving this goal will be illustrated in the next chapter.

Chapter 6

Parallel Implementation of Standard Parabolic Marching Schemes

> *We are on the threshold of a new era in computer architecture...*
> *A number of worthwile ideas have been formulated and implemented to some extent.*
> *Substantial further progress at this point requires results from real programs.*
>
> —G. S. Almasi, in *"Overview of parallel processing"*
> *Parallel Computing, Volume 2 (1985) pp. 191 –203.*
>
> *To the uninitiated the entire area of multiprocessor algorithms research*
> *appears to be a mundane exercise in code implementation, or a "support activity".*
> —Anthony Skjellum, *[119, p. 141]*

We analyse the parallel characteristics of several time-stepping schemes for linear parabolic partial differential equations. We discuss the classical explicit methods (forward Euler, Heun and DuFort-Frankel), three standard implicit methods (the first and second order backward differentiation formulae, and the Crank-Nicolson rule), the line hopscotch technique and the ADI formula of McKee and Mitchell. Three numerical kernels are identified and studied in particular: the explicit update step, the solution of a linear system by means of the multigrid method, and the solution of tridiagonal systems of equations by means of substructured Gaussian elimination. It is shown that numerical efficiency must usually be traded off against parallel efficiency.

6.1   Introduction

The time-accurate numerical solution of parabolic partial differential equations is computationally expensive, even on moderate grid sizes. As a result, considerable effort has been expended in formulating accurate and efficient numerical approximation methods. With the advent of parallel computing systems, efforts have been directed towards the development of new methods with a high inherent degree of parallelism, e.g. [28, 33, 105, 110, 140], and towards the implementation of the classical techniques on the novel hardware, e.g. [65, 77, 82, 109]. For the latter, attention is focused on the efficient implementation of a variety of linear algebra kernels, ranging from matrix vector products to sophisticated multigrid based linear system solvers.

When selecting an appropriate algorithm for a given parabolic problem on a particular machine one is often hampered by the lack of efficiency results for the total computational process. This should not only include the parallel characteristics of the numerical kernels and various other costs such as initialization, coefficient evaluation and matrix setup. The basic questions of numerical stability, consistency and accuracy should also be addressed. A selection based on a literature survey is further complicated by widely varying machine characteristics, the use of different programming languages and operating systems, different complexities of problems being solved, etc... As to our knowledge a systematic comparison on one machine of different parabolic marching schemes was lacking.

In this chapter we would like to present such a comparison of several well-known numerical methods, implemented for solving a fairly general problem class on a multiprocessor. The text is based on [144] and the companion report [143]. In the next section we briefly review some of the mathematical characteristics of the solution methods. We also identify the numerical kernels which serve as the basic building blocks for constructing the parabolic solvers. Starting in section 6.3, we give a detailed description of their parallel implementation. We discuss in particular the parallelization of an *explicit* update step (section 6.4), of the *multigrid method* (section 6.5), and of *tridiagonal system solvers* (section 6.6). Speedup and efficiency figures obtained on the Intel iPSC/2 hypercube are reported in section 6.7. Some numerical examples are presented in section 6.8. We end in section 6.9 with some concluding remarks.

6.2 Problem class and discretization

6.2.1 Problem class

We consider a general, linear, second order, parabolic partial differential equation,

$$\frac{\partial u}{\partial t} = C_{xx}\frac{\partial^2 u}{\partial x^2} + C_{xy}\frac{\partial^2 u}{\partial x \partial y} + C_{yy}\frac{\partial^2 u}{\partial y^2} + C_x\frac{\partial u}{\partial x} + C_y\frac{\partial u}{\partial y} + Cu + f, \qquad (6.1)$$

defined on a rectangular domain $\Omega = [a,b]\times[c,d]$, for $t \in [t_0, t_f]$, with a given initial value of u at time t_0, and with boundary conditions of type

$$u = g \ (\text{ Dirichlet }), \quad \frac{\partial u}{\partial n} + r\,u = s \ (\text{ Mixed }) \quad \text{or} \quad \text{periodic.} \qquad (6.2)$$

($\partial/\partial n$ denotes the derivative in the outside normal direction.) On each side of Ω, we allow one type of boundary condition. The differential equation coefficients and the functions f, g, r and s depend on x, y and t.

6.2.2 Discretization

We discretize (6.1) and (6.2) by using standard second order central finite differences on a rectangular grid, Ω^{h_x,h_y}, with $(I+1)\times(J+1)$ grid points,

$$\Omega^{h_x,h_y} = \left\{ (x_i, y_j) \mid x_i = a + ih_x,\ h_x = \frac{b-a}{I} \text{ and } y_j = c + jh_y,\ h_y = \frac{d-c}{J} \right\}. \quad (6.3)$$

We refer to appendix A for a detailed description of some technical issues related to the construction of the matrix stencils, the treatment of the boundary conditions, and so on. In the sequel we shall assume square mesh sizes, i.e., $h=h_x=h_y$, for notational convenience. Application of the numerical method of lines leads to a linear system of ordinary differential equations,

$$\frac{du^h}{dt} = L^h\,u^h + f^h, \quad (6.4)$$

where u^h is a vector, the components $u^h_{ij}(t)$ of which approximate $u(t, x_i, y_j)$. The vector f^h and the discretized operator L^h are time-dependent. L^h is typically a large matrix with five or nine diagonals, corresponding to the standard five-point and nine-point discretization stencils.

The methods for solving (6.1) or (6.4) are generally classified as either *explicit*, *implicit* or *semi-implicit*, depending on the time-discretization. An overview of some standard marching schemes of the three types is given below, mainly for setting the notation and in order to recall some basic properties. They may be found in standard text-books such as [72, 86]. Throughout this chapter, we assume that the time-increment for the underlying finite difference approximation, denoted by "τ", is a priori known and constant. In the formulae a subscript k will denote function values at time-level $t_k = t_0 + k\tau$. For instance, u^h_k is the set of grid point values $u^h_{k,i,j}$, $0 \leq i \leq I$ and $0 \leq j \leq J$, which approximate the ODE solution values $u^h_{i,j}(t_k)$ or the PDE solution values $u(t_k, x_i, y_j)$.

- A method is classified as *explicit*, if the values at a new time-level may be calculated directly from known values at previous time-levels (without having to solve a system of equations). Well-known is the forward Euler method, which is first order accurate in time,

$$u^h_{n+1} = u^h_n + \tau \left(L^h_n\,u^h_n + f^h_n \right). \quad (6.5)$$

Of higher accuracy is the second order explicit Runge Kutta formula, also known as the explicit method of Heun,

$$u^h_{n+1} = u^h_n + \frac{\tau}{2} \left(L^h_n\,u^h_n + f^h_n + L^h_{n+1}\,u^h_{n+1\,*} + f^h_{n+1} \right), \quad (6.6)$$

in which $u^h_{n+1\,*}$ is the solution obtained by Euler's formula. A careful analysis of (6.6) shows that it can be computed by applying a formula of similar nature as (6.5) twice. Both explicit schemes suffer from a severe stability constraint. The time-step has to be chosen very small in order to restrict excessive error growth. As such, the optimal time-step is restricted by stability, rather than by accuracy. An explicit algorithm that uses the values at two previous time-steps was presented by DuFort and Frankel. They start from the unconditionally unstable Richardson formula,

$$u^h_{n+1} = u^h_{n-1} + 2\tau \left(L^h_n u^h_n + f^h_n\right), \tag{6.7}$$

and derive a new formula by replacing $u^h_{n,i,j}$ by $\frac{1}{2}\{u^h_{n+1,i,j} + u^h_{n-1,i,j}\}$ in the equation at the (i,j)-th grid point, for every i and j. The modified scheme is unconditionally stable. The consistency order is $O(h^2 + \tau^2 + (\tau/h)^2)$ and as such, convergence is only guaranteed if $\lim_{h\to 0}\tau/h = 0$.

- We consider three *implicit* methods. As they are unconditionally stable the time-increment is limited by accuracy requirements only. The A-stable method of order two which supplies the smallest error constant is the trapezoidal rule, also known as the Crank-Nicolson method,

$$\left(1 - \frac{\tau}{2}L^h_{n+1}\right)u^h_{n+1} = \left(1 + \frac{\tau}{2}L^h_n\right)u^h_n + \frac{\tau}{2}\left(f^h_n + f^h_{n+1}\right). \tag{6.8}$$

It is of interest to note that (6.8) may be considered as a combination of the conditionally stable explicit scheme and an unconditionally stable implicit scheme, as expressed by formulae (6.9) and (6.10),

$$u^h_{n+1*} = u^h_n + \frac{\tau}{2}\left(L^h_n u^h_n + f^h_n\right) \tag{6.9}$$

$$\left(1 - \frac{\tau}{2}L^h_{n+1}\right)u^h_{n+1} = u^h_{n+1*} + \frac{\tau}{2}f^h_{n+1}. \tag{6.10}$$

The trapezoidal rule performs less satisfactorily for solving equations in which high frequency components occur in the solution, e.g., when a discontinuity between the initial values and boundary values exists. It may be shown that for adequate damping of these components τ must be of order $O(h)$, [73]. A family of methods based on backward differentiation does not suffer from this problem and allows arbitrarily large time-steps (from the viewpoint of stability). The simplest method in this family is the first order backward differentiation formula, BDF(1), or backward Euler method,

$$\left(1 - \tau L^h_{n+1}\right)u^h_{n+1} = u^h_n + \tau f^h_{n+1}. \tag{6.11}$$

The following second order formula, BDF(2), belongs to the same class,

$$\left(1 - \frac{2}{3}\tau L^h_{n+2}\right)u^h_{n+2} = \frac{4}{3}u^h_{n+1} - \frac{1}{3}u^h_n + \frac{2}{3}\tau f^h_{n+2}. \tag{6.12}$$

- The alternating direction implicit method (ADI) and the line hopscotch technique are classified as *semi-implicit* methods. In the ADI approach, a single cycle of computation requires the solution of two different sets of tridiagonal systems. In a first step an intermediate solution is obtained through a line-by-line solution of tridiagonal systems in the x-direction. The second step involves a line-by-line calculation in the

y-direction. Out of the few ADI formulae that can handle such general problems as (6.1), we implemented a generalized form of the formula of McKee and Mitchell,

$$(1 - \frac{\tau}{2}Lx_{n+1}^h)u_{n+1*}^h = (1 + \frac{\tau}{2}Lx_n^h + \tau\, Ly_{n+1/2}^h + \tau\, Lxy_{n+1/2}^h)u_n^h + \tau f_{n+1/2}^h \quad (6.13)$$

$$(1 - \frac{\tau}{2}Ly_{n+1}^h)\,u_{n+1}^h = u_{n+1*}^h - \frac{\tau}{2}Ly_{n+1}^h u_n^h \quad (6.14)$$

in which the subscript $n+1/2$ denotes the average of the function values at t_n and t_{n+1}. The directional operators Lx^h (Ly^h) and Lxy^h are given by

$$Lx^h = (C_{xx}\frac{\partial^2}{\partial x^2} + C_x\frac{\partial}{\partial x} + \frac{1}{2}C)^h, \quad Lxy^h = (C_{xy}\frac{\partial^2}{\partial x\partial y})^h. \quad (6.15)$$

The method is first order accurate in time and unconditionally stable. Second order accuracy is obtained when C_{xy} is identically zero.

In the line hopscotch method a red-black colouring is applied to the grid lines in either x-direction or y-direction. In a given time-step the forward Euler method (6.5) is used to calculate the new values on, say, the red grid lines. The backward Euler method (6.11) is then applied to compute the values on the black grid lines. One first uses the explicit formula. The implicit step then amounts to solving tridiagonal systems defined along the black grid lines. In the next step the role of red and black grid lines are interchanged. The method is stable but only conditionally consistent, just like the DuFort-Frankel formula. Except from the very first step the explicit evaluation can be done very efficiently. Indeed, the following relation holds on the grid lines along which the explicit formula is to be evaluated,

$$u_{n+1}^h = 2u_n^h - u_{n-1}^h. \quad (6.16)$$

6.2.3 The numerical kernels

We now identify the numerical kernels that are central to the discussed marching schemes.

- In an explicit computation a value at a grid point is calculated as a linear combination of known values at previous time-levels,

$$u_{n+1,i,j}^h = \sum_{k,l=-1}^{1} \{\alpha_{ij,kl}^n u_{n,i+k,j+l}^h\} + \beta_{ij}^n u_{n-1,i,j}^h. \quad (6.17)$$

This linear combination extends across nine grid points (five, if $C_{xy} \equiv 0$). The computational complexity of the linear combination does not depend on the particular method. Indeed, β_{ij}^n vanishes in the Euler method and in the two computational stages of the Heun method, and $\alpha_{ij,00}^n$ is zero by construction in the DuFort-Frankel scheme.

- The implicit methods transform the parabolic problem into a sequence of elliptic partial differential equations to be solved at the successive time-levels. The multigrid method has shown to be well-suited as a general iterative technique for solving these problems. It was applied successfully in the parabolic solvers described in [11, 38, 120,

123]. For a general discussion of the multigrid algorithm we refer to [44]. A short (minimal) introduction was presented in section 3.3.1.

As with any iterative algorithm it is important to supply a good initial approximation in order to restrict the number of iterations necessary for obtaining convergence. In a time-dependent problem the value of the solution at the previous time-level is usually adequate, especially when the time-step is small. If the solution is known to have a smooth time behaviour a low order extrapolation of the solution at previous time-levels may provide an even better approximation, see e.g. [120, 123]. A third possibility is based on the idea of nested iteration or full multigrid (and is usually opted for when solving elliptic partial differential equations). The initial approximation on the fine grid is then determined by interpolation of an approximation of the solution obtained by solving a coarse grid discretization of the problem.

• In the semi-implicit techniques tridiagonal systems of linear equations defined along grid lines are to be solved. In the case of a periodic boundary condition the tridiagonal coefficient matrix is extended with non-zero values in the upper right and lower left corners. The general format of the coefficient matrix for a system with N unknowns is given below in figure 6.1. The well-known tridiagonal Gaussian elimination algorithm, also known by the name of Thomas algorithm, is the fastest known solution method when the matrix is genuinely tridiagonal, i.e., when $c_0 = b_{N-1} = 0$. The computational complexity is then of order $8N$. This simple scheme is however not applicable in the case of periodic boundary conditions. In that case Gaussian elimination becomes about twice as expensive, i.e., order $17N$.

$$
\begin{pmatrix}
a_0 & b_0 & & & & & & c_0 \\
c_1 & a_1 & b_1 & & & & & \\
 & c_2 & a_2 & b_2 & & & & \\
 & & & \ddots & \ddots & \ddots & & \\
 & & & & \ddots & \ddots & \ddots & \\
 & & & & & c_{N-2} & a_{N-2} & b_{N-2} \\
b_{N-1} & & & & & & c_{N-1} & a_{N-1}
\end{pmatrix}
$$

Figure 6.1: A periodic tridiagonal coefficient matrix

6.3 Parallel implementation: preliminaries

6.3.1 The parallel computer model

The algorithms have been implemented on a Intel iPSC/2 hypercube multiprocessor. Our implementation is specialized towards the hypercube topology, and makes extensive use of the properties detailed in chapter 5. However, many of the ideas that are developed in the following sections immediately extend to other *distributed memory,*

MIMD computers. We envision in particular the use of the two-dimensional processor mesh. This processor structure has recently regained importance as the penalty for non-adjacent message exchanges was seriously reduced thanks to various improvements in communication hardware. Consequently, we shall try to keep the discussion as machine independent as possible. In particular, when discussing various communication schemes, e.g., based on rings, arrays or trees, we shall refer to the processors by their logical indices. We shall only refer to the hypercube-specific, i.e., Gray-encoded node numbers, when the hypercube properties allow for an optimized implementation.

Basic to the machines envisaged is that node processors independently execute their own program and that coordination and synchronization is performed by exchanging messages. To each message a communication cost is associated, part of which is independent of the message length, the startup time, and part of which is proportional to the message length. Messages can be sent from any processor to any other processor, but the communication cost, or at least a fraction of it, is proportional to the communication distance. Messages may float concurrently and undisturbed in distinct parts of the network. However, a penalty is to be paid if communication paths intersect. In this model it is of crucial importance to reduce the communication requirements of an algorithm by minimizing the number of messages, the length of each message and the distance between sending and receiving processors. The latter will minimize message travel times and avoid excessive communication channel contention (different messages wanting to use the same physical communication link).

6.3.2 The grid partitioning approach

For distributing the workload to the processors a grid partitioning strategy is to be selected. It should balance the computational workload, in the sense of keeping as many processors as possible computationally occupied at any time and it should minimize the communication overhead. One obvious possibility for achieving the first goal is a static decomposition in which a given processor is assigned the task of all computations associated with mesh points located in some a priori determined subset of the domain. The importance of the second goal is heavily dependent on the problem considered (amount of work to be performed per grid point, number of grid points per processor) and on the machine characteristics (relative cost of message transfer versus arithmetic, communication cost as a function of communication distance, etc.).

For our application we have implemented the standard decomposition that has been used successfully by many authors on a variety of multiprocessors, see e.g. [19, 21, 53, 81, 82, 121]. The processors are arranged in a rectangular array structure and are mapped onto the domain of the partial differential equation, see figure 6.2. Each processor is held responsible for doing all computations on the grid points in the interior of its part of the physical domain and on some of its boundaries (e.g. the south and west boundaries). With the terminology of [54], the processor is said to have the *update right* for the grid points in its subregion. Since the sizes of the subdomains are equal, an almost balanced load distribution will be guaranteed. However, when the number of processors is large and when the number of grid points is small, it may happen that some processors have no grid points in their subdomain. These processors are then called *inactive* or *idle* on that particular grid. The existence of inactive processors may lead to programming difficulties and a performance deterioration, as will be exemplified in the next sections. In particular, the neighbouring active processors are not necessarily adjacent processors in the processor mesh. Furthermore, their logical distances in the mesh may differ depending on the direction, see figure 6.3 for an example.

On the hypercube an optimal processor mapping can be performed following the binary reflected Gray code array embedding technique discussed in section 5.3.3. Each hypercube processor is then physically adjacent to its logically neighbouring processors in the processor mesh. Additionally, by the periodicity property 5.3.6, we know that a processor that is on the border of the mesh is physically connected to the processor that has the same row or column index on the opposite border. This end-around connection is especially important for problems with periodic boundary conditions.

The values at neighbouring grid points are often needed in the computations at a given grid point, e.g., when evaluating a discretization stencil or when calculating an average of grid point values. When that point is situated on the border of a processor's subdomain, values are needed that are treated by neighbouring processors. In order to avoid message passing each time such a value is referenced, we use the classical trick of storing in the processor's local memory a copy of the neighbouring grid line along each interior subdomain border. In the sequel we shall refer to these as the *grid lines in the overlap area*. The overlap area and the local subdomain together constitute the *extended subdomain* of a processor. This data decomposition is graphically represented in figure 6.2. We call a grid *consistent* if the grid points in the overlap area have their correct values, i.e., when they duplicate the values of the corresponding grid lines in the neighbouring processors. An *inconsistent* grid can be made consistent by issuing the appropriate message passing commands which will be discussed in the next section.

The calculations at each time-level can be divided into two separate steps. First, stencil coefficients are evaluated, matrices and right-hand sides are computed. This step requires no communication. Possible loss of full parallel efficiency is due to the load imbalance that may arise from an uneven distribution of grid points or is due to operations that have different arithmetic complexities for different grid points, e.g., in the treatment of boundary conditions. In the second step we either explicitly update the grid point values, or solve a linear system by means of the multigrid method, or solve a number of tridiagonal linear systems. The communication structure needed for each of these numerical kernels will be discussed next.

6.4 The explicit update step

The evaluation of the explicit update formula (6.17) can proceed concurrently at each grid point. No communication is needed provided that one starts with a consistent grid. If a consistent grid is needed for subsequent calculations, the values in the overlap area have to be communicated afterwards. This is not always necessary. E.g., the formula (6.9) is used to calculate the right-hand side in the Crank-Nicolson method. For the parallel solution of the resulting linear system, (6.10), the values in the overlap area of the right-hand side grid are not needed.

The values in the overlap area can be made consistent by exchanging the border

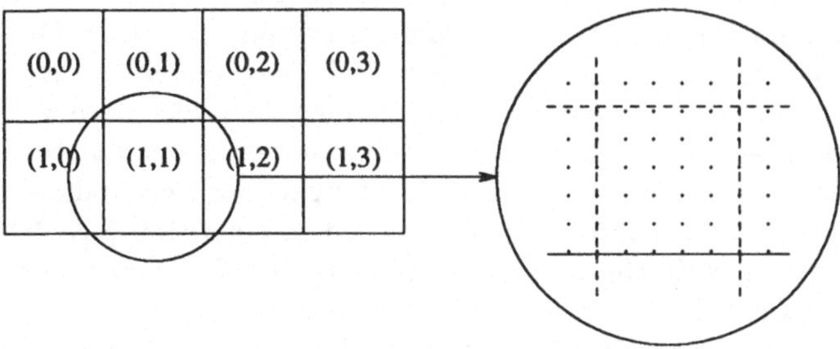

Figure 6.2: Grid partitioning and a processor's subdomain with overlap area.

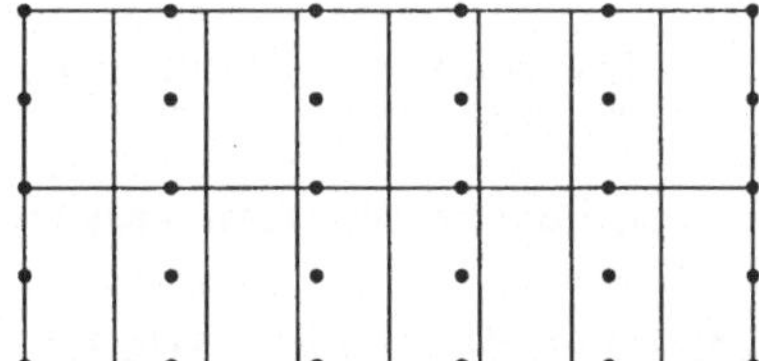

Figure 6.3: Active and inactive processors on a coarse grid; the neighbouring active processors of an active processor are not necessarily logically adjacent.

grid lines. Information is sent to and received from the neighbouring processors in the processor mesh. (End-around connections are used to identify neighbouring processors in the case of periodic boundary conditions.) The exchange of information for a five-point stencil grid is displayed in figure 6.4. The grid points depicted with "•" belong to the interior of the subdomain. The "o" grid points belong to the overlap region. The message exchanges may proceed concurrently in the four directions. Note that the same scheme is used whenever the corner values of the extended domain are not further needed. A different scheme is used in the case of a nine-point stencil update or, more in general, when the corner values of the extended region are needed in subsequent computations. To ensure correctness of the corner points, communication with diagonal neighbours is required, which is achieved with the communication scheme of figure 6.5. A horizontal exchange is followed by an exchange in the vertical direction.

Both communication schemes require a total of four message exchanges, one along each side. With the scheme of figure 6.4 the four messages may overlap and take place at the same time, at least in theory. The scheme of figure 6.5 requires two sequential steps in each of which two messages may proceed concurrently. In practice however — as was noted in section 5.4.2 and in reference [8]— only part of the message cost can be overlapped. In particular, the message startup time, which is often dominant, cannot be overlapped. Consequently, it is expected that the difference in communication time between both schemes will be minimal; this is confirmed by timing experiments.

In our implementation full advantage is taken of communication with calculation overlap by using the nonblocking iPSC/2 communication primitives. We first update the border grid points and immediately send their new values to the neighbouring processors. The calculation of interior grid points may then take place while the values of border grid points are being sent and while the values of the overlap region are being received. For instance, in the case of a five-point stencil explicit update, the series of operations executed by each processor is displayed in algorithm 6.1. Its advantages are threefold. To start with, the communication cost is reduced as part of it becomes "invisible" to the user. Secondly, by issuing the receive commands at such an early stage in the algorithm a possible copy operation of message data from a system buffer into the user message buffer is avoided, see section 5.4.1. Thirdly, when the number of interior grid points is large enough, chances are high that the messages will have been received when their data is needed. This avoids processors having to wait for information to arrive.

Algorithm 6.1: five-point stencil explicit update

```
update :
      -initiate receive operation from north, south, east and west neighbours
      -update the grid points on the local boundaries
      -initiate the send operation to the south, north, west and east neighbours
      -update the interior grid points
      -await the end of the receive operation
      -await the end of the send operation
```

Finally, we would like to mention one further implementation issue. The iPSC/2 communication primitives only support a send or receive operation of a vector stored contiguously in memory. By that, a copy operation transferring the message data to a temporary message buffer may have to take place before the actual send operation can be issued. Similarly, message data may have to be copied from the message buffer into the grid data structure after the receive operation has been completed. Our data structures are such that we store the grid points values by columns. Consequently, copy operations are required for the messages going in the north or south directions.

6.5   The multigrid solver

6.5.1   Introduction

The parallel implementation of multigrid has attracted a great deal of attention in the last decade, and a vast number of papers and reports have been devoted to the subject. This is so because of the ever growing importance of multigrid as a very

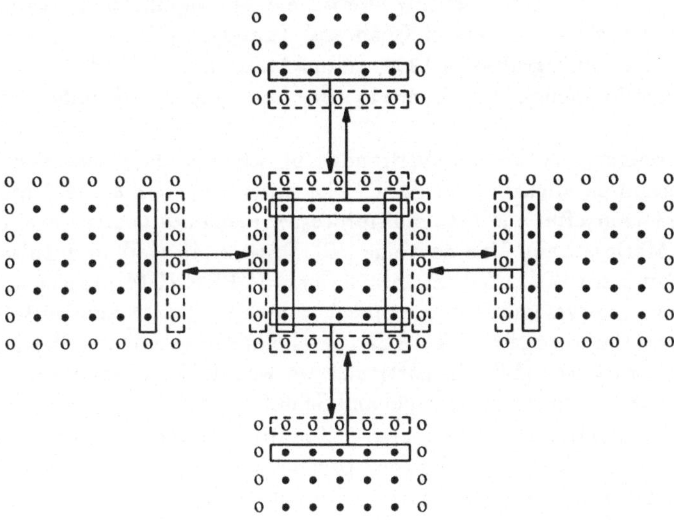

Figure 6.4: Concurrent exchange of local boundaries without update of the overlap corner grid points.

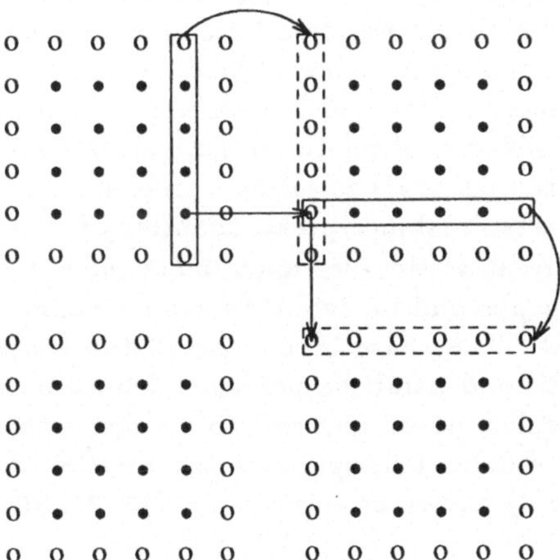

Figure 6.5: Two-step exchange of local boundaries with update of the overlap corner grid points.

efficient technique for solving complicated partial differential equations, and because of the many interesting computer science problems that arise in its implementation. We do not intend to overview the existing literature in any significant detail here. Instead, we refer to the overview papers of Chan and Tuminaro [23], who cover the state of the art in parallel multigrid until 1987, and of McBryan et al., [81], who also address more recent developments. A reference to some basic papers will suffice for our current purposes.

An early reference is the one by Brandt [10], which already discusses many of the essential issues in parallelizing multigrid. The experience with actual implementations was described soon after the first multiprocessor machines became available, e.g. by Thole [127], McBryan and Van de Velde [82], Bassett [4], Naik and Ta'asan [87], and Chan and Tuminaro [21]. Within the German SUPRENUM project substantial effort was put in the systematic development of highly efficient and portable multigrid codes, based on the so-called PARMACS (parallel macros) and the SUPRENUM communications library, see [53]. In particular we would like to mention the successful MGDEMO-code developed by Hempel and Schüller, [54].

The parallel inefficiency of the coarse grid multigrid operations was soon realized to be the main bottleneck in achieving good parallel performance. This inefficiency is due to the imbalance of communication and arithmetic complexities on these grids, and is due to the idling of many processors. This so-called *coarse grid problem* led to the development of various techniques for reducing the coarse grid communication complexity, such as agglomeration and shifting region partitionings [54, 87, 132] or multigrid variants in which some of the communication is skipped [54]. Other researchers developed parallel multigrid algorithms in which all of the processors are kept busy at all times, e.g., by simultaneous relaxations on different grid levels, or by having multiple coarse grids. We mention the concurrent iteration technique of Gannon and van Rosendale [35], the superconvergent multigrid method of Frederickson and McBryan [32], and the residual splitting technique of Chan and Tuminaro [22]. Recently, a lot of effort has been put into the parallelization of multigrid methods for solving complicated problems on irregular domains, such as 3D anisotropic problems, equations on general block structured domains and codes that incorporate adaptivity and irregular grid structures.

The goals of the current section are much more modest, though. We describe the implementation principles and we detail the communication requirements of some basic multigrid operators. They have been integrated in a library for solving linear variable-coefficient elliptic and parabolic problems defined on a rectangular domain, see [131, 132]. In particular, we do not seek to develop a theoretical model for the prediction and analysis of obtainable speedups and parallel efficiencies. Such studies have appeared elsewhere by numerous authors, e.g. [12, 21, 81, 150].

6.5.2 Parallelizing the multigrid components

• **Pre-smoothing** and **post-smoothing** are often performed by pointwise Gauss-Seidel relaxation, which shows good smoothing properties for a large class of isotropic and slightly anisotropic elliptic operators. The scheme that is usually preferred for the parallel implementation uses colouring, see e.g. [10] for an early reference. In the case of a five-point stencil, the grid points are divided into two subsets which are generally referred to as the *red* subset and the *black* subset. This division can be done in such a way that in order to calculate the update to a red point, only values of neighbouring black points are needed and vice versa, see e.g. the left picture in figure 6.6. The order of updating within a given subset is arbitrary; the calculation at one grid point does not influence the calculation at another. Consequently, the points with the same colour can be updated in parallel. A similar colouring scheme with four colours can be used to parallelize the nine-point stencil pointwise Gauss-Seidel relaxation, see the picture to the right. In conclusion, the global relaxation step is divided into two or four partial relaxation steps, which are executed in sequence.

In order to update the values in the overlap region, communication is needed after each partial relaxation step. The schemes depicted in figures 6.4 and 6.5 are used, but only values at grid points with the correct colour are sent. With the red/black scheme a total of eight messages is to be sent per processor, each with a length equal to half the number of grid points along one edge of the subdomain. The communication complexity of the four colour scheme varies according to the colouring strategy and the order of updating. A discussion of possible schemes and optimality results are presented in the papers by Kolp [68] and Alef [1]. We shall illustrate some of their ideas by counting the number of messages in two alternative four-colour strategies. Both algorithms start and end with a consistent grid. Consider figure 6.6 and the update order "+", "x", "o", "•", that is, first the "+" set is updated, then the "x" set, and so on. In this case the complexity is similar to that of the two-colour scheme. Two messages are to be sent in each direction. (We disregard issues of overlapping and synchronization, and only count the number of messages and their lengths.) Alternatively, consider the the ordering "+", "o", "•", "x". Observe that the processor to the north requires the updated values of the overlap "+" points only during the third partial relaxation step. By that, the message containing the "+" values, which would normally be sent after the first step, can be postponed. It can be combined with the message containing "o" grid points , and sent as one large message after the second partial relaxation step. The messages containing "•" and "x" grid points, destined to the southern processor can be grouped in a similar way. Compared to the previous scheme, the number of messages is reduced by two. The total message volume is unchanged. Our implementation of the four-colour relaxation employs the first ordering scheme, i.e., the one with two-colour scheme complexity. It allows for a more straightforward implementation as each boundary is treated similarly.

It has been mentioned before that the iPSC/2 communication primitives only support a send or a receive operation of a vector stored contiguously in memory. As such, the coloured subsets of the local boundaries are to be compressed to and decompressed from a contiguous message buffer before and after a communication step. In our data structures the vertical grid lines are contiguous. Consequently, it may be advantageous to send them as a whole, without first extracting the coloured subset. It is our experience that avoiding the copy operation outweighs the overhead of sending twice the amount of data actually needed.

• The **defect** or **residual** can be calculated in parallel at each grid point. Whether the values in the overlap region are to be communicated depends on the operation that

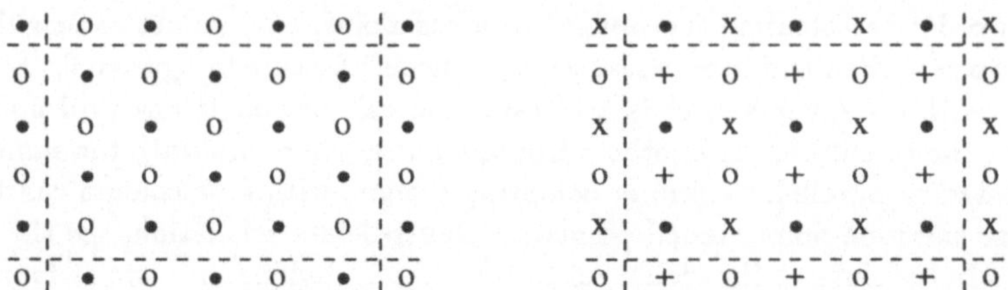

Figure 6.6: Grid colouring with two (left) and four (right) colours.

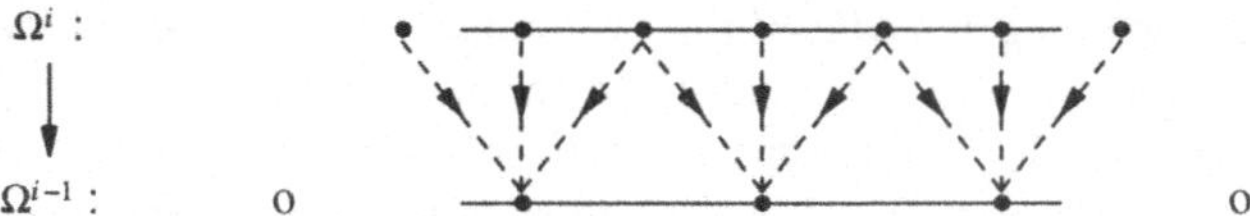

Figure 6.7: Dependence of coarse grid on fine grid values in full weighting.

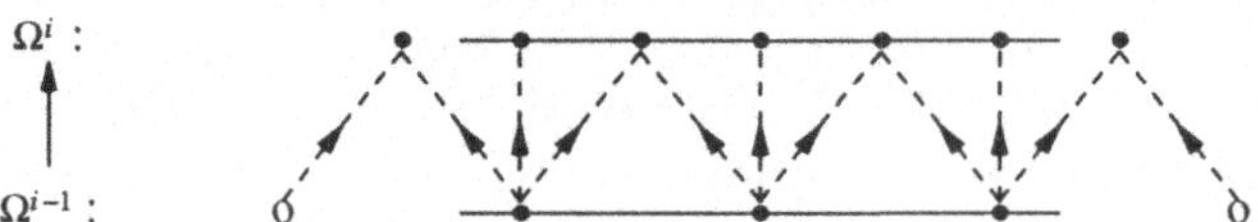

Figure 6.8: Dependence of fine grid on coarse grid values in a linear prolongation.

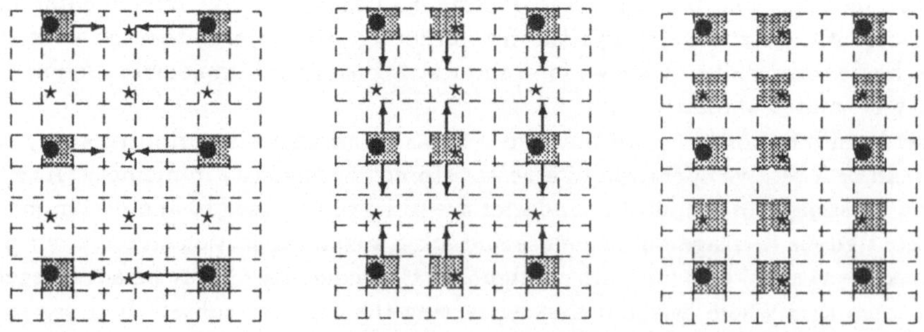

Figure 6.9: The reviving of idle processors in two steps, during the prolongation.

follows, i.e., the nature of the restriction. When the restriction is a full weighting, a consistent grid is required. Then the communication scheme of figure 6.5 is to be used.

- **The restriction** is an inter-grid operation. It transfers information from a fine grid Ω^i to a coarse grid Ω^{i-1}. A one-dimensional processor-subdomain on Ω^i and Ω^{i-1} is drawn by the solid lines in figure 6.7. The grid points that are indicated by a symbol "o" belong to the overlap region. The dependence of coarse on fine grid values in the case of a full weighting restriction operator is shown by means of the arrows. All interior coarse grid points can be calculated in parallel with the use of local information only, i.e., with grid point values that are readily available in memory. When the values in the overlap area are required in subsequent calculations, they must be communicated. This is normally not the case in the standard linear multigrid scheme. In the multigrid full approximation scheme, however, communication is required after the restriction of the fine grid approximation.

- **An** analogous picture, figure 6.8, clarifies the dependence of values of the fine grid points on the values of the coarse grid points in the case of a linear **interpolation** or **prolongation** formula. In this case the extended subdomain associated with the fine grid is completely contained in the extended region of the coarse grid. All prolongated grid point values, interior ones as well as the ones in the overlap region, can be calculated without any message passing.

A complication arises though, when a processor is inactive or idle on the coarse grid, but active on the fine grid. In that case, it has not participated in any of the coarse grid operations, and therefore it lacks the necessary information to calculate the interpolated values. A possible situation is displayed in the picture of figure 6.9. The coarse grid points are indicated by a "•" and the fine grid points that have no coarse grid counterpart by a "⋆". The subdomains are indicated by the squares. The active processors are the ones that are associated with shaded regions. The picture to the left shows the active processors on the coarse grid. Each of these can calculate the prolongated grid point values in their extended subdomains. The idle processors have entered a waiting state in a downward branch of a multigrid cycle, and need to be revived. This is done in two communication steps as is shown in the picture. First, the active processors awaken their "sleeping" neighbours in the horizontal direction. Those can afterwards calculate their interior fine grid points. This step is followed by a communication step in the vertical direction in order to revive the remaining idle processors. Observe that the communication operations are not necessarily of nearest neighbour type, and that the communication distances may vary in different directions.

In the full multigrid algorithm a bicubic interpolation formula is often used to prolongate the coarse grid approximation to the fine grid. Its parallel implementation is not as straightforward as that of the bilinear interpolation. Indeed, application of the bicubic formula requires a minimum number of four by four grid points. This complicates implementation on very coarse grids. For parallelizing this formula we have followed a similar approach as the one of Hempel and Schüller in [54]. When there are too few grid points in a processor's subdomain a biquadratic formula is applied, and unsymmetric formulae are used near processor boundaries. As in the above reference, this is the only place in which the parallel code returns results which are slightly different from those of the sequential code.

6.5.3   Agglomeration/deagglomeration strategies

Basic idea of agglomeration/deagglomeration

In a multigrid application a sequence of fine grids to coarse ones has to be mapped onto the processor structure. Several different mapping schemes have been suggested, such as schemes that map different grid levels on different processors [18, 20, 35] and schemes that map each grid level on all processors. In the latter case a distinction is made between schemes in which each processor is assigned to a fixed part of the domain, independent of the mesh size (so-called *fixed region partitioning* [87]), and schemes in which the subdomain of a processor differs from grid level to grid level (e.g. *agglomeration* [53, 54, 132] or *shifting region partitioning* [87]). In this section we shall discuss an approach based on agglomeration.

In the fixed region partitioning strategy each processor is associated with a fixed region of the computational domain, irrespective of the size of the discrete mesh. This has some definite implementation advantages. For instance, when a coarse grid point belongs to the subdomain of a processor, so will the corresponding grid points that belong to the finer grids. The only communication required in the multigrid operators is then of "intra-grid" type, that is, there is no communication across multigrid levels. With this strategy however some processors may have very few grid points on the coarse grids, or even no grid points at all. By that, this conceptually simple decomposition scheme results in a substantial reduction of performance. Indeed, it is known that small problems cannot be solved efficiently on systems with a large number of processors. The communication and various sequential overheads largely dominate the calculation. In particular, this is due to the fact that the number of messages and the associated message startup cost do not decrease with a decreasing number of grid points. (This is true only as long as the processors have two-colour boundaries in each coordinate direction. In addition on coarse grids a change in communication protocol (on the iPSC/2) may occur, leading to a shorter startup time.)

As a result it may be advantageous to compute the multigrid coarse grid operations with a smaller number of processors than the number of processors used on the fine grid. At a certain stage in the multigrid algorithm it is then necessary to gather the grid point values and reconstruct the grid on a subsystem of processors. The strategy by which the grid is collected is commonly called an *agglomeration strategy*. The opposite operation, by which a grid is expanded or scattered to a system with a larger number of processors, is called *deagglomeration*. The use of agglomeration/deagglomeration generally reduces communication distances and the number of send/receive operations in the remaining part of the multigrid algorithm. As an added advantage, possible programming difficulties associated with processors that do not have any grid points or very few grid points in their domain do not occur. The latter was an important argument for implementing agglomeration subroutines in the study [6]. Agglomeration implies moving to another processor for a large part of the grid points. It gives rise to a large amount of message traffic. This communication cost can be minimized by a careful choice of the active processors and the way they are mapped onto the agglomerated grids. This is particularly easy on a hypercube system, as will be illustrated below.

Some agglomeration/deagglomeration guidelines

We shall first derive some guidelines by looking at a few deficient assignment schemes shown in figure 6.10. The processors of a three-dimensional hypercube, arranged into a four-by-two processor mesh, are mapped onto a discretized domain (picture a). The subdomain each processor is originally assigned to is marked out by a rectangle, in which we have written the binary node number. (Note that we use the Gray code array embedding.) Suppose the considered grid has too few grid points to be efficiently distributed to eight processors. It might then be decided to reduce the number of processors and to continue on a two-by-two subcube. For this, the grid defined on the eight-processor system has to be redistributed to a four-processor system. A first possible assignment is illustrated in picture b. To arrive at this decomposition four nearest neighbour messages are required, initiated by the processors that lose their update rights. The latter send all of their grid point values to the remaining processors $((010)_2 \rightarrow (000)_2, (011)_2 \rightarrow (001)_2, (100)_2 \rightarrow (110)_2$ and $(101)_2 \rightarrow (111)_2)$. These messages can proceed concurrently. The resulting four-processor system however does not have the hypercube topology, a connectivity which we may wish to keep. A scheme which requires the same amount of communication and retains the hypercube nature of the system is displayed in picture c. It has the drawback that the mapping of the processor grid onto the computational grid logically differs from the original one. Consequently, subroutines written for the original partitioning could not be used for the agglomerated partitioning. A scheme which retains both the hypercube nature and the logical processor mapping is presented in picture d. The required number of messages however has increased to six, with a corresponding increase in total communication volume, and communication is no longer of nearest neighbour type.

This example illustrates some conditions which are to be imposed on the selection of a subsystem of processors. The subsystem has to be a hypercube, in order to allow a simple recursive application of agglomeration steps. The subsystem must be mapped onto the computational domain in the same way as the original grid, so as to facilitate code development. To minimize the communication overheads each processor of the agglomerated grid should retain the subdomain that it was originally assigned to. This will reduce the communication volume and the communication distances.

The reduce and exchange type agglomeration strategies

An agglomeration which satisfies the above requirements is stated below as algorithm 6.2. We call it *reduce type* agglomeration, since the dimension of the active cube is reduced by one after each agglomeration step. Note first that we assume a $'k+l'$-dimensional hypercube, the nodes of which are arranged in an array with 2^k rows and 2^l columns. By formula (5.2), the node number of processor (i, j) is then to be interpreted as the concatenation of k bit number $(b_{k-1}b_{k-2}\ldots b_1 b_0)_2$, and an l-bit number, $(a_{l-1}a_{l-2}\ldots a_1 a_0)_2$. Algorithm 6.2 specifies the necessary actions by each processor for an agglomeration in the vertical direction, that is to say, the number of rows in the processor mesh is halved. An agglomeration in the horizontal direction proceeds analogously, though bit a_0 should be considered.

The subcube of processors with bit b_0 equal to zero remains active, whereas the

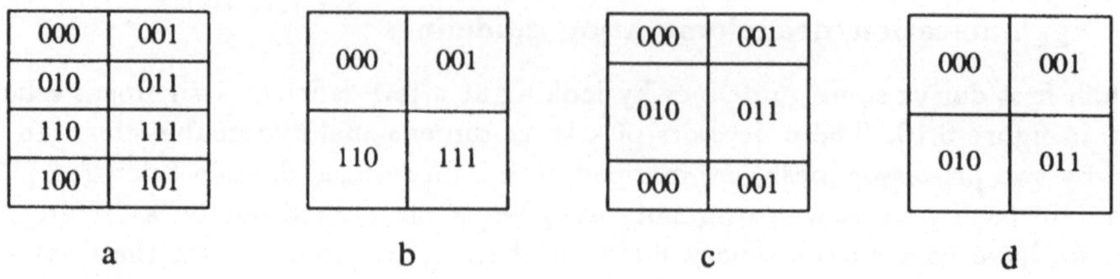

Figure 6.10: Grid agglomeration: different subgrid assignment schemes.

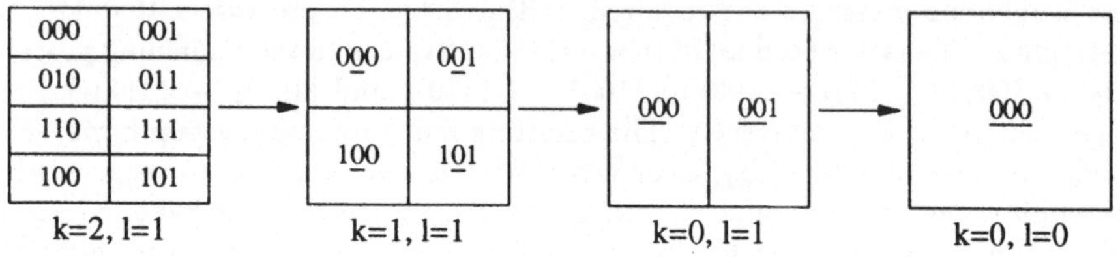

Figure 6.11: Reduce type agglomeration: in each agglomeration step half of the processors go idle; communication is required in the agglomeration and deagglomeration phases. (The underlined bits must be removed from the binary node representation when applying the agglomeration algorithm given in the text.)

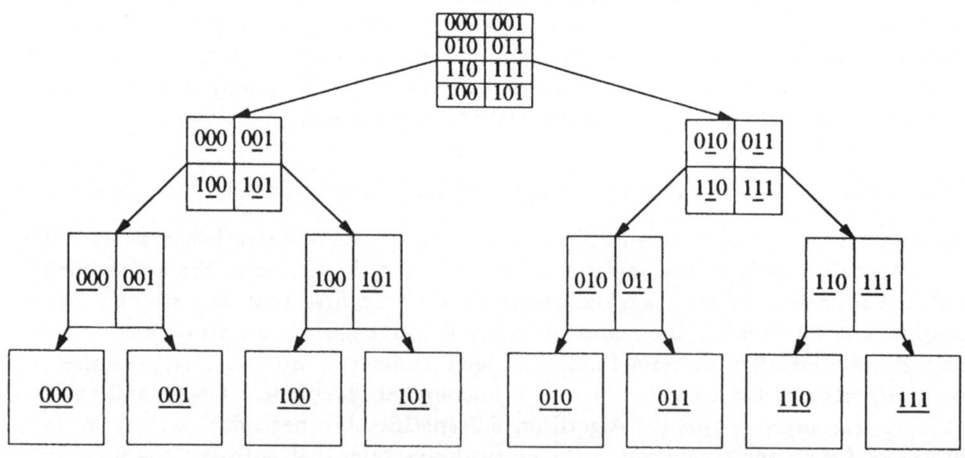

Figure 6.12: Exchange type agglomeration: no processor is idle; computations are duplicated in order to avoid communication in the deagglomeration phase.

Algorithm 6.2: reduce type agglomeration in the vertical direction.

(My node number is $(b_{k-1}\ldots b_1 b_0 a_{l-1}\ldots a_1 a_0)_2$ in a '$k+l$'-dimensional cube.)

if bit $b_0 = 1$ **then**
- send all grid point values to processor $(b_{k-1}\ldots b_1 0 a_{l-1}\ldots a_1 a_0)_2$
- become inactive

else
- receive all grid point values from processor $(b_{k-1}\ldots b_1 1 a_{l-1}\ldots a_1 a_0)_2$
- assemble the agglomerated grid
- continue execution in the '$k+l-1$'-dimensional subcube of processors of which bit b_0 equals 0.

endif

---

complementary subcube goes idle. The latter will be "revived" later on in an ascending branch of a multigrid cycle during a deagglomeration step. It can easily be seen that the communication is of nearest neighbour type, since communicating processors differ in precisely one bit. The total communication volume is equal to half the total number of grid points. Each processor of the "surviving" subcube doubles its subdomain by joining its grid points with those of a neighbouring region. When the zero b_0-bit is *deleted* from the binary node representation of the processors in the active '$k+l-1$'-dimensional hypercube, further agglomeration can be performed with the same algorithm. Such an agglomeration sequence is illustrated in figure 6.11. In this figure we have underlined the "deleted bits", i.e., the bits which must be removed from the binary node representation when applying the agglomeration algorithm.

An alternative agglomeration rule is suggested in the algorithm 6.3. It defines the so-called *exchange type agglomeration*, which is related to the *modified shifting region partitioning* scheme [87], and to a technique mentioned in [150]. Each processor sends all its values to and receives all the values from a companion processor, the binary representation of which differs in only one bit. Afterwards, when the two subcubes have assembled the entire grid, they separately continue with identical computations. As before, further agglomeration to still lower dimensional subcubes is possible, as is illustrated by an example in figure 6.12. Note that with this scheme no communication is necessary in the deagglomeration. Indeed, the local part of the deagglomerated grid is a subset of the local part of the agglomerated grid, with roughly half the number of grid points. So, each processor may locally assemble its part of the deagglomerated grid from values which it has readily available.

To conclude, in each agglomeration/deagglomeration step with exchange type agglomeration one nearest neighbour message exchange is executed per processor. As the communication between different pairs of processors may occur simultaneously, and since the communication paths do not intersect, the total communication cost is precisely the cost of one exchange operation. In the reduce type scheme there is one communication step in the agglomeration phase and one in the deagglomeration phase. This is slightly less efficient as will be shown by the timing results in section 6.7.2.

Algorithm 6.3: exchange type agglomeration in the vertical direction

(My node number is $(b_{k-1} \ldots b_1 b_0 a_{l-1} \ldots a_1 a_0)_2$ in a '$k+l$'-dimensional cube.)

- exchange (send+receive) all the grid point values with the grid point
  values of processor $(b_{k-1} \ldots b_1 \bar{b}_0 a_{l-1} \ldots a_1 a_0)_2$,
- assemble the agglomerated grid
- continue execution in the '$k+l-1$'-dimensional subcube of processors
  that have the same value of the b_0-bit

6.6   The tridiagonal systems solver

6.6.1   Introduction

The solution of tridiagonal systems of linear equations is central to many numerical algorithms. Consequently, the parallel implementation of tridiagonal solvers has received a lot of attention, e.g. in [19, 55, 63, 65, 69, 77, 109, 146]. In the ADI and line-hopscotch methods we are faced with the problem of solving tridiagonal systems defined along horizontal or vertical grid lines. The decomposition of each such system is predetermined by the overall decomposition chosen for the domain of the partial differential equation. From section 6.3.2, one sees that each system is evenly distributed over a linear array of processors. From the discussion about Gray code embedded processor meshes it is clear that (in a hypercube-optimized implementation) each of these linear arrays is actually a *subcube* of the hypercube.

Let P denote the number of processors in the linear array, and let d denote the corresponding subcube dimension, i.e., $P = 2^d$. We set M to the number of tridiagonal systems distributed over the linear array, and N to the number of equations per system. The equations are written as follows[1],

$$eq_i : \quad c_i x_{i-1} + a_i x_i + b_i x_{i+1} = r_i, \quad \text{with} \quad 0 \le i \le N-1 , \tag{6.18}$$

where any unknown x_i with index i for which $i < 0$ or $i > N-1$ is identified with the unknown $x_{i \bmod N}$. The coefficient matrix was given in figure 6.1.

The classical Thomas algorithm is inherently sequential and cannot be parallelized efficiently. (The two-processor case is an exception to this rule, see e.g. [55, 69], a special case which we shall not consider further.) In the multiple system case parallelization can be performed by using pipelining [55, 63], at the cost of a large number of messages, or by using a different decomposition scheme, called *scattered decomposition* [19, 65]. However, neither of the two is adequate in our application. A variety of other methods have been proposed, most of which are based on the ideas of substructuring and multifrontal elimination. The equations are locally condensed and reduced to a small tridiagonal system with one or two equations per processor. This intermediate system is solved in one way or another and the remaining variables are determined in

---

[1]For notational convenience we assume the coefficients a_i, b_i, c_i to be variables in the computer science sense. That is to say, their value may change during the computations.

a backsubstitution step. The algorithm that we have implemented is commonly called the algorithm of Wang [146]. A similar algorithm based on the substructuring idea appeared earlier in a paper by Sameh and Kuck [111]. In the next section we shall review the algorithm of Wang. Note that we consider a more general case than in the above references, since we allow the matrix to be periodic.

6.6.2 Substructured Gaussian elimination

We shall first describe the algorithm as it is applied for solving a *single* tridiagonal system of equations and afterwards make the extension to the case of multiple systems. Throughout the discussion we shall assume that the number of equations, N, is evenly divisible by the number of processors, P. The extension to the more general case does not pose any significant problems. The distribution of the tridiagonal system, derived from the overall domain decomposition, is such that each processor has n contiguous rows, with $n=N/P$. If the processors of the linear array are consecutively numbered from 0 to $P-1$, processor k is assigned to the following equations :

$$eq_i : \qquad c_i x_{i-1} + a_i x_i + b_i x_{i+1} = r_i \quad \text{with} \quad i = kn, \ldots, (k+1)n - 1 \qquad (6.19)$$

The distribution of a sixteen-equation system onto four processors is shown by way of illustration in figure 6.13 (upper left). Each processor is assigned to one contiguous block of equations as indicated by the dashed horizontal lines.

The system is reduced to a tridiagonal system with only one equation per processor in three steps.

- In the first step each processor k eliminates its lower diagonal elements c_i, with $i = kn + 1, \ldots, (k+1)n - 1$, a process which introduces fill-in elements f_i below the main diagonal. The resulting coefficient matrix is shown in figure 6.13 (upper right). The elimination can proceed concurrently on each processor and requires no communication.

- In the second step the upper diagonal elements b_i, with $i = (k+1)n - 3, \ldots, kn$, are eliminated, which is illustrated in the lower left figure. This step is again fully parallel. It introduces the fill-in elements g_i, situated above the main diagonal.

- In the final step the corner elements $b_{(k+1)n-1}$ are eliminated. For this, inter-processor communication is necessary. Each processor sends and receives one equation to and from a processor responsible for an adjacent set of equations. More precisely, processor k sends equation eq_{kn} to its "upper" neighbour, processor $(k-1) \bmod P$, and receives equation $eq_{(k+1)n \bmod P}$ from its "lower" neighbour, processor $(k+1) \bmod P$. The reduced coefficient matrix is shown in figure 6.13 (lower right).

The equations $eq_{(k+1)n-1}$, for $k = 0, \ldots, P-1$, constitute a decoupled linear system with one equation per processor. This system can be solved separately from the other unknowns. We postpone the discussion about how to solve this intermediate system to the next section. After this system has been solved, each processor can determine

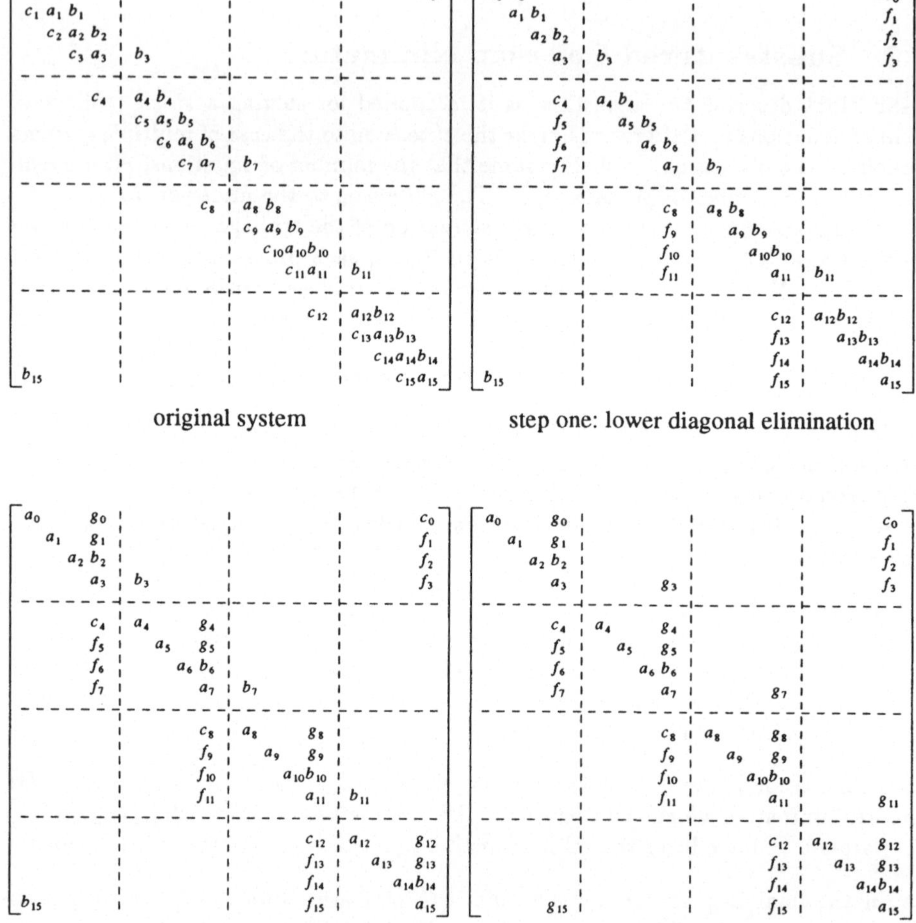

Figure 6.13: Substructured Gaussian elimination.

its remaining unknowns if it knows, besides the value of its local unknown $x_{(k+1)n-1}$, also the value of $x_{(kn-1)\bmod P}$, a value which was calculated by processor $(k-1)\bmod P$. Each processor therefore sends its local value $x_{(k+1)n-1}$ to the lower neighbour processor and receives the corresponding value from the upper neighbour. Backsubstitution for the interior unknowns can then proceed in parallel, without further message passing.

The algorithm immediately extends to the case of multiple tridiagonal systems. To minimize communication overhead each of the steps should be applied to all systems before proceeding to the next step. The message exchanges in the elimination phase can be grouped together and are sent as one large message. The message exchanges in the backsubstitution step can be treated similarly. The substructured Gaussian elimination algorithm is executed on each of the horizontal (or vertical) subcubes of the processor mesh simultaneously, with no communication between processors of different rows (or columns). The consistency of the grid of solution values must be guaranteed afterwards by updating the values in the overlap area. One of the additional border grid lines is already filled out correctly by the communication in the backsubstitution step. Consequently three additional messages are needed.

6.6.3 Solution of the intermediate system

In this section we shall give a brief characterization of three techniques that may be used to solve the intermediate system. They are interesting in that they introduce communication schemes different from the ones discussed previously. The intermediate system is first rewritten as

$$eq_k \; : \quad c_k x_{k-1} + a_k x_k + b_k x_{k+1} = r_k \quad \text{with} \quad k = 0,\ldots,P-1. \tag{6.20}$$

For notational convenience, we have renumbered the unknowns and re-used the symbols a_k, b_k, c_k and r_k. From the discussion in the previous section we recall that equation eq_k is stored by processor k, which is the "$k+1$"-th processor of the linear processor array. We first concentrate on the case $c_0=b_{P-1}=0$. The changes to include periodic boundary conditions are discussed afterwards.

The use of recursive doubling

In the first step of the recursive doubling method each equation eq_k is combined with eq_{k-1} and eq_{k+1} in order to eliminate the values of c_k and b_k. The original system of P equations is then decoupled into two systems of $P/2$ equations each. This is illustrated in figure 6.14. The two systems may be decoupled further by a recursive application of the same procedure. After $\log_2(P)$ steps, (6.20) is reduced to P systems of one equation only, which can be solved immediately. The basic step of the algorithm is mathematically formulated in the following remark.

Remark 6.6.1 In step j, with $j = 1,\ldots,\log_2(P)$, of the recursive doubling algorithm equation eq_k is combined with the equations at distance 2^{j-1}, i.e., with eq_s and with eq_t where $s = k - 2^{j-1}$ and $t = k + 2^{j-1}$, in order to eliminate its coefficients c_k and b_k. (To be formally correct the tridiagonal system is assumed to be extended with equations eq_r for $r < 0$ and $r > P-1$ such that $c_r=b_r=0$ and $a_r=1$.)

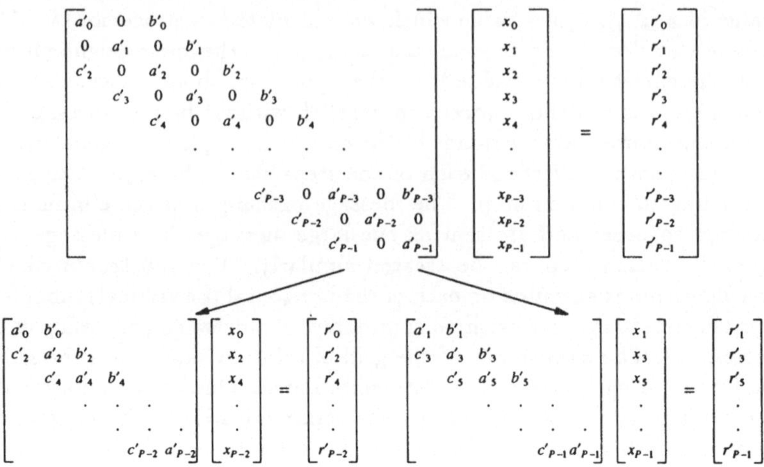

Figure 6.14: Coefficient matrix after the first step of the recursive doubling procedure.

The parallelization of the recursive doubling procedure is straightforward. Each processor is assigned one equation to which it applies all elimination steps. In every step all eliminations are performed simultaneously and a good load balance is achieved. Communication is needed to make the necessary equations available to each processer. In step j processor k needs the equation from the processors $k + 2^{j-1}$ and $k - 2^{j-1}$. In this step the latter two processors need the equation of processor k. Two exchanges are therefore needed per processor in every step. In the last step, however, only two equations remain and one exchange suffices. The extension to the case of M tridiagonal systems of equations is obvious. The number of messages remains the same; their length is multiplied by the factor M.

The overall complexity of a parallel algorithm is determined by the processors that have the largest amount of work in each algorithmic step. In the parallel implementation of recursive doubling the processors responsible for the first and last equation in each system have a somewhat lesser arithmetic and communication complexity than the processors responsible for the other equations. Only the work of the latter processors should therefore be analysed. It can be shown by straightforward counting that the arithmetic cost of an elimination step is 12 floating point operations (flops) in each of the first $\log_2(P)$-1 steps and 5 flops in the final step. The final step reduces the problem to one equation in one unknown, which is solved in one operation. The total computational cost is therefore equal to $(12\log_2(P) - 6)M$ flops. In this operation count we have included the right-hand side computations along with the coefficient matrix elimination. A total of $2\log_2(P) - 1$ messages are exchanged, each of length $4M$ doubles (three equation coefficients and one right-hand side, for each tridiagonal system). From Gray code property 5.3.7, it is clear that the physical distance between the communicating processors is two, except for the first step when the processors are adjacent.

The use of balanced cyclic reduction

The cyclic reduction algorithm for solving (6.20) is very similar to the recursive doubling technique. It differs in that the elimination of coefficients is applied to half of the equations only. In figure 6.15 we have depicted the coefficient matrix after the first elimination step, in which the even equations are used to eliminate the off-diagonal elements of the odd equations. The resulting system is a tridiagonal system, involving the $P/2$ odd variables. It can be further reduced by the same reduction process. The reduction phase terminates when one equation with one unknown remains. A backsubstitution process is needed to determine the remaining unknowns.

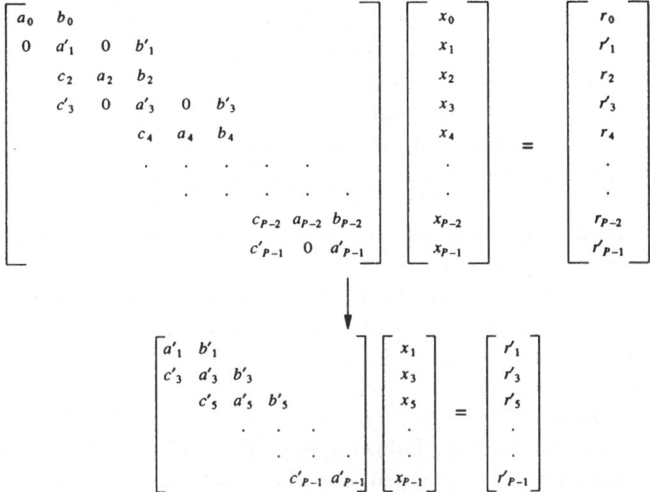

Figure 6.15: Coefficient matrix after the first step of the cyclic reduction procedure.

In every reduction step one may select the odd or the even equations to construct the reduced system. The choices made will eventually determine the equation to which the tridiagonal system converges. Two possible reduction sequences for solving an eight-equation system are presented in figure 6.16. With "$eq_a \rightarrow eq_b$" we denote that equation eq_a is used to eliminate coefficients in equation eq_b. In the first example the reduction process converges to equation seven. In each step the "odd" equations are retained. (In all but the first step the equations should actually be renumbered prior to identifying the even and odd equations.) In the second example the process converges to equation two. The even equations are retained in the first and in the third step, while the odd equations are retained in the second step. Generalizing from the two examples we may state the rule to have the cyclic reduction process converge to an arbitrary equation.

Remark 6.6.2 In each reduction step j, with $j = 1, \ldots, \log_2(P)$, of a cyclic reduction process which converges to eq_ν the off-diagonal elements are eliminated in the equations eq_α where $|\alpha - \nu|$ is a multiple of 2^j. To this end equation eq_α is combined with eq_s, $s = \alpha - 2^{j-1}$, and with equation eq_t, $t = \alpha + 2^{j-1}$.

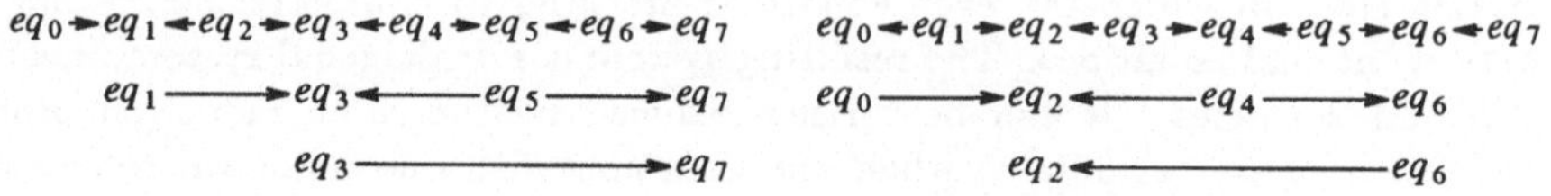

Figure 6.16: Two reduction sequences to reduce an eight-equation system.

When the cyclic reduction algorithm is applied to solve a single tridiagonal system of P equations on P processors, performance will suffer from a severe load imbalance. In each step half of the active processors are turned inactive until one processor remains, namely the one that has the equation to which the process converges. The other processors are reactivated in the backsubstitution step. Only one processor participates in all reduction steps and as such it is the bottle-neck. In the case of multiple tridiagonal systems it is possible to obtain a very good load balance by having different systems converge to different processors. This approach is called *balanced cyclic reduction* in [65] and is opposed to the *naive* cyclic reduction process that is obtained when every system converges to the same equation.

The M systems are divided into P sets, I_k, which contain M/P tridiagonal systems defined on consecutively numbered grid lines. (For simplicity, we only consider the case that M is evenly divisible by P). A first strategy is to have the set I_k converge to equation eq_k. The resulting balanced cyclic reduction process is graphically represented in the left three pictures of figure 6.17, for a configuration with eight processors. The figures should be understood in the following way. The tridiagonal systems extend from the left to the right. They are grouped into the P sets I_k, represented as horizontal strips separated by dotted lines. Each processor k, with $k = 0, \ldots, P-1$, holds precisely one equation of each system, namely equation eq_k, and as such is responsible for doing the computations in one column of the figure.

The reduction of the equation of each set I_k to equation eq_k is displayed in a similar manner as was done in figure 6.16. However, the arrows are not emanating from one single equation but from a set of equations (the dashed rectangles), indicating that all of them are needed in the elimination process of the set of equations to which the arrows point. In this example, three steps are needed to reduce each system to one single equation. The equations that are retained or eliminated in each step for each set can be determined by the rule that was mentioned in remark 6.6.2 [2]. We observe that no processor is idle at any time and that load balance is good. Backsubstitution is carried out afterwards in the reverse order compared to the reduction phase.

---

[2] Before continuing, the reader may wish to verify that the three steps of the two examples of figure 6.16 correspond to the three left-hand pictures (rows I_7 and I_2) in figure 6.17.

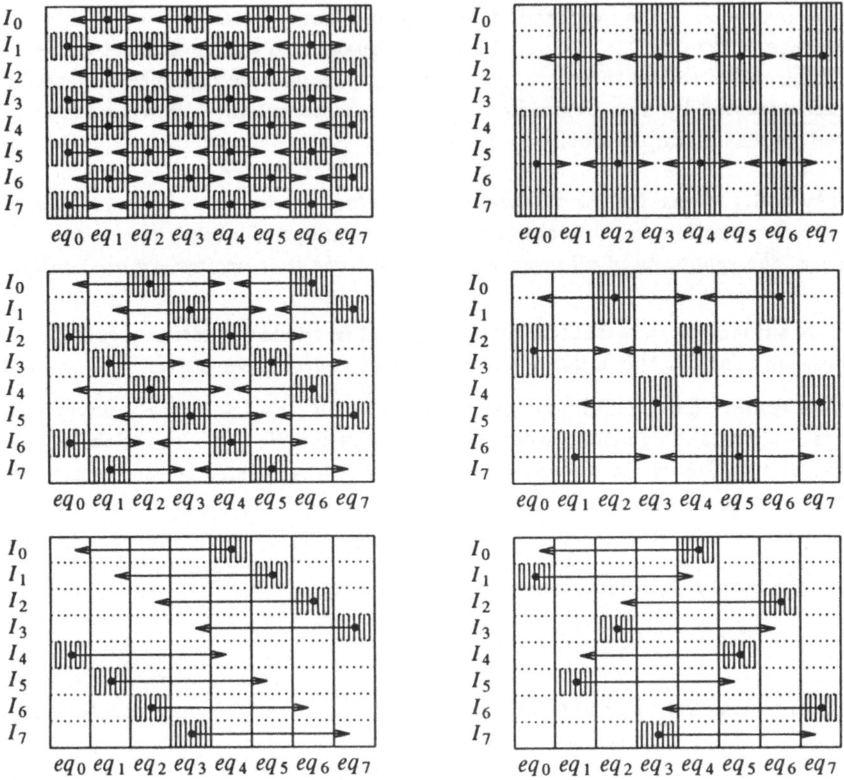

Figure 6.17: Three communication steps with balanced cyclic reduction. The left three pictures illustrate a simple balancing scheme. The scheme in the three pictures to the right is optimized and has minimal data transfer demands.

From figure 6.17 we can identify the necessary communication. By way of illustration we shall look at the communication performed by processor 3 (which handles column eq_3). In the first step it sends its equations that belong to the union set $I_0 \cup I_2 \cup I_4 \cup I_6$ to its left and right neighbours. At the same time it receives from these neighbours the equations eq_2 and eq_4 that belong to $I_1 \cup I_3 \cup I_5 \cup I_7$. In the second step processor 3 sends the equations that belong to $I_1 \cup I_5$ to its neighbours at distance two, i.e., to the processors 1 and 5. From these processors it receives the equations eq_1 and eq_5 that belong to $I_3 \cup I_7$. In the final step processor 3 exchanges its equations of set I_7 for the equations eq_7 of set I_3, with its sole neighbour at distance four, processor 7.

This simple scheme has one definite disadvantage. The equations of the different systems are normally stored contiguously in memory, the one after the other, starting with those of I_0, then those of I_1 and so on. Consequently the sets that are to be sent are not contiguous. As such either one message has to be sent per set I_k or the different sets are to be copied first to a buffer and sent afterwards. Either solution introduces an additional overhead. This overhead of sending additional messages or copying to and from message buffers can be avoided when the equation to which each set converges is chosen differently. In the three graphs on the right-hand side of figure 6.17 we adopt a scheme suggested by Van Driessche in [128], expressed in the following property for which we present a simple proof.

Property 6.6.1 *Consider the balanced cyclic reduction process where set I_k is made to converge to equation eq_p with the binary representation of p being the reverse of that of k. Then, the equations of all messages that are to be sent or received are contiguous.*

Proof . It suffices to show that in every step of the cyclic reduction algorithm every message contains the equations of a union of consecutive sets I_k.

From remark 6.6.2, it follows that in step j, with $1 \leq j \leq \log_2(P)$, any processor p receives sets of equations from the processors $p \pm 2^{j-1}$. These are the equations involved in the cyclic reduction processes which converge to the equations eq_ν where

$$|\nu - p| \quad \text{is a multiple of } 2^j. \tag{6.21}$$

Let the binary representation of p be given by

$$p = (p_{d-1} p_{d-2} \cdots p_1 p_0)_2 \, ,$$

where $d=\log_2(P)$. The ν-values that satisfy (6.21) have the representation

$$\nu = (\nu_{d-1} \nu_{d-2} \cdots \nu_j p_{j-1} \cdots p_1 p_0)_2 \, ,$$

where $\nu_{d-1} \nu_{d-2} \cdots \nu_j$ is an arbitrary sequence of $d - j$ zeroes and ones.

By the rule set out in the proposition, the equations associated to any such value of ν belong to the set I_k, with

$$k = (p_0 p_1 \cdots p_{j-1} \nu_j \cdots \nu_{d-2} \nu_{d-1})_2 \, .$$

Consequently, the equations received by processor p in step j are those belonging to the following union of sets, which is contiguous in memory,

$$\bigcup_{i=0,..,2^{d-j}-1} I_{\alpha+i}, \quad \text{with } \alpha = (p_0 p_1 \cdots p_{j-1} 00 \cdots 00)_2 \, .$$

$$\square$$

Remark 6.6.3 The property is valid for a linear array of processors. It does not rely on any hypercube or Gray code characteristic. However, if the array is Gray code embedded in a hypercube, it can easily be seen by property 5.3.7 that the physical distance between the communicating processors is two, except for the first step, in which it is one.

We can now determine the parallel computational complexity of the balanced cyclic reduction algorithm. In each step j, with $j = 1, \ldots, \log_2(P) - 1$, elimination is performed on $M/2^j$ systems at the cost of 12 operations per system. Five operations are needed per system in the corresponding backsubstitution step. In the final reduction and first backsubstitution step a total of 9 flops are executed on each of M/P systems. This amounts to $17M - 25M/P$ flops, for the algorithm as a whole. In each step of the reduction and the backsubstitution two messages are exchanged, except for one step in which only one message is needed. This leads to a total of $4\log_2(P) - 2$ exchanges. The message length is 4 doubles per system in the reduction step and one double in the backsubstitution. With the number of systems per step as determined above, this amounts to a total of $10M - 15M/P$ doubles.

The Transpose - Gaussian Elimination - Transpose algorithm

Both recursive doubling and cyclic reduction compute the solution in place, i.e., the equations remain in their original location throughout the computation. An alternative is to collect all equations for a single system in one processor, a data movement which is similar to a transpose. The system may then locally be solved by using the sequential Thomas algorithm. A second transpose sends the solution variables back to the processors from which the equations were collected. This scheme was considered in [55, 63, 65, 109] where it is called *algorithm TGET* (Transpose - Gaussian Elimination - Transpose). A similar algorithm, under a different name, was studied in [69] to solve for an intermediate system with two equations per processor, obtained by using a different substructuring technique. The algorithm that we have implemented is based on the *transposition by reflection*, see e.g. [62, 64, 82], and is optimized for use on a hypercube. The process is graphically explained by means of an example in figure 6.18.

The algorithm may be used to transpose matrices of size 2^d by 2^d. In each of d steps successively larger blocks of data, drawn in the double sided boxes, are reflected through one of their corner points, indicated by a "•". For our purposes, the matrix transposition is used to collect M/P intermediate systems to each computing node. Throughout the process each entire column of the matrix is located on one processor. An elementary block of matrix data, denoted in the figure by "$x - y$", contains those equations of processor x that are to be collected by processor y. The necessary routing on a P processor machine is performed in $\log_2(P)$ steps. In each step each processor communicates half of its local data to a processor that is situated successively further away in the linear array. For clarity's sake we have graphically connected the communicating processors by arrows. When the processors are labeled by their Gray code numbers it can be seen that the communicating processors differ in precisely one bit. Thus communication is of nearest neighbour type. For this reason transposition by reflection is perfectly suited for implementation on a hypercube.

The complexity of the algorithm may be determined as follows. After the transposition, each processor solves M/P systems of size P by the Thomas algorithm, which requires $8P - 7$ flops per system. This amounts to a total of $8M - 7M/P$ flops. A total of $2\log_2(P)$ messages are exchanged, half of which are of length $2M$ doubles (first transpose) and half of which are of length $M/2$ doubles (second transpose). The total length of the messages sent by any processor is $5/2\, M \log_2(P)$ doubles.

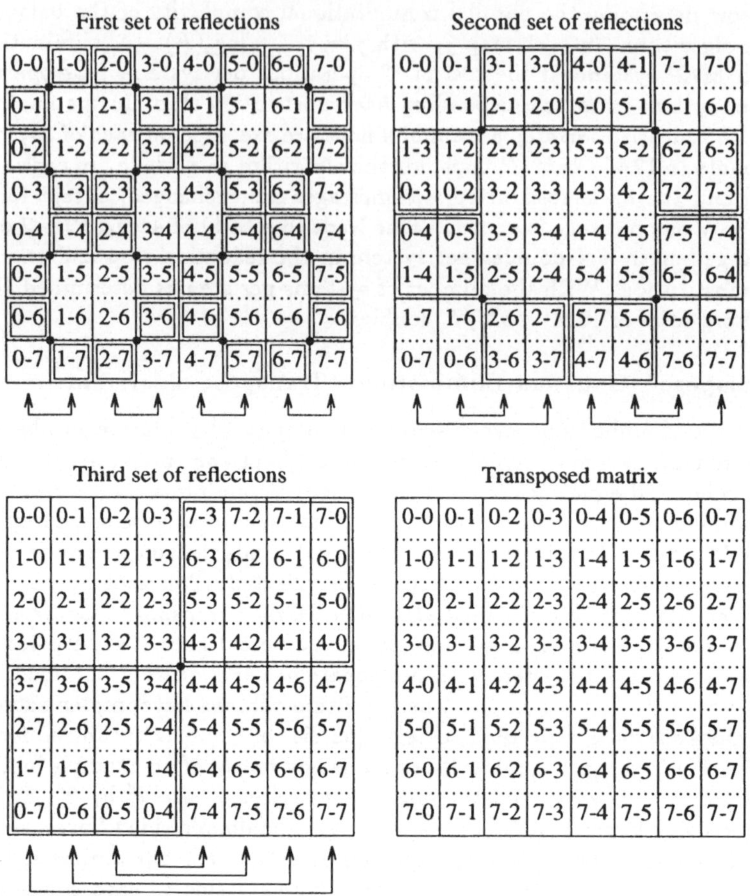

Figure 6.18: Matrix transposition by reflection.

The local data reordering should be considered carefully as it may have an appreciable cost. In each of $\log_2(P)$ steps, each processor sends and receives messages equal to half its local data. Each message has to be copied to a message buffer prior to a send operation in order to make it contiguous, and is copied from the receive buffer afterwards, which amounts to $5M \log_2(P)$ copy operations. This may be an important source of performance degradation and, for large P, will outweigh the arithmetic cost, which is (almost) independent of the number of processors! However, several short-cuts are possible at the cost of some memory for additional communication buffers. The received information need not be copied back into the matrix but may be copied directly into a send buffer if it is part of a future message. The last message of the transpose may be received directly into the data structure of the transposed matrix. A careful count of the number of copy operations then leads to a total of $5/2\ M(\log_2(P) + 1)$.

Some concluding remarks

The complexity of the three intermediate solvers is given in table 6.1.

Table 6.1: Computational cost of intermediate system solvers (no periodic bc).

method	arithmetic	exchanges	total length	local copies
rec.doub.	$(12\log_2(P)-6)M$	$2\log_2(P)-1$	$(8\log_2(P)-4)M$	0
cycl.red.	$17M-25M/P$	$4\log_2(P)-2$	$10M-15M/P$	0
TGET	$8M-7M/P$	$2\log_2(P)$	$5/2\log_2(P)M$	$5/2(\log_2(P)+1)M$

The solvers easily extend to the case of periodic boundary conditions. In the recursive doubling and cyclic reduction algorithms, special attention should be paid to correctly handling the eliminations in which an equation eq_i is involved with index $i<0$ or $i>P-1$. That equation should be identified with $eq_{i\bmod P}$. Similar consideration should be given to the message exchanges. For the above mentioned index values they will cross the borders of the processor array. The computational complexity remains the same on a hypercube multiprocessor, as the end-around connections can be used. The communication in the TGET algorithm is not changed at all. The arithmetic complexity however increases to approximately $17M$ as the simple Thomas algorithm can no longer be used.

Finally, one might wonder whether, perhaps, it would be more appropriate to skip the entire substructuring phase, and to apply any of the three solvers (TGET, cyclic reduction, recursive doubling) to the *original* system instead. This turns out not to be the case. The application of a TGET algorithm to the original set of tridiagonal systems was considered in [69]. There it was shown that the algorithm is inefficient because of the very high communication requirements of the transposition operations. Indeed, without the substructuring phase a very large matrix is to be transposed instead of a very small one. The increase in communication cost exceeds the reduction in arithmetic (which arises from the use of the sequential Thomas algorithm). The application of cyclic reduction to the original system would not reduce the arithmetic complexity (both cyclic reduction and substructuring require about 17 N floating point operations per system). However, it would significantly increase the communication cost. Without substructering, communication is required throughout the computation, whereas with substructuring, communication is restricted to a very limited part of the computation. Similarly, application of recursive doubling to the original system would increase both the arithmetic and the communication costs.

6.7 Timing results on the Intel hypercube

6.7.1 The explicit update step

The speedup of a nine-point stencil explicit update step with different processor numbers is displayed in figure 6.19. The number of grid lines was taken equal in both directions. The following processor configurations were used: a 4×4 mesh, a 4×2 mesh, a 2×2 mesh and a 2×1 mesh.

The left figure shows the speedup that is achieved if the grid is not made consistent after the update step. This is the case, for instance, in the calculation of the right-hand side of the Crank-Nicolson method with formula (6.9). Then, no communication will be needed. However, full parallel efficiency is not always obtained, for various reasons. The load imbalance that arises when the number of grid lines is not evenly divisible by the number of processors is undoubtedly the most important one. Sequential overheads such as procedure calls and loop overheads have a noticeable influence when each processor is allocated very few grid points. A *superlinear speedup* is observed for each of the processor configurations. This is not an uncommon case in highly parallel algorithms and is due to the use of a cache memory in each processor. Programs that get their data from cache generally execute much faster than similar programs that have to fetch their data from the main memory. The larger the number of processors the larger the problem that entirely fits into the fast memory. This also explains why the region of superlinear speedup grows with an increasing number of processors.

The figure to the right supplies the speedups that are obtained with the explicit update formula when communication is applied to update the grid lines in the overlap area. The jagged effect is again due to load imbalance. The large jump in the 16-processor curve is caused by the difference in communication strategy of the iPSC/2 node operating system for short and for long messages. As discussed in section 5.4.2, the change in strategy occurs at message lengths of 100 bytes, which is approximately 13 double precision numbers. A sudden decrease of efficiency will therefore occur for grid sizes that have 13 grid lines per processor in x-direction or y-direction (or at sizes of 11 grid lines if values of the overlap lines are also sent). This is consistent with the single jump observed for the 4×4, 2×2 and 2×1 meshes, and with the two jumps observed in the case of a 4×2 processor mesh.

6.7.2 The multigrid solver

In figure 6.20 we have plotted the speedup versus the problem size for the standard multigrid cycles (V-, W-, F-cycle). The particular cycles that have been timed are characterized by one four-colour pre-smoothing step, one similar post-smoothing step, full-weighting restriction, bilinear prolongation and standard coarsening down to the coarsest grid (one unknown). Agglomeration was used to improve the parallel efficiency on the coarse grid.

The parallel efficiency of the multigrid method depends on the kind of cycle, more precisely on the number of times that the coarse grids are visited. In a V-cycle each grid is visited once in the descending branch and once in the ascending branch. The inefficiency of the coarse grid operations reduces the overall efficiency only slightly. During one W-cycle on a fine grid with l additional coarse grid levels the coarsest grid is visited 2^{l-1} times. The coarse grid operations largely deteriorate the global performance. During each F-cycle the coarse grid is visited l times. The efficiency of the F-cycle therefore takes values between those for the V- and the W-cycles.

Observe that the speedup rapidly increases with an increasing number of grid lines.

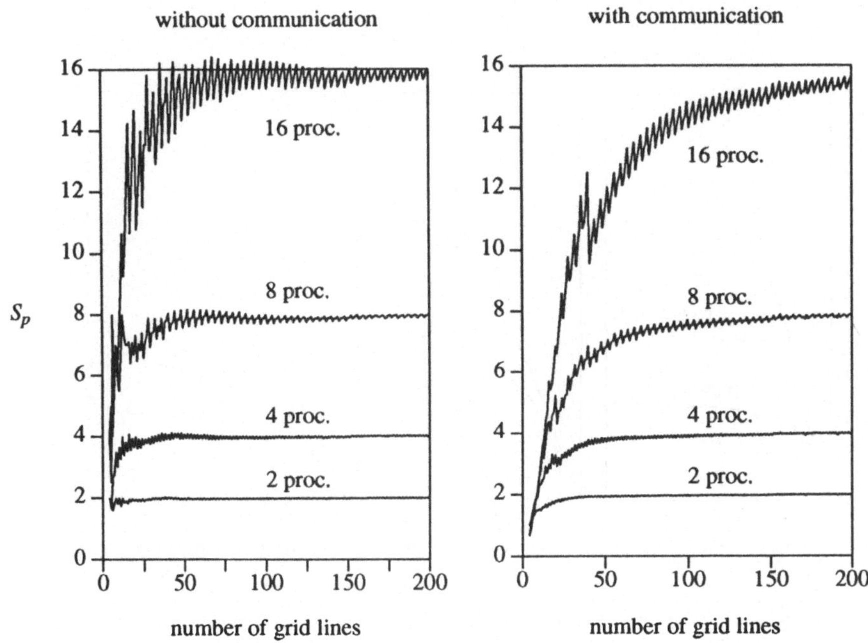

Figure 6.19: Speedup of the nine-point stencil explicit update.

Obviously, the same is true for the parallel efficiency. Next, by a comparison of the measured speedup to the number of processors, it can be derived that the parallel efficiency decreases with an increasing number of processors when the problem size is kept fixed. Both these observations have been made over and over again throughout the parallel computing literature.

The importance of using an agglomeration strategy in order to increase the parallel performance of the coarse grid operations is illustrated in table 6.2 and in figure 6.21. They were taken from our paper [132], in which we consider various multigrid strategies for solving elliptic partial differential equations. The conclusions of the paper however immediately apply to the current case of solving a parabolic problem with a time-stepping method. The timings reported in table 6.2 are the execution times on 16 processors (a 4×4 mesh) of standard multigrid cycles for solving the elliptic problem,

$$x\frac{\partial^2 u}{\partial x^2} + \frac{1}{x}\frac{\partial^2 u}{\partial x\partial y} + \frac{\partial^2 u}{\partial y^2} + (1+\sin(2\pi y))\frac{\partial u}{\partial x} + x\sin(2\pi y)\frac{\partial u}{\partial y} - \frac{\sin(2\pi y)}{x}u = 1$$

on $\Omega = [1,3] \times [0,1]$ with a Dirichlet condition on the western boundary, a mixed boundary condition to the east, and periodicity in north and south direction. (This problem was taken from a report of W. Hackbusch [42].) We used the same standard multigrid operators as above, and coarsening was applied to a grid with 4×2 unknowns. The coarse grid problem is solved by 10 Gauss-Seidel iterations.

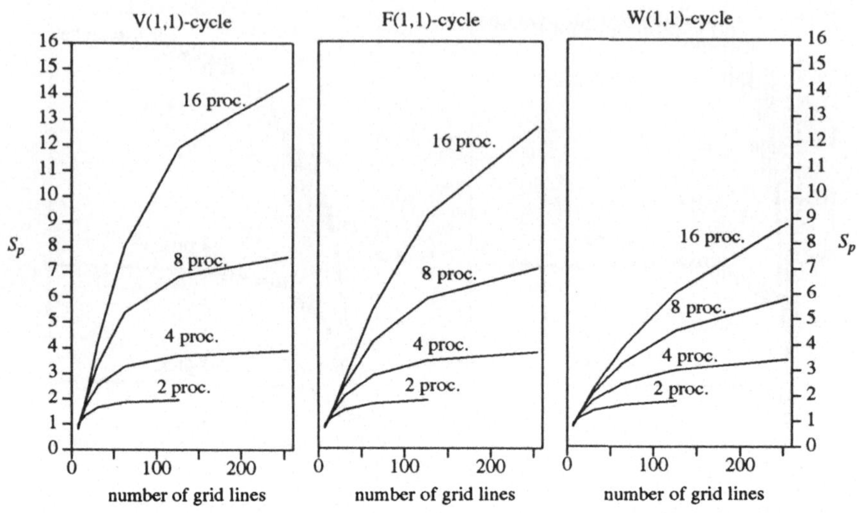

Figure 6.20: Speedup of the standard multigrid cycles for different numbers of processors as a function of the number of grid lines.

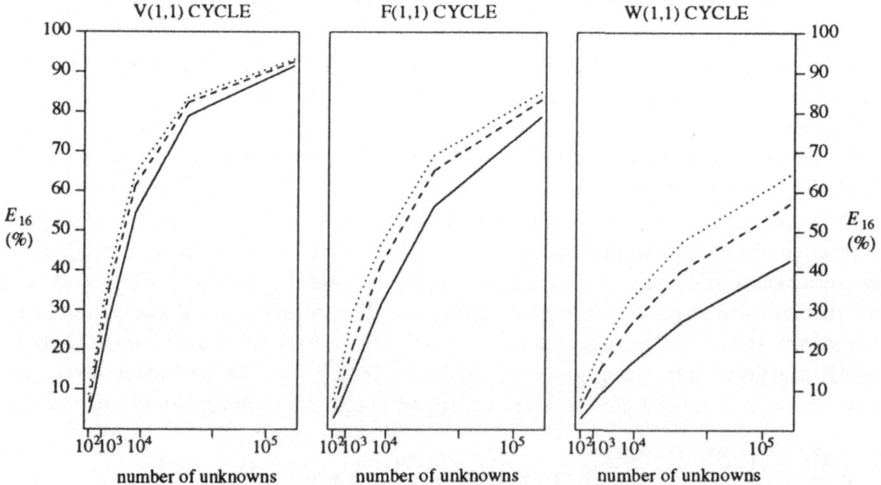

Figure 6.21: Comparison of parallel efficiency for processor idling and agglomeration (solid line: processor idling; dashed line: exchange agglomeration; dotted line: optimized exchange agglomeration).

Table 6.2: Comparison of agglomeration methods on 16 processors; coarse grid problem has 4×2 unknowns, and is solved by 10 Gauss-Seidel iterations.

V(1,1) multigrid cycle (execution time in seconds)				
fine grid	Idling	reduce aggl.	exchange aggl.	optimized aggl.
16×8	0.110	0.068	0.065	0.050
32×16	0.140	0.098	0.095	0.079
64×32	0.202	0.159	0.156	0.140
128×64	0.401	0.357	0.355	0.339
256×128	1.093	1.050	1.047	1.033
512×256	3.750	3.713	3.710	3.688

F(1,1) multigrid cycle (execution time in seconds)				
fine grid	Idling	reduce aggl.	exchange aggl.	optimized aggl.
16×8	0.200	0.108	0.105	0.075
32×16	0.339	0.205	0.200	0.153
64×32	0.540	0.359	0.354	0.292
128×64	0.938	0.714	0.706	0.628
256×128	2.021	1.750	1.740	1.650
512×256	5.734	5.430	5.417	5.296

W(1,1) multigrid cycle (execution time in seconds)				
fine grid	Idling	reduce aggl.	exchange aggl.	optimized aggl.
16×8	0.200	0.108	0.105	0.075
32×16	0.428	0.245	0.239	0.177
64×32	0.918	0.549	0.538	0.416
128×64	2.034	1.293	1.273	1.025
256×128	4.752	3.263	3.224	2.741
512×256	12.148	9.183	9.111	8.104

The first column with timing results displays the execution time of a multigrid cycle in which no agglomeration is used. The grid partitioning is fixed, and some processors go idle on the coarsest grid. The second and third column illustrate the improvement when an agglomeration strategy is used. In this particular case the coarse grid problem is solved on a single processor, the second coarsest grid on a 2×1 processer mesh, while the remaining grids are treated by all 16 processors. Note that the decrease in execution time is almost independent of the number of grid points in the case of a V-cycle; it increases linearly in the case of an F-cycle, and it doubles from row to row in the case of a W-cycle. This observation is consistent with the remark made above on the number of times the coarse grid is visited per cycle. Note further that exchange agglomeration is always somewhat faster than the reduce type agglomeration, albeit that the actual difference is negligible. In one V-cycle on a four-dimensional hypercube four agglomeration steps are to be executed. From the timings in table 6.2, we may conclude that one exchange agglomeration step is about 0.75 msec. faster than one reduce agglomeration step. This is due both to the difference in communication time between a send-receive pair and an exchange, see section 5.4.2, and to a somewhat smaller internal data transfer demand (less copying to and from message buffers is necessary in the exchange versus the reduce type communication). The final column reports the execution time of an optimized exchange agglomeration scheme. As the coarse grid problem is executed on a single processor, we need not use the four-colour Gauss-Seidel method which was used to obtain the results in the other columns. Instead of applying the same subroutine on the coarse grid as on the other grids, we applied an optimized sequential lexicographic scheme. By that we avoid the overhead associated with colouring as well as various other overheads related to the introduction of parallelism in the code (e.g. first updating boundary grid points, checking whether there are neighbours to which messages have to be sent, etc.). As the results illustrate, the additional gain is considerable, especially, as expected, for the W-cycle. Figure 6.21 further illustrates the above in a graphical way.

6.7.3 The tridiagonal system solver

The three solvers for the intermediate tridiagonal systems with one equation per processor have been integrated into the substructured Gaussian elimination solver. A few changes are needed to the analysis of section 6.6.3. To start with, they must cater for the fact that the intermediate systems are not in the desired format for further processing after the first step of Wang's algorithm has been executed. Additionally, the solutions have to be copied back into the grid data structure afterwards. In the recursive doubling and cyclic reduction method $5M$ local copy operations are required to collect the intermediate equations in a contiguous workspace (which is necessary to facilitate the sending of the messages) and to get the solutions back afterwards. In the TGET algorithm the equations can be copied immediately to the send buffers from the equation data structure and the solutions can be copied directly from the receive buffers into the grid data structure. As such, no local copies are required additional to the ones already counted in section 6.6.3.

The solution of the intermediate systems is followed by one communication step to get the solution values which were calculated by the upper neighbour and which are needed in the backsubstitution step. This can be avoided in the TGET algorithm by a small extension to the second transpose (the transpose that sends the solution values back to the processors the corresponding equations were collected from). Each block of data "$x - y$" in figure 6.18 is then enlarged so that it also contains the data "$x - (y-1)$", which are precisely the additional values that are needed in the backsubstitution. By this extension one message of M doubles is avoided at the cost of a more expensive transpose. The total message length is increased with $M/2\log_2(P)$ doubles and the number of local copies increases with $1/2(\log_2(P)+1)M$.

In figure 6.22 we consider the case of Dirichlet or mixed boundary conditions. We compare the execution time of the three solvers on 4, 8 and 16 processors. The timings include the copying of the equations from the equation data structure, the solution of the intermediate system and the communication necessary prior to the start of the backsubstitution. The cost of the remaining parts of the substructured Gaussian elimination algorithm is identical for each of the three methods.

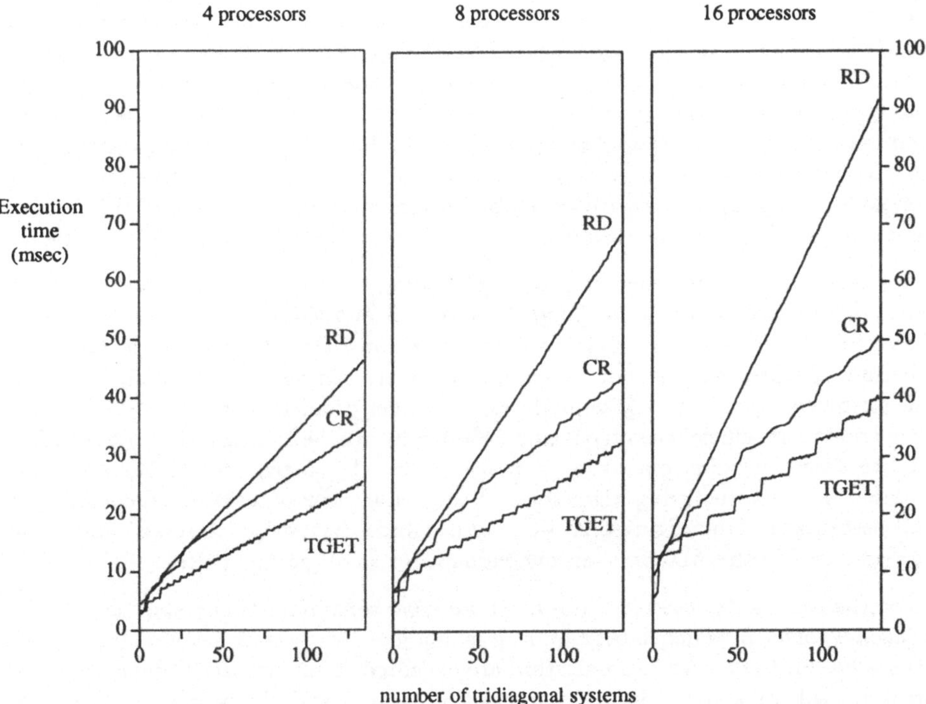

Figure 6.22: Comparison of three intermediate system solvers: recursive doubling, balanced cyclic reduction and "transpose-Gaussian elimination-transpose".

Parallel recursive doubling is especially uncompetitive if the number of systems is large in comparison to the number of processors. The method suffers from a high arithmetic cost, proportional to $\log_2(P)$. The arithmetic cost of cyclic reduction and TGET, on the contrary, is essentially independent of P. The cost of TGET is, however, only half of the cost of cyclic reduction. A cyclic reduction step requires approximately twice as many message startups as a step of the transpose algorithm. However, each step requires transferring only half as much data as the previous step, in contrast to the transpose where the amount of data transferred is the same for all steps. From our experimental results we may conclude that the TGET algorithm performs best, at least for the processor numbers that we consider. For higher processor numbers TGET might be outperformed by the cyclic reduction method because of its "$\log_2(P)M$"-dependence of the local data movement and the message length. This is, however, not obvious from figure 6.22. Due to its superior performance we shall use TGET as the intermediate system solver throughout the rest of this chapter.

The complexity of substructured Gaussian elimination is approximately $17N$ operations per tridiagonal system, where N is the number of equations. The best sequential algorithm in the case of Dirichlet or mixed boundary conditions, which is the Thomas algorithm, requires only $8N$ operations. The obtainable efficiency is therefore restricted to 8/17, or 47%. This low upper limit raises the question of how to decompose the data. For the line hopscotch algorithm, in which tridiagonal systems have to be solved in one coordinate direction only, a strip decomposition is clearly beneficial. The optimal sequential algorithm may then be used. In the ADI method a choice has to be made between square or strip decomposition. In the former case the substructured algorithm is used in both directions. The efficiency is then limited to 47%. In the latter case half of the systems are solved by the Thomas algorithm, half of the systems by substructuring. The upper limit to the efficiency is then $(8+8)/(17+8)$, or 64%.

We have measured the execution time for solving the tridiagonal systems that arise in the ADI algorithm on a N by N grid. A 16×1 and a 4×4 processor mesh were used, leading to a strip and a square decomposition. The efficiencies are presented in figure 6.23. Observe that both decompositions are competitive for small problems, while strip decomposition is to be preferred for larger problem sizes. The theoretically derived maximum efficiencies of 47% and 64% are not reached. This is partly due to fact that two unary negation operators $(a = -b)$ are usually disregarded in the operation count for the substructuring algorithm. They reduce the obtainable efficiency by a small percentage. From figure 6.23 we may conclude that strip decomposition is also to be preferred in the ADI case, except perhaps for small problem sizes.

For the results depicted in figure 6.24 we have separately timed that set of ADI tridiagonal solves in which substructuring is applied. The second set of solves as well as the solves in the hopscotch algorithm are executed at almost 100% efficiency, if we disregard load imbalance. Strip decomposition is used and the number of grid lines is equal in both directions. In the case of periodic boundary conditions, the best sequential algorithm has a complexity similar to the complexity of the substructuring algorithm. The obtainable speedup, shown in the left figure, is then limited by the number of processors only. In the case of Dirichlet or mixed boundary conditions the speedup is limited to 0.47 times the number of processors, as was explained above. This is illustrated in the figure to the right.

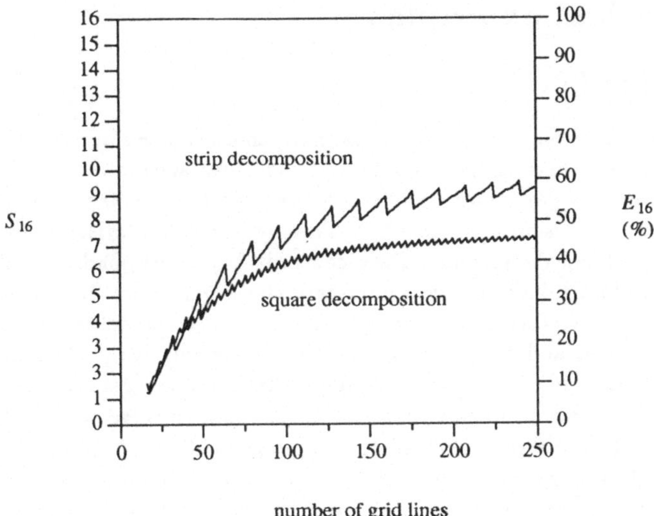

Figure 6.23: Strip decomposition versus square decomposition (on 16 processors).

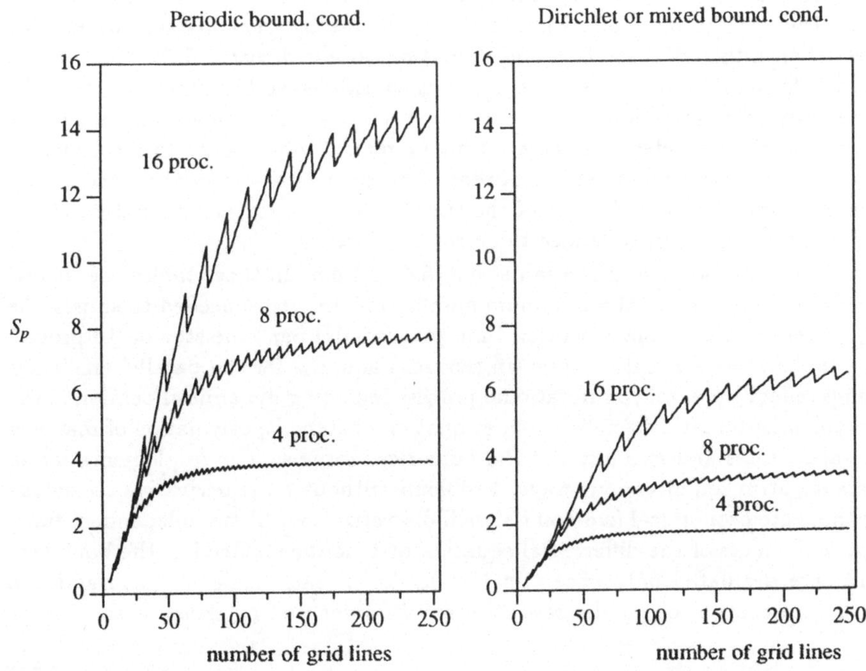

Figure 6.24: Speedup of substructered Gaussian elimination.

6.8 Numerical examples

6.8.1 Introduction

The numerical quality of the discretization method, characterized by its stability, consistency and convergence properties, should be taken into account when choosing an appropriate method for a particular problem. Some general guidelines may be derived from the theoretical discussion in section 6.2. The precise characteristics, however, are to a large extent problem dependent. They can only be determined by a theoretical analysis of the specific problem or by numerical experimentation. In this section we shall solve three non-trivial examples. They were carefully chosen to illustrate some mathematical properties and to give some insight in the relative performance of the eight methods discussed in section 6.2. We hope that they are general enough to be representative for the other problems that one might want to solve.

Each problem is discretized and solved on meshes with 17 by 17, 33 by 33 and 65 by 65 grid lines, distributed over 16 processors. Strip decomposition is used for the ADI and hopscotch methods, square decomposition otherwise. For each method and each grid size we have determined the time-increment, τ, in the following way. First, the solution of the system of ODEs (6.4) is computed by solving the equation with a second order method using very small time-steps. This is the theoretical solution to which all of the methods converge if τ is very small. The difference between this ODE solution and the solution of the parabolic PDE is the *error due to space discretization*. The difference between the ODE solution and a solution obtained with a finite time-step is the *error due to time discretization*. A solution is considered sufficiently accurate if the error due to time discretization is dominated by the error due to space discretization. To be more specific, we select the largest time-increment which leads to a solution of which the former error component is an order of magnitude less than the latter. Both errors are measured at the end point of the time-interval. A further restriction of the time-step will not significantly reduce the error any further.

The results are tabulated in the tables 6.3, 6.4 and 6.5. In those tables the following symbols are used: n_t for the minimum number of time-steps needed to satisfy the time-step criterion, t_{16} for the execution time (in seconds) per time-step on 16 processors, tot_{16} for the total execution time (in seconds) and E_{16} for the parallel efficiency. The timings reflect the total computational process including dynamic allocation of the data structures, automatic discretization, evaluation of stencils, calculation of matrices and right-hand sides and execution of the numerical kernels. The implementation of the solvers is optimized in various ways. Problems without cross derivative are solved by using the more economical five-point stencil discretization. Time-independent functions, e.g. coefficients of the differential equation or functions defined in the boundary conditions, are evaluated only once. In the header of each table we have added a measure of the accuracy of the discrete solution. It is denoted by *error*, with

$$error = \max_{i,j} |u(t_f, x_i, y_j) - u_{K,i,j}^h| \qquad (6.22)$$

where K is the index of the final time-level, i.e., $t_f = t_0 + K\tau$.

6.8.2   Test problem one

The first example is a parabolic equation with variable, but time-independent coefficients and with a mixed derivative,

$$\frac{11x^3 - 10x^2 + 2}{x^3 - 2x^2 + 1}\frac{\partial u}{\partial t} = \frac{3}{2}x^2\frac{\partial^2 u}{\partial x^2} - \frac{5}{2}xy\frac{\partial^2 u}{\partial x \partial y} + \frac{5}{3}y^2\frac{\partial^2 u}{\partial y^2} - \frac{9x^3 - 6x^2}{x^3 - 2x^2 + 1}u - \frac{5}{3}x^2y^2e^{-t}$$

$$(6.23)$$

defined on the unit square, for $t \in [0,1]$, and with the following boundary conditions along the north (N), east (E), south (S) and west (W) sides:

$$N: \quad u(t,x,1) = x^3 - 2x^2 + 1 \qquad\qquad E: \quad u(t,1,y) = y(1-y)(1-y^2)e^{-t}$$

$$S: \quad \tfrac{\partial u}{\partial y}(t,x,0) - u(t,x,0) = -x^3 + 2x^2 - 1 + x^2 e^{-t} \quad W: \quad \tfrac{\partial u}{\partial x}(t,0,y) = 0$$

Its solution is equal to

$$u(t,x,y) = x^3 - 2x^2 + 1 + x^2 y(1 - x^2 y^2)(1 - y)e^{-t} \, . \qquad (6.24)$$

The problem is discretized with central finite differences for spatial discretization and with each of the eight time-stepping methods. The timing and efficiency results are tabulated in table 6.3. Correctness of the discretization may be verified by looking at the evolution of the number of time-steps needed to satisfy the error criterion for an increasing number of grid lines. When the spatial mesh size is halved the value of n_t is multiplied by 4 in the case of a first order method, and multiplied by 2 otherwise.

The Euler and Heun methods are severely restricted by stability constraints and need very small values of τ. No advantage can be taken of the second order accuracy of Heun's method. Both schemes are highly parallel, even for the coarsest discretization.

A small time-step is to be selected for the BDF(1) method for reason of accuracy. At each time-level we use a linear extrapolation of the solution at two previous time-levels as the initial approximation that is input to the multigrid algorithm. This is a valid and frequently used approach in a time-stepping scheme if the solution of the parabolic differential equation is known to have a smooth time behaviour and if the time-step is small, see e.g. [120, 123]. The resulting initial approximation is very close to the true solution and consequently, one V-cycle with one pre-smoothing and one post-smoothing step is sufficient to solve the linear system. Multigrid V(1,1)-cycles also showed to be the most effective for solving the linear system in the second order implicit methods. Due to the larger time-steps three cycles are necessary in the case of the 17 by 17 mesh and four cycles in the case of the 33 by 33 and 65 by 65 mesh. Low efficiency values are measured for the implicit methods. This is due to the inefficient multigrid operations on the coarse grids. One time-step of BDF(1) on the 17 by 17 grid with 16 processors is ten times more expensive than an explicit Euler step, although it is only three times more expensive on a single processor. The comparison is even worse with Crank-Nicolson and BDF(2), as their parallel efficiency is lower due to the larger number of multigrid cycles.

The unconditionally stable DuFort-Frankel and hopscotch methods need very little time per time-step and can be parallelized well. As such they outperform the other methods on the coarse 17 by 17 grid. They are only conditionally consistent though, and a small time-step is needed to achieve the desired accuracy on the finer grids. The relatively low efficiency of the hopscotch method for the 17 by 17 grid may be explained by load imbalance. In each of the two steps of the hopscotch algorithm an operation is to be performed on half of the grid lines only. Consequently, at any given time only half the number of processors are active.

The ADI formula is of first order accuracy for problems with a mixed derivative. (This is verified by the different values of n_t.) For this reason it is substantially slower than Crank-Nicolson and BDF(2), notwithstanding the much lower cost per time-step and the higher parallel efficiency.

We plotted the parallel efficiencies of the eight solvers as a function of problem size and number of processors in figure 6.25. Each of the methods may be parallelized up to almost 100% efficiency when the problem is large. The only exception is ADI. The parallel efficiency of its kernel is restricted to 64%. The overall efficiency may be somewhat larger. This is due to the set-up of the tridiagonal systems and right-hand sides, operations which are fully parallel.

6.8.3　Test problem two

The second example has time-dependent coefficients but no mixed derivative,

$$\frac{\partial u}{\partial t} = \frac{t}{4(x+1)^2}\frac{\partial^2 u}{\partial x^2} + \frac{t}{4(y+1)^2}\frac{\partial^2 u}{\partial y^2} - \frac{t}{4(x+1)^3}\frac{\partial u}{\partial x} - \frac{t}{4(y+1)^3}\frac{\partial u}{\partial y}. \tag{6.25}$$

It is again defined on the unit square, for $t \in [0,1]$, and has four Dirichlet boundary conditions. The solution is equal to

$$u(t,x,y) = \sin((x+1)^2 + (y+1)^2)e^{-t^2}. \tag{6.26}$$

The problem is solved with a five-point discretization. The timing results are given in table 6.4. This example is such that the stability constraint is more easily dealt with than the accuracy requirement. This explains why the performance of the Heun method is superior to the performance of the Euler method. It also explains why the number of time-steps for explicit Euler and for BDF(1) are approximately equal.

The linear system in BDF(1) is again solved by using one V(1,1)-cycle. In the case of the second order implicit schemes we derived the initial approximation at each new time-level by the full multigrid or nested iteration technique. In this case the initial approximation on the fine grid is derived by interpolation from an approximate solution on the next coarser grid. Full multigrid significantly reduces the parallel efficiency because of the frequent visits to the coarse grid. However, for this problem, it still proved to be faster than the use of straightforward multigrid cycling. Line hopscotch and DuFort-Frankel are very effective on the 17 by 17 mesh. They are however surpassed on the finer grids. The ADI method is second order accurate for this example. Some more time-steps are needed to satisfy the error criterion than the number of time-steps that are needed when Crank-Nicolson is used. However, the ADI time-steps are less costly and more parallel, which is why ADI is faster than the second order implicit methods. The relative difference between the methods will however decrease when the grid is further refined.

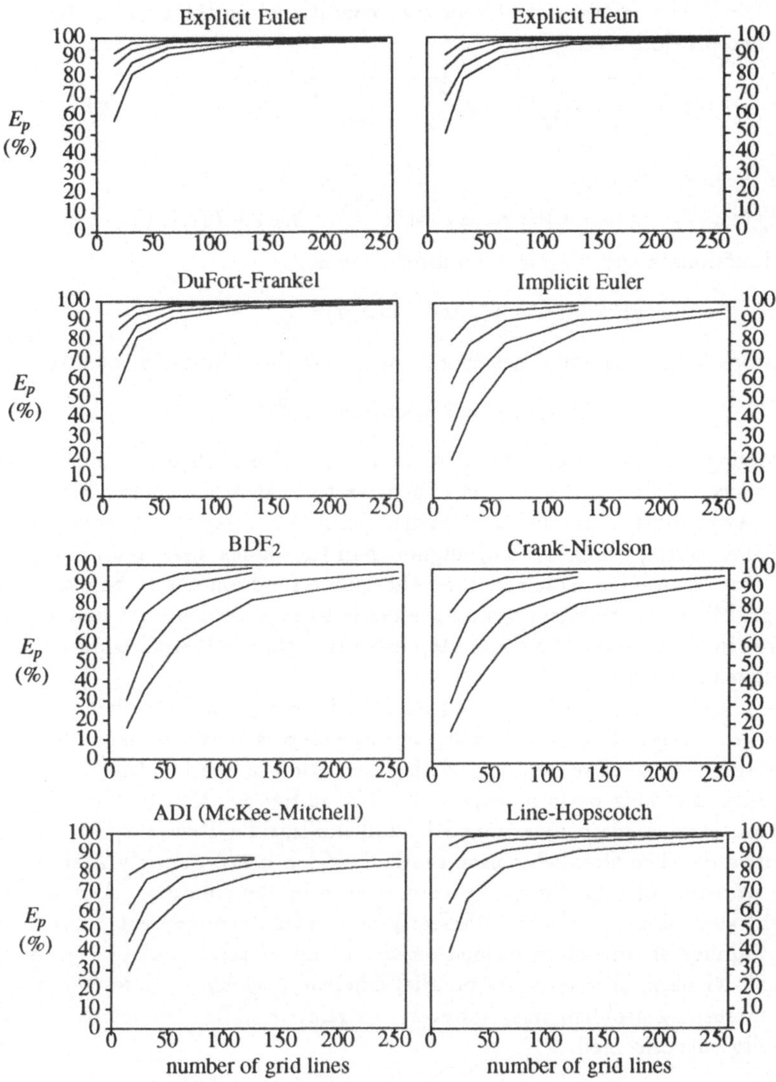

Figure 6.25: Efficiency of the eight parabolic solvers as a function of the problem size. The number of processors is equal to 2, 4, 8 or 16 (top curve to bottom curve).

6.8.4　Test problem three

As a final problem we consider the numerical solution of the general heat equation, with two Dirichlet and two mixed boundary conditions, on the same problem domain as in the two previous examples,

$$a(x,y)\frac{\partial T}{\partial t} = \frac{\partial}{\partial x}(k(x,y)\frac{\partial T}{\partial x}) + \frac{\partial}{\partial y}(k(x,y)\frac{\partial T}{\partial y}) + Q(t,x,y) \tag{6.27}$$

$$
\begin{aligned}
N: \quad & T(t,x,1) = g(t,x) & E: \quad & T(t,1,y) = h(t,y) \\
S: \quad & \frac{\partial T}{\partial y}(t,x,0) + T(t,x,0) = p(t,x) & W: \quad & \frac{\partial T}{\partial x}(t,0,y) + T(t,0,y) = q(t,y).
\end{aligned} \tag{6.28}
$$

The two functions a and k are chosen arbitrarily as,

$$a(x,y) = 1 + x + y \quad \text{and} \quad k(x,y) = e^{4(x-1/2)^2 + 4(y-1/2)^2}. \tag{6.29}$$

The functions Q, g, h, p and q are chosen such that the solution is equal to

$$T(t,x,y) = 2 + \sin(5xy)\,e^{-2t(x+y)}. \tag{6.30}$$

The timing results are presented in table 6.5. The problem is characterized by a source function $Q(x,y,t)$ which is expensive to evaluate. The evaluation of this function may proceed in parallel at each grid point. This explains the execution times and efficiency values, which are both higher than the values measured for the previous test problems. In the hopscotch method the source function has to be evaluated only at those grid lines where a tridiagonal system is to be solved (i.e. at half the number of grid lines in each step). The single step execution time is therefore low compared to that of other methods.

The explicit methods, BDF(1), and hopscotch need a large number of time-steps. They are not competitive as the cost per time-step is dominated not by the complexity of the numerical kernel, but by the evaluation of the function Q. The second order implicit methods perform very well. The initial solution to the linear system in each time-step is again determined by full multigrid. Two V(2,1)-cycles are applied afterwards. The Crank-Nicolson method, which is not strongly A-stable, suffers from the presence of high frequency components in the solution. A minimal number of time-steps is necessary to sufficiently reduce these components, which explains the high number of time-steps needed on the 17 by 17 mesh. ADI is initially faster than Crank-Nicolson thanks to its parallel efficiency, which is up to 3 times higher. With an increasing problem size, however, the relative difference decreases, and ADI is eventually outperformed.

6.9　Concluding remarks

In this chapter we first briefly recalled some of the standard parabolic marching schemes. The overview was by no means exhaustive as we excluded some important subjects such as the L.O.D. techniques, methods based on extrapolation and local or global combination, and so on. In addition we did not pursue some matters that are of great importance to the development of robust software. By assuming a constant time-increment, we evaded the question of how to select step-sizes so as to attain a desired accuracy. We experimentally determined an optimal multigrid parameter choice for each of the problems and each of the grid sizes. This should be determined automatically in any robust solver. However, most of the constituent components of the solvers currently in use can be related to the schemes and numerical kernels discussed. The results should therefore also be of practical use to anyone developing fully automatic software.

Each of the time-marching methods considered can be parallelized with close to 100% efficiency, if the problem is large enough. The only exception is the ADI method, as the efficiency of a parallel implementation of its numerical kernel is restricted to 64%. For large problems the performance of each method is entirely determined by its numerical properties. The introduction of parallelism does not alter their ranking. As such the unconditionally stable second order methods largely outperform the methods with more limited stability or consistency characteristics.

For "small" problems (small with respect to the number of processors) the requirement of high parallel efficiency contradicts the requirement of high numerical quality. Differences in execution time, which may be large on one processor, are reduced and formerly uncompetitive methods may become competitive. The performance of the ADI and of the implicit methods is seriously degraded by their parallel inefficiency. The degradation is especially prominent for the methods using a multigrid based linear system solver. The explicit methods as well as the hopscotch method remain highly parallel. As such, they may surpass the second order methods on coarse grids.

As the number of processors in commercially available multiprocessor systems is constantly increasing, the number of interesting problems that may be classified as "small" continuously grows. Consequently, new parallel algorithms are needed to tackle the problems with a limited number of grid points per processor. These methods should either increase the parallel efficiency of the fast second order implicit methods, or improve the numerical quality of the explicit methods. The latter could be achieved by extending the stability region and thus reducing the time-step constraint of the classical techniques, see e.g. [28, 105]. The former can be obtained by calculating the solution on several or on all time-levels at once, e.g., by using *windowed relaxation* techniques or *waveform relaxation* [110, 140]. The parallel efficiency of the latter will be the main subject of the next chapter.

Table 6.3: Timing results for test problem one (in seconds).

Problem 1: 17 by 17 discretization (16 by 16 unknowns), *error* = 1.7e-4

method	n_t	t_1	t_{16}	tot_1	tot_{16}	E_{16}
Forward Euler	533	0.071	0.008	38.2	4.14	57.7%
Heun	533	0.107	0.013	57.1	7.06	50.6%
DuFort-Frankel	155	0.079	0.008	12.3	1.31	58.7%
BDF(1)	167	0.237	0.084	39.8	14.00	17.8%
BDF(2)	12	0.495	0.213	6.1	2.58	14.9%
Crank-Nicolson	7	0.522	0.218	3.9	1.56	15.5%
ADI	210	0.139	0.029	29.4	6.15	29.9%
Hopscotch	112	0.064	0.010	7.3	1.16	39.7%

Problem 1: 33 by 33 discretization (32 by 32 unknowns), *error* = 4.4e-5

method	n_t	t_1	t_{16}	tot_1	tot_{16}	E_{16}
Forward Euler	2475	0.271	0.021	671.8	51.6	81.4%
Heun	2475	0.410	0.032	1014.7	80.3	79.0%
DuFort-Frankel	615	0.298	0.023	183.9	14.0	82.0%
BDF(1)	820	0.856	0.144	702.3	118.5	37.0%
BDF(2)	25	2.288	0.463	58.0	11.6	31.1%
Crank-Nicolson	13	2.402	0.476	32.0	6.3	32.0%
ADI	955	0.532	0.066	508.8	62.9	50.5%
Hopscotch	460	0.234	0.022	108.5	10.1	67.3%

Problem 1: 65 by 65 discretization (64 by 64 unknowns), *error* = 1.1e-5

method	n_t	t_1	t_{16}	tot_1	tot_{16}	E_{16}
Forward Euler	10700	1.060	0.072	11349	775	91.5%
Heun	10700	1.610	0.112	17234	1202	89.6%
DuFort-Frankel	2400	1.163	0.079	2793	190	91.8%
BDF(1)	3460	3.268	0.326	11309	1127	62.7%
BDF(2)	50	8.901	1.004	448	50	55.6%
Crank-Nicolson	25	9.340	1.045	237	26	56.1%
ADI	3950	2.099	0.194	8295	768	67.5%
Hopscotch	1800	0.893	0.071	1611	127	79.0%

Table 6.4: Timing results for test problem two (in seconds).

Problem 2: 17 by 17 discretization (15 by 15 unknowns), $error = 1.7e\text{-}3$

method	n_t	t_1	t_{16}	tot_1	tot_{16}	E_{16}
Forward Euler	691	0.086	0.009	59.2	6.18	60.3%
Heun	144	0.108	0.014	15.6	2.00	48.9%
DuFort-Frankel	71	0.091	0.009	6.5	0.67	60.8%
BDF(1)	687	0.178	0.054	122.5	36.89	20.8%
BDF(2)	52	0.220	0.099	11.5	5.17	13.9%
Crank-Nicolson	26	0.233	0.102	6.2	2.65	14.6%
ADI	31	0.167	0.033	5.3	1.05	31.6%
Hopscotch	64	0.062	0.010	4.1	0.66	38.6%

Problem 2: 33 by 33 discretization (31 by 31 unknowns), $error = 4.3e\text{-}4$

method	n_t	t_1	t_{16}	tot_1	tot_{16}	E_{16}
Forward Euler	2735	0.343	0.026	938.5	71.8	81.7%
Heun	735	0.432	0.035	317.7	25.8	77.0%
DuFort-Frankel	280	0.368	0.028	103.1	7.8	82.6%
BDF(1)	2725	0.716	0.102	1952.6	279.1	43.7%
BDF(2)	104	0.858	0.192	89.5	19.9	28.0%
Crank-Nicolson	52	0.913	0.197	48.0	10.2	29.2%
ADI	61	0.669	0.079	41.3	4.9	53.0%
Hopscotch	211	0.242	0.022	51.3	4.7	68.0%

Problem 2: 65 by 65 discretization (63 by 63 unknowns), $error = 1.1e\text{-}4$

method	n_t	t_1	t_{16}	tot_1	tot_{16}	E_{16}
Forward Euler	10850	1.382	0.095	14999	1029	91.1%
Heun	3365	1.748	0.123	5882	415	88.5%
DuFort-Frankel	1103	1.484	0.101	1638	112	91.6%
BDF(1)	10800	2.902	0.263	31346	2838	69.0%
BDF(2)	207	3.421	0.431	709	89	49.6%
Crank-Nicolson	104	3.649	0.448	382	47	51.0%
ADI	122	2.705	0.234	332	29	72.2%
Hopscotch	800	0.957	0.072	766	58	83.1%

Table 6.5: Timing results for test problem three (in seconds).

Problem 3: 17 by 17 discretization (16 by 16 unknowns), *error* = 8.5e-4						
method	n_t	t_1	t_{16}	tot_1	tot_{16}	E_{16}
Forward Euler	5535	0.216	0.016	1196.8	90.90	82.3%
Heun	5535	0.241	0.021	1336.7	116.71	71.6%
DuFort-Frankel	299	0.222	0.017	66.6	5.07	82.2%
BDF(1)	142	0.309	0.066	44.2	9.35	29.5%
BDF(2)	13	0.476	0.191	6.5	2.35	16.1%
Crank-Nicolson	30	0.503	0.204	15.6	6.17	15.8%
ADI	82	0.286	0.039	23.9	3.22	46.4%
Hopscotch	209	0.129	0.014	27.4	2.93	58.5%

Problem 3: 33 by 33 discretization (32 by 32 unknowns), *error* = 2.1e-4						
method	n_t	t_1	t_{16}	tot_1	tot_{16}	E_{16}
Forward Euler	24000	0.846	0.057	20315.2	1357.7	93.5%
Heun	24000	0.942	0.066	22603.1	1572.6	89.8%
DuFort-Frankel	1190	0.869	0.058	1035.4	69.0	93.8%
BDF(1)	540	1.178	0.133	637.3	71.9	55.4%
BDF(2)	25	1.776	0.367	45.7	9.3	30.7%
Crank-Nicolson	36	1.856	0.387	68.7	14.1	30.5%
ADI	165	1.118	0.102	186.1	17.0	68.3%
Hopscotch	840	0.497	0.038	418.8	32.0	81.8%

Problem 3: 65 by 65 discretization (64 by 64 unknowns), *error* = 5.0e-5						
method	n_t	t_1	t_{16}	tot_1	tot_{16}	E_{16}
Forward Euler	104000	3.359	0.217	349000	22548	96.8%
Heun	104000	3.740	0.246	389000	25546	95.1%
DuFort-Frankel	4750	3.447	0.222	16377	1055	97.0%
BDF(1)	2140	4.615	0.366	9880	783	78.9%
BDF(2)	48	6.906	0.828	337	40	52.5%
Crank-Nicolson	50	7.192	0.865	367	44	52.5%
ADI	330	4.436	0.342	1471	113	81.1%
Hopscotch	3360	1.953	0.134	6568	450	91.3%

Chapter 7

Computational Complexity of Multigrid Waveform Relaxation

...we give below one-equation or one-line descriptions of a variety of computers.

$$C(CRAY - 1) = Iv^{12}[12E_p^{12} - 16M^{50}]_r; 12E_p = \{3F_{p64}, 9B\}$$
$$C(CYBER - 205) = Iv^{12}[4\bar{F}_{p64} - 512M_{16K*32}^{80K}]$$
$$C(HYPERCUBE) = I[2^4\bar{C}]; \bar{C} = C(2 * INTEL8080)$$

More complex structures may require the extension of the formula into the second dimension, in the manner of a structural chemical formula.

—R.W. Hockney and C.R. Jesshope, *[56]*.

We detail the arithmetic complexity of the multigrid waveform relaxation method. It is shown that the complexity is comparable to that of the best sequential solvers in the case of initial boundary value problems. It is better by a factor of 2.5 in the case of time-periodic problems. The communication complexity of a parallel implementation based on a spatial grid partitioning approach is analysed and compared to that of a similar implementation of standard initial value and time-periodic solvers. Finally, we discuss the vectorization of the waveform relaxation method.

7.1   Introduction

A theoretical analysis of multigrid waveform relaxation was given in chapters 3 and 4. It was shown that the method is a rapidly converging iterative procedure. We also compared its performance to the performance of standard waveform relaxation methods (Gauss-Seidel, Jacobi, and SOR). In order to compute the solution to a parabolic partial differential equation multigrid waveform relaxation requires only a marginal fraction of the work of the standard waveform relaxation methods. In the current chapter we shall complete the picture, and motivate the real computational effectiveness of multigrid waveform relaxation. That is, we shall analyse its computational complexity, and compare our findings to similar complexity estimates for standard parabolic solvers.

We restrict the analysis to the case where *constant global time-steps* are used. We do not consider the exploitation of multi-rate integration. If we did, it would be very difficult to derive *general* quantitative results. Indeed, the effectiveness of multi-rate integration strongly depends on the characteristics of the partial differential equation, and in particular on the smoothness of the solution. We shall also assume that the standard solvers use similar time-steps and similar time-discretization formulae as the waveform relaxation methods. That is to say, they derive a solution on the same *space-time grid*. Such a grid is displayed in figure 7.1. It can be interpreted as a set of discretized functions located at the grid points of the discrete spatial domain. This is the viewpoint taken by the waveform relaxation methods. Or, it can be interpreted as a sequence of spatial grids defined on successive time-levels. This corresponds to the viewpoint of the time-marching schemes, and was pictured in figure 1.1, chapter 1.

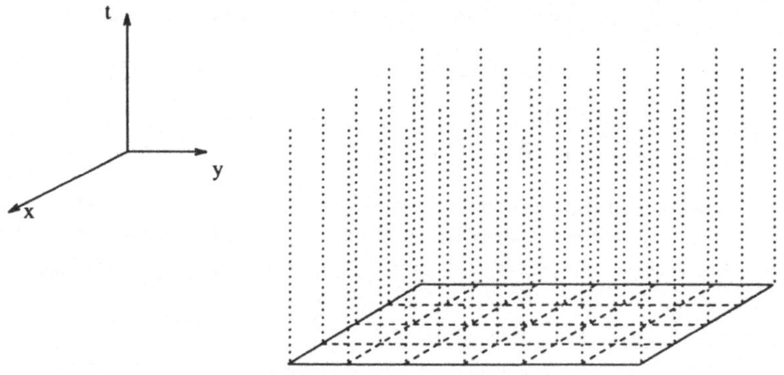

Figure 7.1: A space-time grid, considered as a set of vectors located at the grid points of the discrete spatial mesh.

We determine the arithmetic complexity of the initial value and time-periodic waveform methods in section 7.2. The parallel implementation and communication complexity are discussed in section 7.3. Finally, vectorization is dealt with in section 7.4.

7.2 Arithmetic complexity

7.2.1 Initial boundary value problems

The basic operation in the waveform relaxation smoothing step corresponds to calculating the solution to a linear ordinary differential equation in a single unknown,

$$\frac{dx}{dt} + v\,x = w, \quad x(0) = x_0 . \tag{7.1}$$

x is an unknown function located at a grid point. The right-hand side w is a linear combination of known functions, either located at the same grid point as x, or at neighbouring grid points. v is derived from the discretization of the PDE coefficients. Depending on the problem, v may either be a constant, or a function of time.

With the use of a linear multistep method for time-discretization, (7.1) becomes,

$$\tfrac{1}{\tau}\sum_{j=0}^{k}\alpha_j x_{n-k+j} + \sum_{j=0}^{k}\beta_j v_{n-k+j} x_{n-k+j} = \sum_{j=0}^{k}\beta_j w_{n-k+j}, \quad n = 0,1,2,\ldots \quad (7.2)$$

We assume that k starting values (denoted by negative indices) are given. (7.2) can be rewritten into a k-th order recurrence relation, and used for step-by-step determination of the discrete solution. For instance, with the trapezoidal rule we get the following first order recurrence,

$$x_n = a_n x_{n-1} + r_n \quad \text{with} \quad a_n = \frac{1 - v_{n-1}\tau/2}{1 + v_n\tau/2} \quad \text{and} \quad r_n = \frac{(w_n + w_{n-1})\tau/2}{1 + v_n\tau/2}. \quad (7.3)$$

The remaining discrete-time multigrid waveform relaxation operators (restriction, prolongation, correction, and defect calculation) are linear combination operators which operate on discretized functions, i.e., on vectors. For instance, the discrete-time equivalent to the continuous-time full weighting formula is given by,

$$u^{H}_{n,I,J} = \frac{1}{16}\begin{bmatrix} 1 & 2 & 1 \\ 2 & 4 & 2 \\ 1 & 2 & 1 \end{bmatrix} u^{h}_{n,i,j}, \quad n = 0,1,2,\ldots \quad (7.4)$$

Before we can estimate the cost of a method, we have to specify precisely what problems that we consider. Indeed, the number of operations strongly depends on the characteristics of the differential equation, on the discretization, and on the solution procedure. For the analysis below we consider two-dimensional parabolic PDEs with variable but time-independent coefficients. Spatial discretization is by standard 5-point or 9-point stencils, and time-discretization is by the trapezoidal rule or Crank-Nicolson method. We use standard multigrid parameters: multi-colour smoothing (red/black in the case of 5-point stencils, 4-colour in the case of 9-point stencils), bilinear interpolation and full-weighting. The hierarchy of grids is derived by standard coarsening from a regular fine grid Ω^{h_k},

$$\Omega^{h_0} \subset \Omega^{h_1} \subset \ldots \subset \Omega^{h_i} \subset \ldots \subset \Omega^{h_k} \quad \text{with} \quad h_i = 2^{k-i}h_k.$$

We assume that the cost of solving the coarse grid (Ω^{h_0}) problem can be neglected.

Floating point operation counts

The number of floating point operations (*flops*) involved in each discrete-time multigrid waveform relaxation operation is proportional to the number of grid points in the space-time grid. Let the proportionality constants be denoted by C_S (smoothing), C_D (defect calculation), C_R (restriction), C_P (prolongation), and C_C (correction). A count of the number of operations leads to the numbers specified in table 7.1 [1].

---

[1] These values correspond to the operators as they are implemented in our programs used for getting the timing results in chapter 8. The implementation is fairly optimized.

Table 7.1: Number of floating point operations (averaged over the number of grid points) for discrete-time multigrid waveform relaxation operators.

stencil	C_S	C_D	C_R	C_P	C_C
5-point	12	4	1.5	1.5	1
9-point	20	8	2.25	1.5	1

By way of illustration we shall determine the numbers for the smoothing and the defect calculation steps. The calculation of the discrete function w in (7.1) requires 8 flops for a 5-point stencil and 16 flops for 9-point stencil (per grid point). Calculation of each r_n in (7.3) requires 2 flops. More precisely, it requires one addition and one multiplication (by the factor $(\tau/2)/(1 + v_n\tau/2)$). This factor is a constant and has to be calculated only once, since v is a constant. (The PDE coefficients do not depend on t.) Note that also a_n is a constant. Two additional flops are required for each x_n. Summing up leads to the counts 12 and 20 for C_S. We shall now calculate C_D. In the case of a 5-point stencil the defect to be calculated after a red/black smoothing step is zero at half the number of the grid points. The non-zero defects can be calculated at a cost of 8 flops per grid point. This leads to an average cost of 4 flops ($C_D = 4$). In the case of a 9-point stencil, the defect is zero in one fourth of the grid points only. The determination of the non-zero defects requires 7 flops at one half of the grid points, and 15 flops at the remaining one fourth. In addition, at 3/4 of the grid points two successive smoothing iterates are to be subtracted. This leads to an average of 8 flops.

Given the above cost factors one can easily estimate the cost of the multigrid algorithm, based on the formulae 4.3.6 and 5.3.1 in the book by Hackbusch [44]. Assume the multigrid algorithm performs a total of α_{WR} V-(1,1)-cycles, α_{WR} W(1,1)-cycles, or 1 full multigrid waveform cycle using 1 V(1,1)-cycle at each grid level. The average cost per grid point on the fine (space-time) grid is determined as follows[2].

$$C_V \simeq 4/3\,(2C_S + C_D + C_R + C_P + C_C)\,\alpha_{WR} \simeq 43\,\alpha_{WR} \quad (5pt)$$

$$C_W \simeq 2\,(2C_S + C_D + C_R + C_P + C_C)\,\alpha_{WR} \simeq 64\,\alpha_{WR} \quad (5pt)$$

$$C_{FMG} \simeq 16/9\,(2C_S + C_D + C_R + C_P + C_C) \simeq 57 \quad (5pt)$$

$$C_{FMG} \simeq 16/9\,(2C_S + C_D + C_R + C_P + C_C) \simeq 94 \quad (9pt)$$

Comparison with standard time-stepping

In each time-step with a time-marching scheme a discretized elliptic partial differential equation is solved by using multigrid. The cost of the elliptic problem set-up, i.e., right-hand side vector calculation, is proportional to the number of fine-grid grid points at the time-level. (Note that the coefficient matrix has to be calculated only once.) The value of the proportionality constant (C_E) is given in table 7.2. This table also displays the values of proportionality constants for the (elliptic) multigrid operators.

---

[2]We neglect the costs of the full multigrid interpolation and of the coarse grid solver.

Table 7.2: Number of floating point operations (averaged over the number of grid points) for Crank-Nicolson time-stepping multigrid operators.

stencil	C_E	C_S	C_D	C_R	C_P	C_C
5-point	14	9	5	1.5	1.5	1
9-point	22	17	13.5	2.25	1.5	1

For simplicity's sake we consider a Crank-Nicolson method which applies the same number of multigrid cycles at each time-level (e.g., one FMG step, or two V(1,1)-cycles). Under this assumption (which is not unrealistic) we can derive estimates for the total number of floating point operations executed in the algorithm. Obviously, it is equal to the cost per time-level multiplied by the number of time-levels. Below we consider the time-stepping method with use of α_E elliptic multigrid V(1,1)- or W(1,1)-cycles applied to an initial approximation equal to the value obtained in the previous time-step, and with use of full multigrid with 1 V(1,1)-cycle on each grid level. The numbers given are the average cost per grid point on the whole of the space-time grid.

$$C_V \quad \simeq \quad C_E + 4/3 \, (2C_S + C_D + C_R + C_P + C_C) \, \alpha_E \quad \simeq \quad 14 + 36 \, \alpha_E \quad (5pt)$$

$$C_W \quad \simeq \quad C_E + 2 \, (2C_S + C_D + C_R + C_P + C_C) \, \alpha_E \quad \simeq \quad 14 + 54 \, \alpha_E \quad (5pt)$$

$$C_{FMG} \simeq \quad C_E + 16/9 \, (2C_S + C_D + C_R + C_P + C_C) \quad \simeq \quad 62 \qquad (5pt)$$

$$C_{FMG} \simeq \quad C_E + 16/9 \, (2C_S + C_D + C_R + C_P + C_C) \quad \simeq \quad 115 \qquad (9pt)$$

It will be illustrated in chapter 8 that the use of the full multigrid procedure in both the waveform relaxation and the Crank-Nicolson method often leads to an approximation with algebraic error smaller than the discretization error. In that case additional iterations do not reduce the actual error any further. When the use of one full multigrid step is satisfactory, waveform relaxation is about 10% to 20% faster than the Crank-Nicolson method. Note, however, that the estimates do not include the cost of PDE-coefficient and PDE right-hand side evaluation. This cost does not depend on the choice of the particular algorithm (waveform or time-stepping) and may constitute an important fraction of the total computation time. The relative difference in execution time between waveform relaxation and time-stepping will correspondingly be diminished.

When a small algebraic error is wanted, or when pure multigrid cycling is used, Crank-Nicolson will be more efficient. The convergence factor of Crank-Nicolson is usually smaller, as it corresponds to the time-level convergence factor of the first time-level in the multigrid waveform relaxation method. In addition, using information from previous time-levels, good initial approximations may be obtained with small initial error. As such we can expect α_E to be smaller than α_{WR}.

7.2.2 Time-periodic problems

The only difference between the initial value and the time-periodic waveform algorithms is the ODE solver used in the smoother. While an initial value problem is to be solved in the former case, the latter requires the solution of a boundary value problem,

$$\frac{dx}{dt} + v\,x = w, \quad x(0) = x(T) \,. \tag{7.5}$$

This equation can be solved with standard techniques such as shooting, discretization or dynamic simulation. In the case of discretization with a linear multistep method and a constant time-increment $\tau = T/N$, we get a linear system of N equations of the form (7.2). (As before, we extend the discrete solution by N-periodicity.)

In the sequel, we shall consider the special case of a discretization by the trapezoidal rule. Then, we get the equations,

$$a_{n,n-1}x_{n-1} + a_{n,n}x_n = b_n, \quad n = 0,\ldots,N-1 \,,$$

$$\text{with} \quad a_{n,n-1} = -\tfrac{2}{\tau} + v_{n-1}, \quad a_{n,n} = \tfrac{2}{\tau} + v_n, \quad b_n = w_{n-1} + w_n \,.$$

With $x := [x_0\ x_1\ \ldots\ x_{N-1}]^t$, $b := [b_0\ b_1\ \ldots\ b_{N-1}]^t$, and $a_{0,N-1} := a_{0,-1}$, these equations can be assembled into the form $Ax = b$, with

$$A = \begin{pmatrix} a_{0,0} & & & & & a_{0,N-1} \\ a_{1,0} & a_{1,1} & & & & \\ & a_{2,1} & a_{2,2} & & & \\ & & & \ddots & & \\ & & & & \ddots & \\ & & & & a_{N-1,N-2} & a_{N-1,N-1} \end{pmatrix} \,.$$

This is a bidiagonal circulant matrix, if v is a constant. In the case of a k-step multistep method the resulting matrix will be of similar form, yet with $k+1$ diagonals. The particular matrix displayed above allows a straightforward LU-decomposition, i.e., $A = L.U$, with L and U having the format shown below.

$$L = \begin{pmatrix} 1 & & & & \\ l_1 & 1 & & & \\ & l_2 & 1 & & \\ & & \ddots & \ddots & \\ & & & l_{N-1} & 1 \end{pmatrix}, \quad U = \begin{pmatrix} u_0 & & & & e_0 \\ & u_1 & & & e_1 \\ & & u_2 & & \ddots \\ & & & \ddots & e_{N-2} \\ & & & & u_{N-1} \end{pmatrix} \,.$$

The coefficients l_i, u_i and e_i can be calculated as follows.

$$\begin{cases} i = 0 & : \quad u_0 = a_{0,0}; \quad e_0 = a_{0,N-1} \\ i = 1,\ldots,N-2 & : \quad l_i = a_{i,i-1}/u_{i-1}; \quad u_i = a_{i,i}; \quad e_i = -l_i e_{i-1} \\ i = N-1 & : \quad l_{N-1} = a_{N-1,N-2}/u_{N-2}; \quad u_{N-1} = a_{N-1,N-1} - l_{N-1}e_{N-2} \end{cases}$$

In order to solve the system $Ax = b$, the LU-decomposition step is followed by the usual two-stage back-substitution step, i.e., solve $Ly = b$ and then $Ux = y$. The necessary arithmetic is detailed below.

$$\begin{cases} y = L^{-1}b & : \quad y_0 = b_0; \quad y_i = b_i - l_i y_{i-1}, \quad i = 1,\ldots,N-1 \\ x = U^{-1}y & : \quad x_{N-1} = y_{N-1}/u_{N-1}; \quad x_i = (y_i - e_i x_{N-1})/u_i, \quad i = 0,\ldots,N-2 \end{cases}$$

A count of the floating-point operations reveals that the complexity of solving the system is about $7N$ if we include the cost of the LU-decomposition, and $5N$ if we don't. When v is a constant, these counts become $6N$ and $5N$. However, this is only the cost of the core of the ODE-integrator. If the other operations are taken into account (construction of the right-hand side w, of the "A"-matrix coefficients and of the right-hand side "b"), we get a total of $14N$ (or 15 with LU-decomposition). This leads to the constant $C_S = 14$ (or, $C_S = 15$). This should be compared with the value $C_S=12$ derived in the previous section.

These counts are for five-point stencils and variable but time-independent PDE coefficients. In the case of more general problems the counts for the time-periodic and for the initial value waveform relaxation smoothers are relatively even closer.

Finally, if the costs of restriction, prolongation, correction, defect calculation, initialization, PDE-coefficient and right-hand side evaluation (which are method independent) are taken into account, we may safely state that the cost of one time-periodic multigrid waveform cycle is (almost) equal to that of an initial value cycle. Since the convergence properties of both algorithms are very similar, we can conclude that the cost of solving a time-periodic problem is (almost) equal to that of solving the corresponding initial value problem. This conclusion holds also for other time-discretizations (e.g., more general multistep methods), as long as a method is used that solves (7.5) at a cost similar to that of solving only a few initial value problems.

Comparison with standard time-periodic solvers

We already argumented in chapter 4 that the cost of solving a time-periodic problem by dynamic simulation, by global discretization and the use of a direct sparse solver, or by shooting is substantially higher than the cost of solving an initial boundary value problem. It remains to consider the multigrid method of the second kind.

To this end we introduce some additional notation. Let $W_i(IBVP)$ and $W_i(TPP)$ denote the cost of solving an initial boundary value problem and a time-periodic problem on Ω^{h_i}. Let α_M be the number of cycles of the multigrid method of the second kind required for solving a time-periodic problem on Ω^{h_k}. In the multigrid method of the second kind several initial boundary value problems are to be solved at each grid level (these are the calculations of type $Y^{h_i} := T^{h_i}Y^{h_i} + K^{h_i}$ in algorithm 4.1). Since the algorithm is based on a W-cycle, every grid Ω^{h_i} is visited precisely 2^{k-i} times in a descending W-cycle branch during each cycle on Ω^{h_k}. In half of these visits two initial boundary value problems are to be solved (one in the smoothing and one in the defect calculation step). In the other half the problem in the smoothing step can be skipped. Indeed, when the initial condition is zero, the smoothing step simplifies to $Y^{h_i} := K^{h_i}$. The total number of initial boundary problems, n_i, to be solved on grid Ω^{h_i} ($0 < i < k$) in one cycle of the multigrid method of the second kind on Ω^{h_k}, is therefore equal to

$$n_i = \tfrac{3}{2}\, 2^{k-i} . \tag{7.6}$$

On Ω^{h_0} a time-periodic problem is to be solved. This is to be done 2^{k-1} times.

If in the multigrid method of the second kind a time-discretization method is used with sufficient smoothing behaviour, e.g. a backward differentiation formula, the number of time-steps may be decreased in going from one grid level to a coarser one. In

our implementation we have considered the BDF(2) formula with $\Delta t = $ const. h. That
is, the number of time-steps is halved from one grid level to the next. Let $n_{t,i}$ denote
the number of time-levels on grid Ω^{h_i}. Then,

$$n_{t,i} = 2^{i-k} n_{t,k} \quad (0 \le i \le k) . \tag{7.7}$$

Consequently, the total number of grid points (in space and time) differs by a factor
of (about) 8 between successive grids. From this, it follows that we may take

$$W_i(IBVP) = 8^{i-k} \, W_k(IBVP) \quad (0 \le i \le k) . \tag{7.8}$$

when an algorithm is used the cost of which is proportional to the number of grid
points. We can now calculate the cost of the multigrid method of the second kind. In
the derivation below we neglect the costs of restriction, prolongation and correction,
and the cost of the coarse grid solver [3].

$$
\begin{aligned}
W_k(TPP) &= \left(\, 2 \, W_k(IBVP) + \sum_{i=1}^{k-1} n_i W_i(IBVP) \, \right) \alpha_M \\
&= \left(\, 2 + \sum_{i=1}^{k-1} \tfrac{3}{2} \, 2^{k-i} \, 8^{i-k} \, \right) W_k(IBVP) \, \alpha_M \\
&= \left(\, \tfrac{5}{2} - 2 \cdot 4^{-k} \, \right) W_k(IBVP) \, \alpha_M \\
&\simeq \, 2.5 \, W_k(IBVP) \, \alpha_M
\end{aligned}
$$

Experience shows that $\alpha_M = 1$ is often sufficient to solve the time-periodic problem
to discretization accuracy. In that case, the cost of the multigrid method of the second
kind is about 2.5 times the cost of solving a single initial boundary value problem.
This is substantially better than the other standard time-periodic solvers. Yet, it is 2.5
times slower than the full multigrid time-periodic waveform relaxation method, which
calculates the solution to the time-periodic problem to similar accuracy.

7.3   Parallel implementation

7.3.1   Grid partitioning

The parallelization of the multigrid waveform relaxation method is based on an identical grid partitioning strategy as discussed in chapter 6. The spatial grid is divided
into subgrids which are distributed over the processors. Each processor has the update
right for all the grid point *functions* which are located in its subdomain. As such,
each processor is in fact responsible for doing the calculations on a *box* of grid points
of the space-time grid. Note that even for relatively coarse spatial meshes the number of space-time grid points per processor (and, associated with this, the arithmetic
complexity per processor) may be fairly large.

When function values at neighbouring grid points are needed, they are obtained
by message passing. The communication structures are identical to the ones used in
the parallelization of the elliptic multigrid method, see chapter 6. The only difference
is the length of each message. Instead of transferring sets of scalars, sets of discrete
functions are moved from one processor to another.

---

[3]If we include the cost of the coarse grid solver and assume that $W_0(TPP) = W_0(IBVP)$ then we
get $W_k(TPP) = (\tfrac{5}{2} - \tfrac{3}{2} 4^{-k}) W_k(IBVP) \alpha_M$.

7.3.2 Communication complexity

Initial boundary value problem solvers

Let M_i denote the *total number of messages* sent by each processor in *one multigrid cycle* (V, W, FMG) applied to an *elliptic problem* on grid Ω^{h_i} (e.g., at one time-level of the space-time grid). If M_i differs from processor to processor, we take the maximum value. Let L_i denote the corresponding *total message length*. We set $TM_i(IBVP)$ and $TL_i(IBVP)$ to denote the total number of messages and the total message length in the algorithm as a whole for solving a parabolic problem on Ω^{h_i}.

The communication complexity of the *time-stepping* methods is easy to determine. It equals the complexity per time-level ($M_i \alpha_E$ and $L_i \alpha_E$) multiplied by the number of time-levels $n_{t,i}$. The number of messages sent in one multigrid *waveform relaxation* cycle is precisely equal to the number of messages sent in one elliptic multigrid cycle. The length of the messages in one waveform cycle is equal to the length in one elliptic cycle multiplied by $n_{t,i}$, the length of a discretized function. The cost of both methods for solving a problem on the fine grid, Ω^{h_k}, is reported in table 7.3.

Table 7.3: Communication complexity of initial boundary value problem solvers

	time-stepping algorithm	multigrid waveform relaxation
$TM_k(IBVP)$	$n_{t,k} M_k \alpha_E$	$M_k \alpha_{WR}$
$TL_k(IBVP)$	$n_{t,k} L_k \alpha_E$	$n_{t,k} L_k \alpha_{WR}$

In the case of the full multigrid approach a value of 1 for α_E and α_{WR} is often satisfactory. The total message length in both solvers is then identical, whereas the number of messages differs by a factor of $n_{t,k}$.

Time-periodic problem solvers

The communication complexity of a time-periodic multigrid waveform relaxation cycle is identical to that of an initial value multigrid waveform relaxation cycle. The communication cost is reported in table 7.4, where we have used $\bar{\alpha}_{WR}$ to denote the number of time-periodic multigrid waveform relaxation cycles.

The derivation of the communication complexity of the multigrid method of the second kind is much more involved. Some further assumptions are helpful. We assume in particular that the length of the messages in each of the multigrid operators is proportional to $1/h$, and that the number of messages is independent of h. These are very natural assumptions for 2D problems. In addition, we assume that no communication is required for solving the coarse grid problem, i.e., $M_0=0$ and $L_0=0$. (For instance, we assume that the coarse problem is solved on a single processor, and we neglect certain anomalies on the very coarse grids.) Furthermore, we restrict the analysis to multigrid V-cycles. This leads to,

$$M_i = \frac{i}{k} M_k \ . \tag{7.9}$$

The total message length per cycle can be determined by considering the following second order recurrence relation, $L_{i+1} - L_i = 2\,(L_i - L_{i-1})$ or $L_{i+1} - 3L_i + 2L_{i-1} = 0$. Its solution, satisfying $L_0 = 0$, is easily found to be,

$$L_i = \frac{2^i - 1}{2^k - 1} L_k \quad (\;\simeq\; 2^{i-k} L_k \quad \text{if } i \text{ and } k \quad \text{are sufficiently large }) . \tag{7.10}$$

The total length $TL_k(TPP)$ and the total number of messages $TM_k(TPP)$ for solving a time-periodic problem by α_M cycles of the multigrid method of the second kind can now be calculated.

$$\begin{aligned}
TM_k(TPP) &= (2\ TM_k(IBVP) + \textstyle\sum_{i=0}^{k-1} n_i\ TM_i(IBVP))\ \alpha_M \\
&= (2\,n_{t,k}\ M_k\ \alpha_E + \textstyle\sum_{i=0}^{k-1} n_i\ n_{t,i}\ M_i\ \alpha_E)\ \alpha_M \\
&= (2\,n_{t,k}\ M_k\ \alpha_E + \textstyle\sum_{i=0}^{k-1} \frac{3}{2}2^{k-i}\,2^{i-k}n_{t,k}\,\frac{i}{k}\ M_k\ \alpha_E)\ \alpha_M \\
&= (2 + \textstyle\sum_{i=0}^{k-1} \frac{3}{2}\,\frac{i}{k})\ n_{t,k}\ M_k\ \alpha_E\ \alpha_M \\
&= \tfrac{5+3k}{4}\ n_{t,k}\ M_k\ \alpha_E\ \alpha_M
\end{aligned}$$

$$\begin{aligned}
TL_k(TPP) &= (2\ TL_k(IBVP) + \textstyle\sum_{i=0}^{k-1} n_i\ TL_i(IBVP))\ \alpha_M \\
&= (2\,n_{t,k}\ L_k\ \alpha_E + \textstyle\sum_{i=0}^{k-1} n_i\ n_{t,i}\ L_i\ \alpha_E)\ \alpha_M \\
&= (2\,n_{t,k}\ L_k\ \alpha_E + \textstyle\sum_{i=0}^{k-1} \frac{3}{2}2^{k-i}\,2^{i-k}n_{t,k}\,\frac{2^i-1}{2^k-1} L_k\ \alpha_E)\ \alpha_M \\
&= (2 + \textstyle\sum_{i=0}^{k-1} \frac{3}{2}\,\frac{2^i-1}{2^k-1})\ n_{t,k}\ L_k\ \alpha_E\ \alpha_M \\
&= (2 + \tfrac{3}{2}\,(1 - \tfrac{k}{2^k-1}))\ n_{t,k}\ L_k\ \alpha_E\ \alpha_M \\
&\simeq \tfrac{7}{2}\ n_{t,k}\ L_k\ \alpha_E\ \alpha_M
\end{aligned}$$

Because of its importance in actual implementations we also derive the communication complexity of the full multigrid approach. Note that no *time-coarsening* is used in the full multigrid waveform relaxation method. The length of each vector (function) is independent of the grid level and equal to $n_{t,k}$.

$$TM_k(TPP) = \textstyle\sum_{i=0}^{k} M_i = \textstyle\sum_{i=0}^{k} \frac{i}{k}M_k = \frac{k+1}{2}M_k \tag{7.11}$$

$$TL_k(TPP) = \textstyle\sum_{i=0}^{k} n_{t,k}\ L_i = (2 - \frac{k}{2^k-1})n_{t,k}\ L_k \;\simeq\; 2\,n_{t,k}\ L_k \tag{7.12}$$

The results are collected in table 7.4. Note that the communication complexity of the periodic multigrid waveform relaxation algorithm is much smaller than that of the multigrid method of the second kind. Consider e.g. the following typical set of parameters, taken from one of the case studies in chapter 8: $k=4$, $n_{t,4} = 32$, $\alpha_E=2$, $\alpha_M=1$. With these parameters the number of messages required by the multigrid method of the second kind is larger than that of the full multigrid waveform method by a factor of more than 100. In addition, the length differs by a factor of 3.5.

Table 7.4: Communication complexity of time-periodic problem solvers.

	multigrid second kind	waveform multigrid	waveform fmg
$TM_k(TPP)$	$(5 + 3k)/4n_{t,k}M_k\alpha_E\alpha_M$	$M_k\bar{\alpha}_{WR}$	$(k + 1)/2M_k$
$TL_k(TPP)$	$7/2n_{t,k}L_k\alpha_E\alpha_M$	$n_{t,k}L_k\,\bar{\alpha}_{WR}$	$2n_{t,k}L_k$

7.4 Vectorization

The use of vector processors for executing the multigrid waveform relaxation algorithm results in an important reduction in computing time, as we first illustrated in [134]. Indeed, as mentioned in section 7.2 most waveform relaxation operators are linear combination operators which operate on discretized functions or vectors. Consequently, waveform relaxation methods naturally *vectorize in the time-direction*. This contrasts to the standard vectorization approach, which *vectorizes across space*. (The vectors contain values of grid points that belong to the same time-level).

With use of waveform vectorization the vector speedup of the arithmetic strongly depends on the number of time-levels. It is virtually independent of the size of the spatial grid, the number of multigrid levels, the multigrid cycle used, and the number of processors. This is verified by extensive timing experiments obtained on an Intel iPSC/2 VX hypercube, reported in chapter 8. This grid size independence contrasts sharply to standard multigrid vectorization results, see e.g. McBryan et al. [81] which discusses vectorized multigrid implementations on the Intel hypercube and the SUPRENUM, and Lemke [76], which deals with multigrid on vector-supercomputers, CDC 205, CRAY X-MP and Fujitsu VP200. Standard vectorization does not lead to any speedup unless the number of points per processor is very large. The use of standard spatial vectorization is therefore of limited use on a large-scale parallel processor.

A second advantage of waveform vectorization is the ease of implementation. As the vector operations at a grid point involve the vectors at neighbouring grid points only, no complicated grid restructuring (as in the standard approach) is required.

The only operation in the waveform relaxation method which is not perfectly vectorizable is the core of the ODE integrator used in the smoothing step. It consists of one first order recurrence relation in the case of an initial boundary value problem, and two first order recurrence relations in the case of a time-periodic problem. Since these recurrence relations are inherently sequential they diminish the achievable vector speedup. We can easily calculate their influence on the highest possible gain through vectorization. The latter value is expressed by Amdahl's law as the inverse of the sequential fraction, or,

$$S_{vector} = \frac{n_{vector} + n_{sequential}}{n_{sequential}}.$$

Here, n_{vector} denotes the total number of operations that can be executed in vector-mode. The remaining number of operations is given by $n_{sequential}$. Assume, for instance, that the non-vectorizable fraction of the algorithm is about 5 to 10%. By Amdahl's law

S_{vector} is then a value between 10 and 20. The actual speedup of an implementation will be smaller due to the finite cost of a vector operation, due to vector-pipe startup costs, and various other sequential overheads.

It can be seen that the cost of the recurrence relation makes out only a small fraction of the total computation cost. By way of illustration we consider again the Crank-Nicolson discretization. The number of floating operations involved in the recurrence can be counted. It equals 2 (IBVP) or 3 (TPP) per grid point. This should be compared to the number of flops required by the vectorizable parts of the algorithms. For instance, with the values of table 7.1 the non-vectorizable fraction is given by 4/32 (5pt, IBVP) or 4/53 (9pt, IBVP) for any cycle with two smoothing steps. Note that most of the initialization operations, such as the PDE right-hand side evaluation, are perfectly vectorizable. They further decrease the sequential fraction.

Finally, we would like to mention that the vector speedup can be improved even further. For instance, one could vectorize the recurrence relation by techniques based on cyclic reduction. One could also increase the vector lengths by combining the waveform vectorization with spatial vectorization.

7.5 Concluding remarks

We have shown that the arithmetic complexity of the waveform relaxation method for solving an initial boundary value problem is essentially similar to that of the standard time-stepping schemes. Waveform relaxation for solving time-periodic problems is substantially faster than any other method. These statements are true as long as the multigrid acceleration leads to a rapidly convergent waveform relaxation algorithm. We may have to reconsider the ranking of the methods when the multigrid acceleration is not as successful, e.g. in the case of strongly anisotropic elliptic operators.

Waveform relaxation generally leads to much higher parallel efficiencies than the parallel implementation of standard techniques. This is due to the much lower number of messages and, correspondingly, the tremendous reduction in communication startup costs. As such, waveform relaxation will especially be successful for use on parallel machines that are characterized by a high value of $t_{startup}$ (see chapter 5), the Intel machine being a notorious example. We may also mention that waveform relaxation allows a better exploitation of communication with calculation overlap. This is due to the large message lengths. (The message lengths are equal to the message lengths in standard time-stepping methods multiplied by the number of time-levels.)

Finally, we would like to direct the reader's attention to a paper by P. Worley [151]. He suggests using a parallelized cyclic reduction method for evaluating the recurrence relations in the ODE solvers. This would lead to a space-time grid partitioning in both the space- and the time-direction, and would allow efficient implementation of multigrid waveform relaxation on truly massively parallel systems.

Chapter 8

Case Studies

Suppose you want to teach the "cat" concept to a very young child.
Do you explain that a cat is a relatively small, primarily carnivorous mammal
with retractile claws, a distinctive sonic output, etc.?   I bet not.
You probably show the kid a lot of different cats, saying "kitty" each time,
until it gets the idea. To put it more generally,
generalizations are best made by abstraction from experience.

—R.P. Boas, *"Can we make mathematics intelligible"*
(Am. Math. Monthly 10, p. 727, Dec. 1981).

Although their work differs from the experimental research associated with, say, test
tubes and noxious chemicals, mathematicians, like chemists and other researchers,
often collect piles of data – whether prime numbers or diagrams of knots – before they
can begin to extract and abstract the principles that account for their observations.

—I. Peterson, *"Searching for new mathematics"*,
(SIAM Review Vol. 33, No. 1, pp. 37-42, March 1991).

We present a number of non-trivial, linear and nonlinear examples of initial boundary value and time-periodic parabolic partial differential equations. Each problem is solved with the appropriate variant of the multigrid waveform relaxation method as well as with "the best" standard parabolic solver. The differences in performance are explained and the theoretical results obtained in the previous chapters are illustrated. It is shown that the waveform relaxation methods are competitive on sequential processors, and that they outperform the standard techniques on parallel machines. In particular we illustrate that on a 16-processor vector hypercube waveform relaxation can be faster than any of the standard approaches by a factor of ten up to forty.

8.1   Introduction

In the previous chapter we analysed the parallel efficiency of the waveform relaxation method and we discussed the potential for vectorization. It was evidenced that the use of waveform relaxation generally leads to higher parallel and vector speedups than the use of the corresponding standard time-marching methods. High processor usages, however, are not what the practitioner of numerical simulation aims at. Instead, he wants to obtain the solution to his particular problem in as short a time as possible. Consequently, numerical characteristics have to be taken into account when evaluating the waveform approach as a means for solving partial differential equations on sequential and on parallel computers.

In the present chapter we intend to complete the view of the waveform methods by exemplifying their numerical performance. This is done by means of a careful selection of case studies which are chosen so as to illustrate the wide applicability of the methods. The numerical results allow to verify a number of properties previously mentioned or analysed, such as the sequential and parallel complexities, the effects of linearization and windowing, the speedup due to vectorization, and so on. The main emphasis will be put on the comparison of the waveform relaxation algorithms with the standard parabolic problem solvers. To this end we have implemented for each problem class a good, if not the best, standard solution method. In particular, in the case of initial boundary value problems we compare the waveform approach with the Crank-Nicolson method; in the case of time-periodic problems, the comparison is with the multigrid method of the second kind and with the dynamic simulation method, see section 4.2.

In the next two sections, we briefly talk about some important issues that were taken into account when implementing the programs, and we discuss the presentation of the results. Section 8.4 contains an extensive comparison of the linear initial boundary value solvers. Two examples are reported: one with and one without a cross derivative term; both have Dirichlet and mixed boundary conditions. Two nonlinear initial boundary value problems one of which is a system of equations are considered in section 8.5. Section 8.6 deals with linear time-periodic equations, where the time-periodicity is either due to a time-periodic PDE right-hand side or due to a time-periodic boundary condition. A nonlinear time-periodic system of two equations constitutes the final example and is given in section 8.7. We end in section 8.8 where we briefly discuss the class of problems that can be solved efficiently with waveform relaxation.

8.2　Programming considerations

In an experimental comparison one inevitably compares particular implementations on particular machines. A comparison of different *algorithms* is therefore bounded to fail unless the programs are very carefully implemented. The following precautions were taken to ensure the fairness of our study.

- The programs are such that they solve exactly the *same class of problems*. For instance, all of our linear solvers are made to handle the class of linear second order partial differential equations with variable but time-independent coefficients, on a two-dimensional rectangular domain with boundary conditions of Dirichlet, mixed or periodic type. Note that the programs used for the comparison of the standard parabolic marching schemes in chapter 6 allow the more general time-dependent coefficient case.

- We use a *consistent programming style*, with comparable optimizations and a similar program complexity throughout. The code optimizations are the usual ones that any programmer is expected to perform. Any unusual optimizations that are expected to be absent in a standard implementation are avoided. Sometimes, this results in a somewhat different implementation style for different algorithms. For instance, in a time-stepping method it is standard to evaluate the PDE right-hand side time-level per time-level, although often some savings could be made if the right-hand side were evaluated for all time-levels simultaneously. The latter is the natural "standard" approach in a waveform relaxation context.

- We compare codes that solve a same problem to a *similar accuracy*. To this end the same discretization techniques and parameters (time-step, spatial mesh size, etc.) are applied. We have shown in previous chapters that in that case waveform relaxation and time-stepping are merely two different techniques for solving the same system of equations. The two solutions are then identical, and the residual of approximate solutions may fairly be compared.

Within the framework sketched above, we have tried to select for each method good and possibly optimal parameters. Some of these parameters must be selected depending on the particular hardware platform on which the programs are executed. For instance, the choice of when and how to agglomerate is highly machine- (and algorithm-) dependent. Other parameters have to do with the numerical characteristics of the methods, such as the type of multigrid cycle, the required number of cycles, the number of smoothing steps, and so on. It is very difficult, if not impossible, to determine a priori what selection will lead to the smallest execution time for solving a particular problem to a particular accuracy. This is a problem on a sequential machine, and it is even more problematic on a parallel processor. To do justice to each algorithm, we shall therefore always present results for several different parameter sets. As an added advantage, this will allow the reader to judge the sensitivity of the execution time and the accuracy with regard to the parameters.

8.3 Representation of the results

The results of the experiments are displayed in the figures of the following sections. Some of these figures show diagrams where the *accuracy* of an approximation is plotted *versus the execution time* needed for its computation. This "accuracy" is defined as follows. Let u, u_τ^h, and $\bar{u}_\tau^h$ denote respectively the true solution of the PDE, the exact solution of the discretized problem, and an approximation of the discrete solution. When u is known, or when a very good approximation is available (e.g. obtained on a very fine mesh), the accuracy will be given by some norm of the *error*, $u - \bar{u}_\tau^h$. E.g.,

$$\| \text{ error } \|_\infty = \max_{i,j,k} |u(t_k, x_i, y_j) - \bar{u}_{k,i,j}^h| \qquad \text{(max. norm)}$$

$$\| \text{ error } \|_2 = \left(\sum_{i,j,k} |u(t_k, x_i, y_j) - \bar{u}_{k,i,j}^h|^2 \right)^{1/2} \qquad \text{(Eucl. norm)}$$

where the indices i, j, k run through each of the $n_x \times n_y \times n_t$ unknowns. n_x and n_y denote the number of unknowns along the x- and y-axis, and n_t is the number of time-steps. In the case of a time-periodic problem standard methods aim at finding the solution at one time-level only, e.g., $t = 0$ or $t = T$. They solve a problem with $n_x \times n_y$ variables. In that case, the k-index should be deleted from the above formulae.

When the true solution is not known, "accuracy" will either indicate the norm of the *discrete residual*, or the norm of the *algebraic error*. The former is obtained upon substitution of the approximation in the discrete set of equations. The latter is defined as the difference between the true *discrete* solution and the approximation, i.e., $u_\tau^h - \bar{u}_\tau^h$.

Finally, we want to explain how the results are presented. With the waveform relaxation method an initial approximation is defined along the whole time-interval and is iteratively updated by applying multigrid cycles. Each iterate achieves a certain accuracy and its computation requires a certain amount of computing time. Every such iterate is represented by an "o"-symbol in the diagrams. Successive iterates are connected by lines. These lines are annotated with, e.g., "WR, V(1,1) FMG", which signifies the use of waveform relaxation with V-cycles with one pre-smoothing and one post-smoothing step, and the full multigrid procedure to determine the initial approximation. When a constant initial profile is selected, equal to the value of the initial condition, the letters "FMG" are omitted.

The results obtained with the standard time-stepping schemes show up as discrete points. With each parameter set the solution is advanced time-step per time-step, in a total of t seconds. The resulting approximation is represented by a "+"-symbol at the coordinate (t,accuracy) in the diagram. The annotation is as follows. E.g., "CN, 2 V(1,1)" means the use of the Crank-Nicolson method, where the problem at each time-level is solved by applying 2 V(1,1)-cycles to an initial approximation. That initial approximation is either calculated by extrapolation of the solution at one or at two previous time-levels, or it is calculated by using the full multigrid method. In the latter case the letters "FMG" are added to the annotation. E.g., with "MGM 2nd, BDF(2), 1 V(1,1) FMG" we indicate the use of the multigrid method of the second kind; the time-integrations in the algorithm are performed by the second order backward differentiation method; in each time-step the initial approximation is determined by the full multigrid method and corrected by one V(1,1)-cycle.

8.4   Linear initial boundary value problems

8.4.1   Example 1

Our first example is a linear problem with a cross derivative term,

$$\frac{\partial u}{\partial t} = \frac{\partial^2 u}{\partial x^2} + xy\frac{\partial^2 u}{\partial x \partial y} + \frac{\partial^2 u}{\partial y^2} + f \; , \quad (x,y) \in [0,1] \times [0,1], \; t \in [0, 0.5] \; . \qquad (8.1)$$

This equation is supplemented with Dirichlet conditions on the north (N), east (E) and south (S) boundaries, and a Neumann condition on the boundary to the west (W),

$$N : \quad u(t,x,1) = \sin(5x + 1 + 10t)\, e^{-4t} \qquad E : \quad u(t,1,y) = \sin(5 + y + 10t)\, e^{-4t}$$

$$S : \quad u(t,x,0) = \sin(5x + 10t)\, e^{-4t} \qquad W : \quad \tfrac{\partial u}{\partial x}(t,0,y) = 5\cos(y + 10t)\, e^{-4t}$$

The function $f(t,x,y)$ and the initial condition are such that the solution of (8.1) equals $u(t,x,y) = \sin(5x + y + 10)\, e^{-4t}$.

The problem will be solved for several grid sizes. The spatial mesh is always chosen equidistant and with equal mesh size, h, in x- and y-direction. Three values of h were considered, namely 1/16, 1/32 and 1/64. The time-step, τ, is set to 0.01, independent of the spatial discretization. This leads to a vector length n_t of 50. Because of the Neumann boundary condition, there is one more grid line with unknows parallel to the y-axis than there are parallel to the x-axis, see appendix A. Therefore, the values of n_x and n_y are given by 16, 32, 64 and 15, 31, 63. Spatial discretization is done with second order central differences. This leads to a general *nine-point stencil*.

We shall compare the performance of multigrid waveform relaxation with that of the Crank-Nicolson method. In both approaches the multigrid method uses a four-colour nine-point Gauss-Seidel smoother, standard coarsening down to a coarse grid with mesh size $h = 1/2$ (that is with $n_x = 2$ and $n_y = 1$), full-weighting restriction, bilinear interpolation and a coarse grid solver that performs two Gauss-Seidel iterations. In order to guarantee identical solutions the trapezoidal rule is used for time-integration in the waveform technique. The timing results are depicted in the graphs of figure 8.1. They were obtained on a 16-processor Intel iPSC/2-VX hypercube with *vector*-nodes. Load distribution was performed by means of a two-dimensional partitioning with 4 processors in each coordinate direction. Two sets of results are given for the waveform method. The solid line indicates timings obtained in *scalar* execution mode, while the dashed lines represent the results in *vector* mode.

For the largest problem, an error of the order of the discretization error [1] is obtained with 4 WR V(1,1)-cycles or 3 WR F(1,1)-cycles. If full multigrid is used to determine the starting solution, only one additional V(1,1)- or F(1,1)-cycle is needed. In the Crank-Nicolson method, the initial approximation at each time-level is determined either by the full multigrid method, or as a linear extrapolation of the solution at two previous time-levels. In the former case only one additional cycle is needed at each grid level. In the latter case a fairly accurate initial approximation is also obtained and consequently two additional V(1,1)- or F(1,1)-cycles suffice to solve the linear system at each time step. For the $h=1/16$ discretization, a single V-cycle or F-cycle at each time-level turns out to be sufficient. This is due to the small time-step (too small compared to the spatial mesh size) the very smooth time behaviour of the solution and the use of the extrapolation procedure. Note that in practice it is difficult to know in advance how many cycles are needed at each time-level. One will therefore often do several cycles in excess, to be on the safe side, or use the full multigrid approach with one or two cycles on the fine grid. [2]

On the 16-processor vector hypercube, waveform relaxation turns out to be faster than the Crank-Nicolson method by a factor of about 7 (and even higher, up to a factor of 10, when the corresponding full multigrid approaches are compared). This is due to the *smaller arithmetic complexity* of the waveform method, its *superior parallel characteristics* and the use of *vectorization*. We shall briefly analyse each of these.

---

[1] Additional iterations do not reduce the actual error, $u - \bar{u}_\tau^h$, any further. They only reduce the algebraic error, $u_\tau^h - \bar{u}_\tau^h$ and the residual.

[2] We refer to section 8.5 for two examples where the residual in each time-level is driven to zero by executing many multigrid cycles.

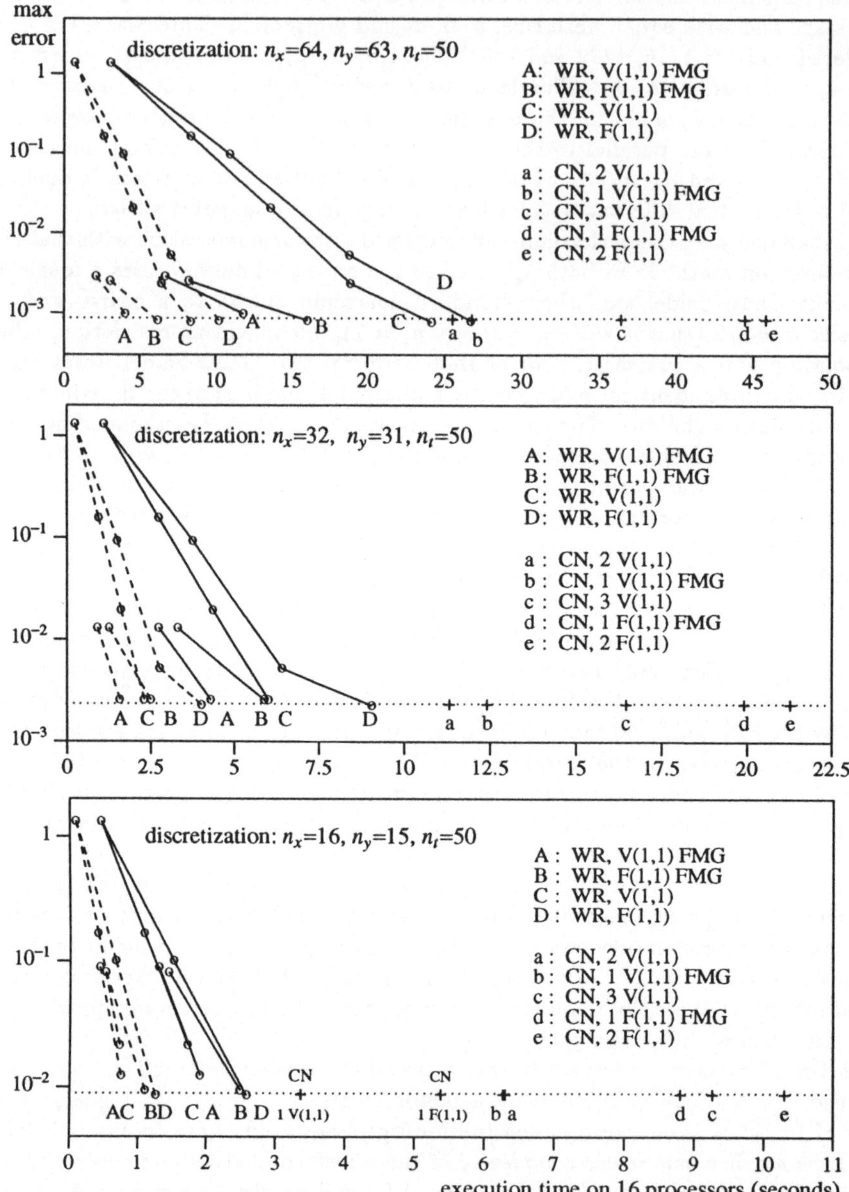

Figure 8.1: Example 1: comparison of Crank-Nicolson time-stepping and linear multigrid waveform relaxation, on 16 processors (solid lines: without vectorization, dashed lines: with vectorization). The horizontal dotted line indicates the discretization error, i.e., the error of the fully converged discrete solution

Table 8.1: Example 1: execution time, in seconds, of the full multigrid waveform relaxation method and of the Crank-Nicolson method with full multigrid solution of the linear problem at each time-level (resp. "A" and "b" in figure 8.1).

method	$n_x = 16,\ n_y = 15,\ n_t = 50$			$n_x = 32,\ n_y = 31,\ n_t = 50$		
	1 proc.	16 proc.	S_{16}	1 proc.	16 proc.	S_{16}
Waveform Relaxation	9.53	1.91	5.0	36.00	4.30	8.4
Crank-Nicolson	12.90	6.32	2.0	48.29	12.52	3.9

Table 8.2: Example 1: execution time, in seconds, of multigrid waveform relaxation (three V(1,1)-cycles) and of Crank-Nicolson with three V(1,1)-cycles at each time-level.

method	$n_x = 16,\ n_y = 15,\ n_t = 50$			$n_x = 32,\ n_y = 31,\ n_t = 50$		
	1 proc.	16 proc.	S_{16}	1 proc.	16 proc.	S_{16}
Waveform Relaxation	15.64	2.36	6.6	62.05	6.00	10.3
Crank-Nicolson	21.49	9.30	2.3	82.41	16.55	5.0

In table 8.1 we have tabulated the execution time of the full multigrid solvers with one V(1,1)-cycle at each grid level, on 1 and on 16 processors. In table 8.2 we report similar results obtained when 3 V(1,1)-cycles are applied. In both tables we have added the parallel speedup, S_{16}. On a single processor our implementation of the waveform method shows to be faster than the implementation of the standard approach by a factor of about 1.33. This is partly due to the lower arithmetic complexity of the numerical kernels, as was discussed in chapter 7. But, it is also due to particular implementation aspects, which, as we feel, are inherent in the waveform approach. E.g., there is the smaller initialization cost in evaluating the PDE right-hand side f and the boundary conditions. In the waveform approach these evaluations appear grid-point-function per grid-point-function. Some intermediate results, e.g. expressions involving the x, y-variables only, may be retained and need not be re-evaluated at every time-level. This is not possible when the evaluations proceed time-step per time-step. Additionally, there is the much lower computational overhead associated with program control, such as loop overhead, procedure call overhead, and, most importantly, indexing overhead. In the Crank-Nicolson method the data structures are two-dimensional grids. To access a variable two indexing operations are required. With waveform relaxation the data structures are two-dimensional grids of vectors. Two indexing operations are required to access a vector, but only one further indexing operation is required to access a particular element. On the average two indexing operations accompany every floating point operation in the Crank-Nicolson method. In the waveform approach calculations are performed with functions (or vectors). Therefore only slightly more than one index operation is required per floating point operation. (The cost of accessing the vector is amortized over the cost of accessing the n_t vector elements.) The importance of the above can be appreciated when it is realized that the cost of a floating point operation lies in the range of 4 to 6 μ-seconds, while the cost of an index operation is about 1 μ-second.

The higher parallel efficiency of the waveform relaxation method results in a supplementary speedup factor of 2 to 2.5. This immediately follows from the S_{16}-values in tables 8.1 and 8.2. The speedup of a multigrid V(1,1)-cycle, executed on a 4 by 4 processor mesh, is displayed in figure 8.2 as a function of the number of time-levels, n_t. As expected, speedup is an increasing function of n_t. For sufficiently large function lengths an asymptotic S_{16} value is approached, the magnitude of which is mainly determined by the effect of load imbalance and coarse mesh processor idling. Also indicated in the figure is the speedup obtained with a parallel implementation of the Crank-Nicolson algorithm. This speedup is, of course, independent of the number of time-levels. As explained in chapter 7, the Crank-Nicolson speedup is much smaller, because of the much higher communication cost.

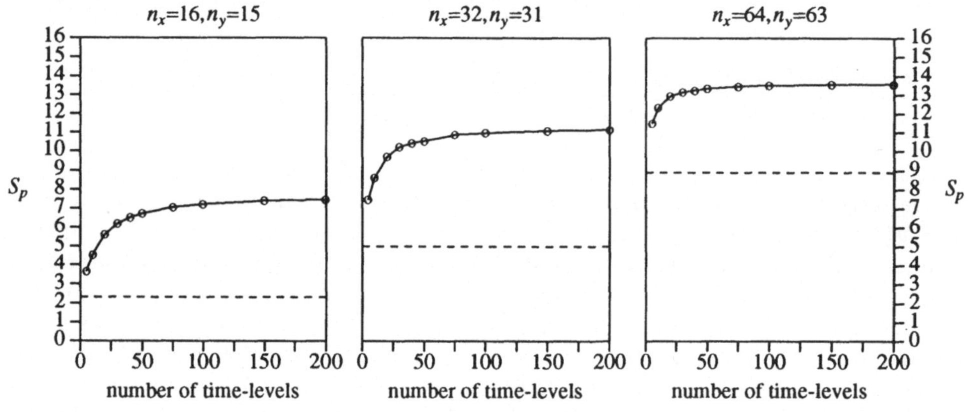

Figure 8.2: Example 1: speedup of a multigrid waveform V(1,1)-cycle on a 16 processors. The dashed horizontal line indicates the speedup of a V(1,1)-cycle in the Crank-Nicolson time-stepping scheme.

The remaining performance difference is due to vectorization. Note first that for the problem sizes and the number of processors considered here, vectorization in the Crank-Nicolson method would not lead to any speedup. In table 8.3 we give the execution times for a small problem, a discretization with 9 by 9 grid lines, solved on a single processor in scalar and vector mode. The values represent the time in seconds for the execution of the full multigrid solver with V- or F-cycles, and the cost of a single V- and F-cycle applied to the fine grid. The low values of the vector speedup are due to the high startup time of the vector operations on the iPSC/2. The speedup would be higher on a processor with more specialized vector hardware. The dependence of the vector speedup on the vector length is obvious. As expected from the discussion in chapter 7, the vector speedup turns out to be virtually independent of the multigrid cycle type.

The full multigrid speedups are higher than the single cycle vector speedups. This is because the initialization cost of the ODE system, a highly vectorizable operation, was included in the full multigrid execution times.

In tables 8.4 and 8.5 we report the execution times of the full multigrid waveform solver on 16 and on 32 processors. (The letters "na" indicate values that are not available, as they could not be obtained due to lack of memory on the vector board.) Note that the vector speedup is lower than in the single processor case. This was to be expected. Indeed, vectorization only speeds up the arithmetic part of the computation, while the communication part remains unaffected.

Table 8.3: Example 1: single processor execution time, in seconds, of different vectorized multigrid waveform relaxation cycles ($n_x = 8$, $n_y = 7$).

n_t	WR FMG V(1,1)			WR FMG F(1,1)			WR V(1,1)			WR F(1,1)		
	scalar	vector	Sp	scalar	vector	Sp	scalar	vector	Sp	scalar	vector	Sp
100	5.41	1.09	5.0	5.78	1.18	4.9	1.86	0.45	4.1	2.27	0.54	4.1
50	2.75	0.67	4.1	2.94	0.73	4.0	0.95	0.28	3.4	1.14	0.34	3.3
25	1.43	0.48	3.0	1.53	0.52	3.0	0.49	0.20	2.5	0.59	0.24	2.5
10	0.63	0.36	1.8	0.68	0.39	1.8	0.22	0.14	1.6	0.27	0.18	1.5

Table 8.4: Example 1: execution time, in seconds, of the vectorized full multigrid waveform relaxation method on 16 processors.

n_t	$n_x = 16, n_y = 15$			$n_x = 32, n_y = 31$			$n_x = 64, n_y = 63$		
	scalar	vector	Sp	scalar	vector	Sp	scalar	vector	Sp
100	3.54	1.12	3.16	8.20	2.42	3.39	(na)	(na)	(na)
50	1.91	0.76	2.51	4.30	1.59	2.70	11.88	4.02	2.96
25	1.10	0.59	1.86	2.43	1.20	2.03	6.53	2.91	2.24
10	0.62	0.49	1.27	1.29	0.97	1.33	3.21	2.25	1.43

Table 8.5: Example 1: execution time, in seconds, of the vectorized full multigrid waveform relaxation method on 32 processors.

n_t	$n_x = 16, n_y = 15$			$n_x = 32, n_y = 31$			$n_x = 64, n_y = 63$		
	scalar	vector	Sp	scalar	vector	Sp	scalar	vector	Sp
100	3.05	0.99	3.08	6.00	1.89	3.18	14.50	4.17	3.48
50	1.66	0.68	2.44	3.22	1.27	2.53	7.61	2.71	2.80
25	0.98	0.54	1.80	1.85	0.99	1.87	4.20	2.03	2.07
10	0.58	0.46	1.25	1.04	0.82	1.27	2.18	1.61	1.35

8.4.2 Example 2

We consider the following, variable coefficient heat flow equation,

$$a(x,y)\frac{\partial u}{\partial t} \;=\; \frac{\partial}{\partial x}k(x,y)\frac{\partial u}{\partial x} + \frac{\partial}{\partial y}k(x,y)\frac{\partial u}{\partial y} + Q \;, \quad (x,y) \in [0,1] \times [0,1] \;, \quad (8.2)$$

where the material properties $a(x,y)$ and $k(x,y)$ denote the heat capacity of the material per unit volume and the thermal conductivity. They were arbitrarily chosen as follows,

$$a(x,y) = 1 + x + y \quad \text{and} \quad k(x,y) = e^{4(x-0.5)^2+4(y-0.5)^2} \;.$$

We impose Dirichlet conditions on the north and east boundaries, and mixed conditions on the boundaries to the south and to the west,

$$N: \quad u(t,x,1) = 2 + \sin(5x)\,e^{-2t(x+1)} \qquad E: \quad u(t,1,y) = 2 + \sin(5y)\,e^{-2t(1+y)}$$

$$S: \quad \frac{\partial u}{\partial y}(t,x,0) + u(t,x,0) = 2 + 5x\,e^{-2tx} \quad W: \quad \frac{\partial u}{\partial x}(t,0,y) + u(t,0,y) = 2 + 5y\,e^{-2ty}$$

The initial condition and the source function $Q(t,x,y)$ are chosen in such a way that the solution to the problem becomes,

$$u(t,x,y) = 2 + \sin(5xy)\,e^{-2t(x+y)} \;. \tag{8.3}$$

The problem is discretized on a regular grid for two different values of the mesh size, namely $h= 1/16$ and $h = 1/128$. The time-increment is taken equal to 0.04 for the coarse problem, and 0.01 for the fine problem. In both cases 25 time-steps are performed. Discretization of the spatial operator leads to a *five-point stencil*. To determine the solution we selected the following standard multigrid operators: coarsening to a grid with size $h=1/2$ ($n_x=n_y=2$), red/black smoothing, half-weighting, bilinear interpolation, and a coarse grid solver that applies two Gauss-Seidel relaxations.

The timing results are presented in figure 8.3. From this figure we can read the appropriate multigrid parameter choices. Observe, for instance, that the full multigrid approach is sufficient to obtain an error of the order of the discretization error, in both the waveform relaxation and the Crank-Nicolson methods. It can also be seen that the the multigrid waveform cycling process is rapidly convergent. The averaged V-cycle convergence factor was measured to be around 0.09. When full multigrid was not used in the Crank-Nicolson method, the initial approximation at each time-level was taken equal to the solution at the previous time-level, i.e., we used a zero-th order extrapolation. For this particular problem this turned out to be more efficient than a first order extrapolation.

Execution times for the full multigrid solvers are reported in table 8.6. The single processor values between brackets are values which were not actually measured due to a lack of computer memory. Instead, they were determined by extrapolation of smaller sized results. The four rows in the table represent different algorithms, different optimizations and different execution modes. The first row corresponds to the standard Crank-Nicolson time-stepping method. The second row represents waveform relaxation results. In determining these values, we did not fully exploit the potential of the

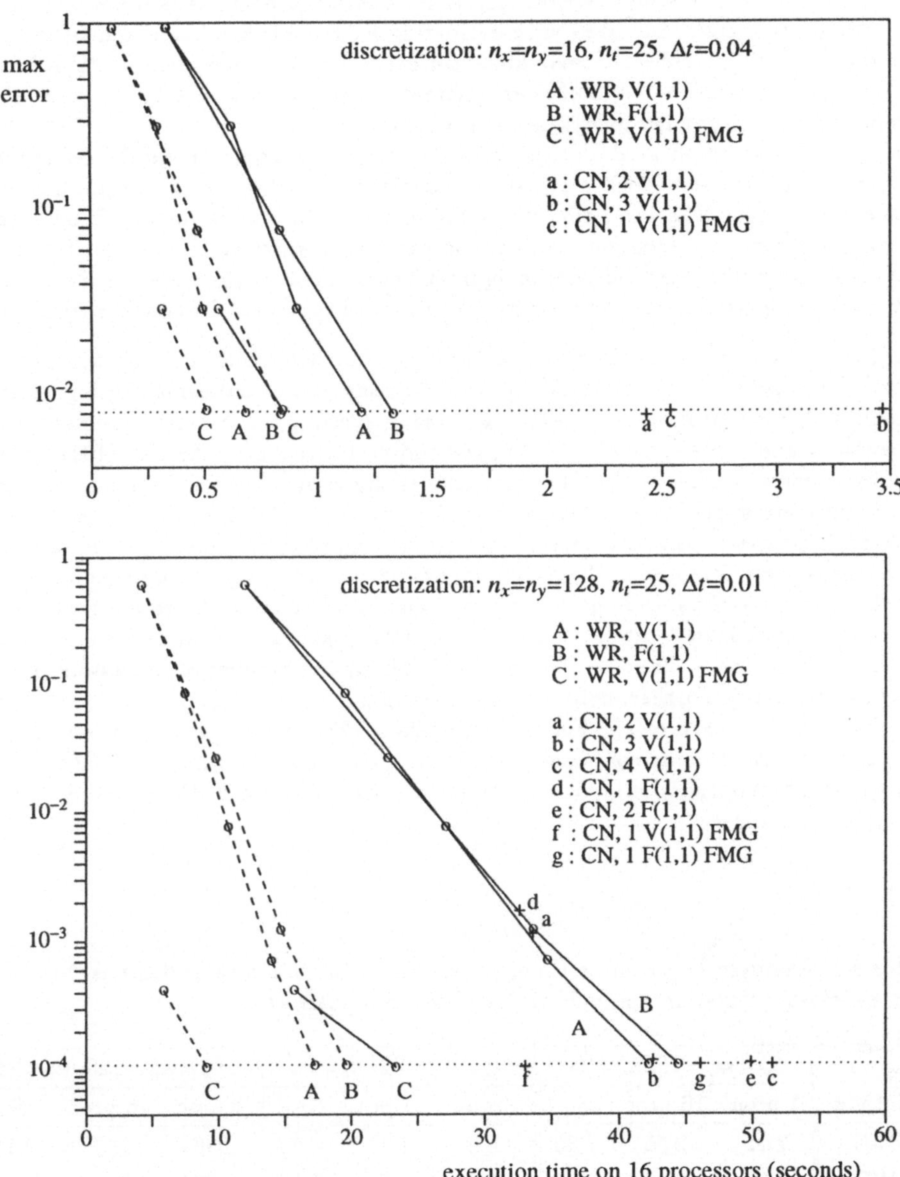

Figure 8.3: Example 2: comparison of Crank-Nicolson time-stepping and linear multi-grid waveform relaxation, on 16 processors (solid lines: without vectorization, dashed lines: with vectorization). The horizontal dotted line indicates the discretization error, i.e., the error of the fully converged discrete solution

waveform relaxation method. We did not store intermediate results in order to reduce the initialization cost associated with evaluating PDE right-hand side and boundary conditions. The differences in single processor execution times between the "C-N" and "WR" rows – approximately 20% – are therefore mainly due to the lower arithmetic complexity and the smaller indexing-overhead of the waveform kernels. The PDE right-hand side Q is a complicated function, costly to evaluate. Here, substantial economies are possible in the waveform method by transferring t-independent expressions out of the inner loops. The third row shows the effect of this optimization. Finally, the fourth row gives the execution times of the optimized waveform code, run in vector mode. (Remark that the solid lines in figure 8.3 correspond to the values in the second row, and the dashed lines correspond to the optimized vector results of the fourth row.)

Table 8.6 enables us to illustrate a variety of further topics as, e.g., the dependence of the speedup on the problem size. The larger the problems the higher the speedups. It is interesting to note the decrease in speedup for the same problem sizes in rows 2, 3 and 4. As the code is optimized and vectorized, only the cost of the arithmetic decreases. Consequently, the communication overhead, which remains constant in absolute terms, becomes relatively more important, and the parallel efficiency deteriorates. The speedups due to parallelization are higher than those reported in table 8.1. This is partly because of the expensive evaluation of the right-hand side function, which is included in the timings and which is an entirely parallel operation that needs no communication. In addition, it is due to a better load balance, since the number of unknowns in x- and y-direction is evenly divisible by the number of processors (four) in those directions. Further, note the low speedup due to vectorization. This is due to the short vector length of 25. For the large 128×128 problem, waveform relaxation loses much of its advantage of being more parallel, since also the time-stepping method has become highly (parallel) efficient. However, the waveform method retains any of its other advantages.

Table 8.6: Example 2: execution time, in seconds, of the full multigrid Crank-Nicolson and waveform relaxation methods on 1 and on 16 processors.

method	$n_x = n_y = 16,\ n_t = 25$			$n_x = n_y = 32,\ n_t = 25$			$n_x = n_y = 128,\ n_t = 25$		
	1 proc	16 proc	S_{16}	1 proc	16 proc	S_{16}	1 proc	16 proc	S_{16}
C.-N.	7.11	2.55	2.80	26.4	5.11	5.17	396	33.0	12.0
WR	5.70	0.85	6.73	21.8	2.20	9.91	(328)	23.4	(14.1)
WR opt	4.04	0.69	5.83	15.4	1.71	9.02	(232)	16.9	(13.7)
WR vec	2.15	0.51	4.25	8.0	1.13	7.08	(120)	9.0	(13.3)

8.5 Nonlinear initial boundary value problems

8.5.1 Example 3

A solid body with initial temperature u_0, is submerged into the environment with temperature u_∞ and starts losing energy through radiation along its sides. This heat flow problem is governed by a quasi-linear parabolic partial differential equation,

$$\rho\, c\, \frac{\partial u}{\partial t} \;=\; \mathrm{div}(\, k(u)\, \mathrm{grad}(u)\,)\,, \tag{8.4}$$

and is supplemented by the nonlinear radiation boundary condition,

$$k(u)\, \frac{\partial u}{\partial n} \;=\; e\, \epsilon\, (\, u_\infty^4 - u^4\,)\,, \tag{8.5}$$

where u is the temperature distribution [K], ρ is the density [kg/m^3], c is the specific heat [J/kg K], k is the thermal conductivity [W/m K], e is the emissivity [-] and ϵ is the Stefan-Boltzmann constant ($5.75 10^{-8} W/m^2 K^4$), see e.g. [97]. With $\partial/\partial n$ we denote the outside normal derivation operator.

We shall numerically determine the evolution of the temperature for $t \in [0, T]$ of a square two-dimensional surface of size l, that is $(x, y) \in [-l/2, l/2] \times [-l/2, l/2]$, which satisfies (8.4) and (8.5). Thanks to symmetry the computational domain may be restricted to $[0, l/2] \times [0, l/2]$, with homogeneous flux boundary condition, $\partial u/\partial n = 0$, imposed along the sides $x=0$ and $y=0$. The following values are chosen for the problem parameters: $u_0=473$, $u_\infty=3$, $\rho=7000$, $c=400$, $e=1$, $k(u)=50(1-0.0008u)$, $l=0.5$ and $T=1545$ seconds. The problem is discretized on an equidistant grid with 17 or 33 grid lines in the x- and y-direction ($n_x=n_y$). The time-discretization will be specified further on.

In figure 8.4 we illustrate the convergence behaviour of waveform relaxation. The figure to the left shows the time-profile of successive approximations $u^{(k)}$ evaluated at a corner point of the domain, $(x, y)=(l/2, l/2)$, and generated by the waveform Gauss-Seidel algorithm. For all grid points ($n_x=n_y=17$ in this case) a constant time-profile equal to the initial condition was used as the starting approximation. As can be seen from the figure the convergence of the algorithm is very slow. The time interval where the approximation is satisfactory only gradually extends as more relaxations are applied. Corresponding results obtained with multigrid waveform relaxation are shown in the middle figure, which clearly illustrates a dramatic improvement in convergence. Multigrid V-cycles are used with one red-black pre-smoothing step and one similar post-smoothing step. Standard coarsening is performed down to a coarse level with mesh size $h=1/2$, i.e., 3×3 unknowns. The differential equation at each grid point is integrated in the smoothing step by solving the linear equation that results after applying one Newton linearization. The computational cost of one such V-cycle is approximately equal to 8/3 work-units, with one work-unit being the cost of one waveform Gauss-Seidel iteration. The observed averaged convergence rate corresponds to the value 0.09. The figure to the right shows the starting approximation on the fine grid as it is obtained by the full multigrid procedure. The cost of this initial step is approximately 8/9 work-units. The subsequent approximations cannot be distinguished graphically from the initial one.

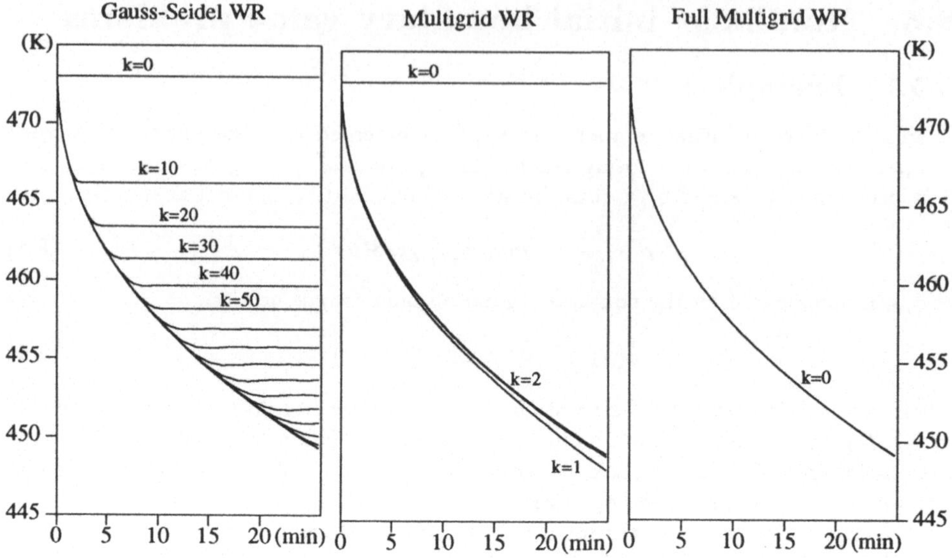

Figure 8.4: Example 3: waveform relaxation iterates $u^{(k)}(t, l/2, l/2)$.

The figures demonstrate the savings that could be achieved by using a *windowing* strategy (especially in the case of the Gauss-Seidel relaxation). In the initial iterations the calculations associated with time-levels for high t values are largely wasted. Later on, the time-levels for low t values have converged to the discrete solution. Further computations at these time-levels may be skipped, since additional iterations only improve the approximation at the end of the time-interval. A windowing strategy should take this typical convergence behaviour into account by limiting the length of the window and by gradually moving it forward as the computation advances. Observe that the use of windowing is of lesser importance in the case of multigrid cycling, and it is even less promising in combination with full multigrid. (However, example 4 will illustrate a situation where windowing is crucial even within the multigrid approach.)

In figure 8.5 the performance of the nonlinear multigrid waveform relaxation method and that of the Crank-Nicolson method are compared. The latter applies the multigrid full approximation scheme (FAS) to solve the nonlinear elliptic problems at each time-level. In this example a variable step-size time-integration is used with Δt_i $(= t_i - t_{i-1})$ determined a priori. It is chosen equal to 5 seconds for the first 10 time-steps and it is doubled every 10 steps. A total of 50 time-steps is applied. The accuracy, in casu the largest residual with respect to the discrete equations, is plotted versus the execution time on 1 and on 16 processors; no vectorization was used. We have applied the Crank-Nicolson method in two different (but both common) ways. In one series of experiments a fixed number of V-cycles is applied to the initial approximation at each time-level, which, here, is chosen equal to the solution obtained in the previous time-step. The results are denoted by "a" (1 V-cycle), "b" (2 V-cycles), "c" (3 V-cycles), "d" (4 V-cycles) or "e" (5 V-cycles). In the second series of experiments an *accommodative*

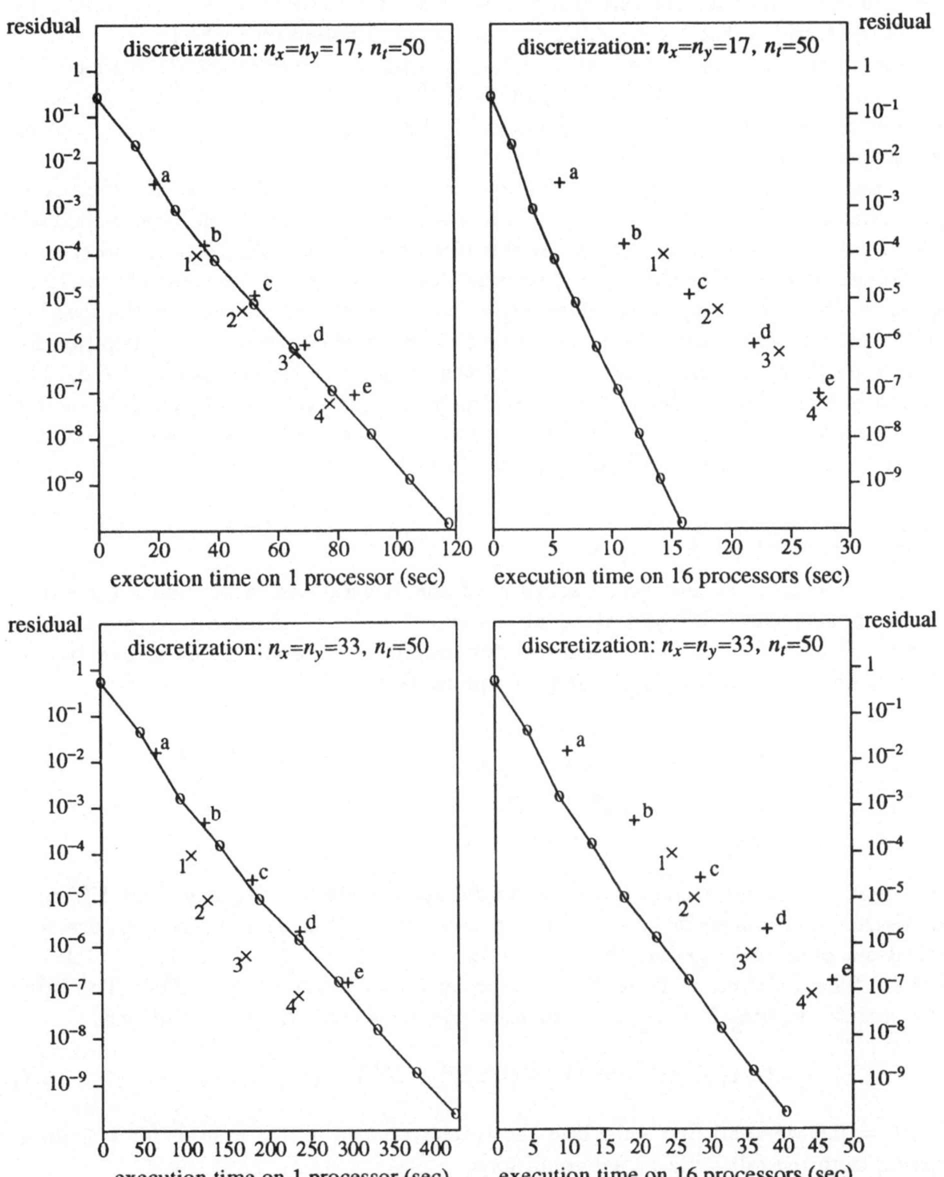

Figure 8.5: Example 3: maximum norm of the residual versus the execution time on 1 and on 16 processors with multigrid waveform relaxation using V-cycles (solid lines) and with the Crank-Nicolson method with fixed number of V-cycles (+) or with full multigrid and a variable number of V-cycles (×).

scheme is implemented. The initial approximation at each time-level is determined by the full multigrid method and multigrid V-cycles are applied until the largest residual is below a certain specified tolerance. The corresponding results are denoted by "1" (tol. $= 10^{-4}$), "2" (tol. $= 10^{-5}$), "3" (tol. $= 10^{-6}$) and "4" (tol. $= 10^{-7}$). Note that a measure of the largest residual may be calculated at negligible cost, e.g. by calculating the residual on a subgrid of the fine grid.

On one processor multigrid waveform relaxation turns out to be as efficient as Crank-Nicolson when the latter uses a fixed number of V-cycles. The Crank-Nicolson method with the accommodative scheme is somewhat faster. On 16 processors, however, waveform relaxation clearly outperforms Crank-Nicolson. In the case of the 17×17 discretization the speedup of waveform algorithm is about 7.5, whereas the the speedup of Crank-Nicolson is only 3. A speedup of 11 is achieved by waveform relaxation for the 33×33 problem, while the speedup of the standard approach is about 6.5. The speedup of the accommodative time-stepping scheme is even worse, which is mainly due to parallel overheads in the full multigrid phase. The latter consists of operations on coarse grids, which are difficult to parallelize efficiently.

8.5.2 Example 4

We shall calculate the numerical solution of the Brusselator, a nonlinear system of two parabolic partial differential equations which we have already encountered in a previous chapter, section 4.7. The two-dimensional Brusselator is described by the following equations defined over the unit square, $\Omega = [0,1] \times [0,1]$,

$$
\begin{cases}
\dfrac{\partial X}{\partial t} = \dfrac{D_X}{L^2} \left(\dfrac{\partial^2 X}{\partial x^2} + \dfrac{\partial^2 X}{\partial y^2} \right) - (B+1)X + X^2 Y + A \\[2ex]
\dfrac{\partial Y}{\partial t} = \dfrac{D_Y}{L^2} \left(\dfrac{\partial^2 Y}{\partial x^2} + \dfrac{\partial^2 Y}{\partial y^2} \right) - X^2 Y + BX
\end{cases}
\tag{8.6}
$$

The equations model a chemical reaction-diffusion process. $X(t,x,y)$ and $Y(t,x,y)$ denote chemical concentrations of reaction products. $A(t,x,y)$ and $B(t,x,y)$ are concentrations of input reagents, which in our experiments are taken constant, $A=2.0$ and $B=5.45$. The diffusion coefficients are given by $D_X=0.004$ and $D_Y=0.008$. The value of the reactor length, L, is specified further. We consider Dirichlet conditions,

$$
X(t,x,y) = A \quad \text{with} \quad Y(t,x,y) = B/A, \quad (x,y) \in \partial\Omega .
\tag{8.7}
$$

For $L = 0.9$ we shall first calculate the time-evolution of the system for $t \in [0,3]$, starting from the following initial conditions,

$$
X(t,x,y) = A + 0.9\sin(\pi x)\sin(\pi y) \quad \text{and} \quad Y(t,x,y) = B/A - 0.9\sin(\pi x)\sin(\pi y) .
$$

Figure 8.6 illustrates the waveform relaxation convergence behaviour for a discretization with $h=1/16$. The multigrid results are obtained using V(1,1)-cycles, full-weighting, bilinear interpolation, and standard coarsening to the mesh with $h=1/2$. The graphs are to be interpreted in a similar way as those in figure 8.4. They illustrate

Brusselator X-component evaluated at (0.5,0.5)

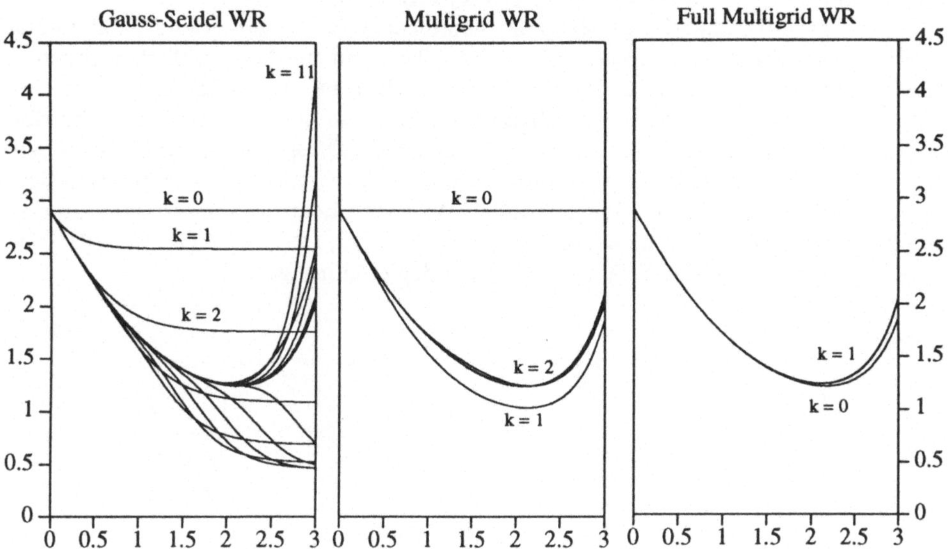

Brusselator Y-component evaluated at (0.5,0.5)

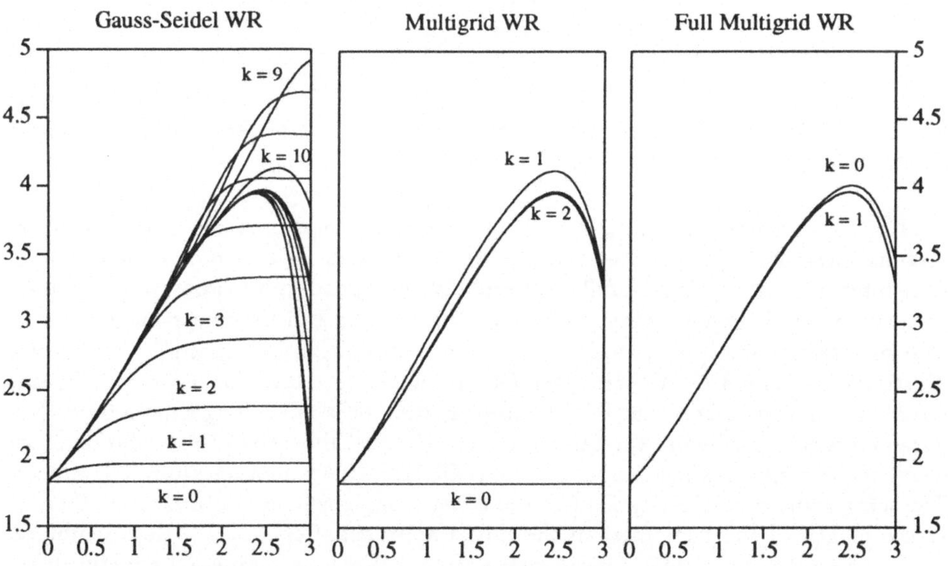

Figure 8.6: Example 4: waveform relaxation for time-integration of the Brusselator problem; the figure shows the profile of approximations $X^{(k)}(t, 1/2, 1/2)$ and $Y^{(k)}(t, 1/2, 1/2)$.

the typical phenomenon that the largest error occurs at the end of the integration interval and they indicate the possible gain of windowing.

In the waveform smoothing operation a nonlinear system of two ordinary differential equations is to be solved at each grid point. This can be done by first applying a Newton linearization around an approximation of the solution profile and then solving the resulting linear system of two ODEs (i.e., using a blockwise waveform Newton method). If necessary this process can be repeated. Often, though, the use of one linearization suffices and does not deteriorate the convergence of the outer iteration, e.g. the multigrid or Gauss-Seidel iteration. We have illustrated this Newton process on a 3×3 grid, i.e., with one set of unknowns, $X(t, 1/2, 1/2)$ and $Y(t, 1/2, 1/2)$. The successive Newton iterates $Y^{(i)}(t, 1/2, 1/2)$ are depicted in figure 8.7.

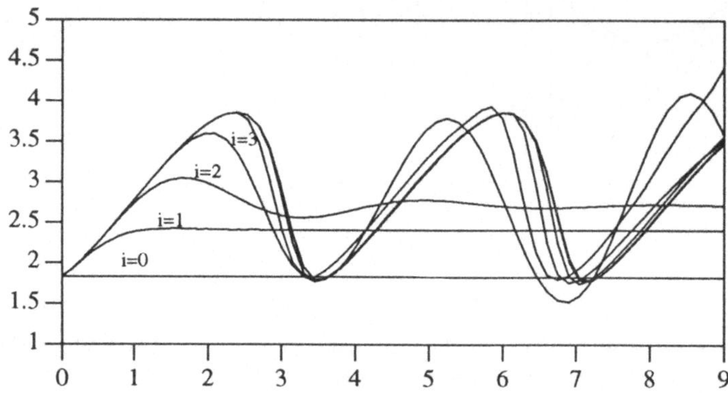

Figure 8.7: Example 4: successive Newton iterates, Y-component of the Brusselator on a $h=1/2$ grid.

The Newton linearization is a rapidly converging iteration when a good approximation is available. This is, e.g., the case when problems are solved on relatively short time intervals with the use of full multigrid. When a good approximation is not available numerical divergence may result, e.g., the solution can become arbitrarily large, or it may take many iterates before convergence attains its full "speed". This effect is illustrated in table 8.7. We computed Gauss-Seidel waveform iterates for the Brusselator starting from a constant initial approximation on a fairly long time-interval (the parameters are specified in the table header). The first line shows the evolution of the norm of the residual when the (discretized) ODEs in the smoother are solved exactly. The second line characterizes the effect of using a single Newton linearization. Because of the length of the time-interval the initial approximation bears little resemblance to the true solution profile. Linearization around this profile cannot be justified; the iteration diverges. The effect of applying more than one Newton step at each grid point is illustrated in the remaining two lines. If the successive solution profiles were plotted, it could be seen that the approximations are erroneous especially at the end of the time-window, where very large solution values are found. In order to avoid this

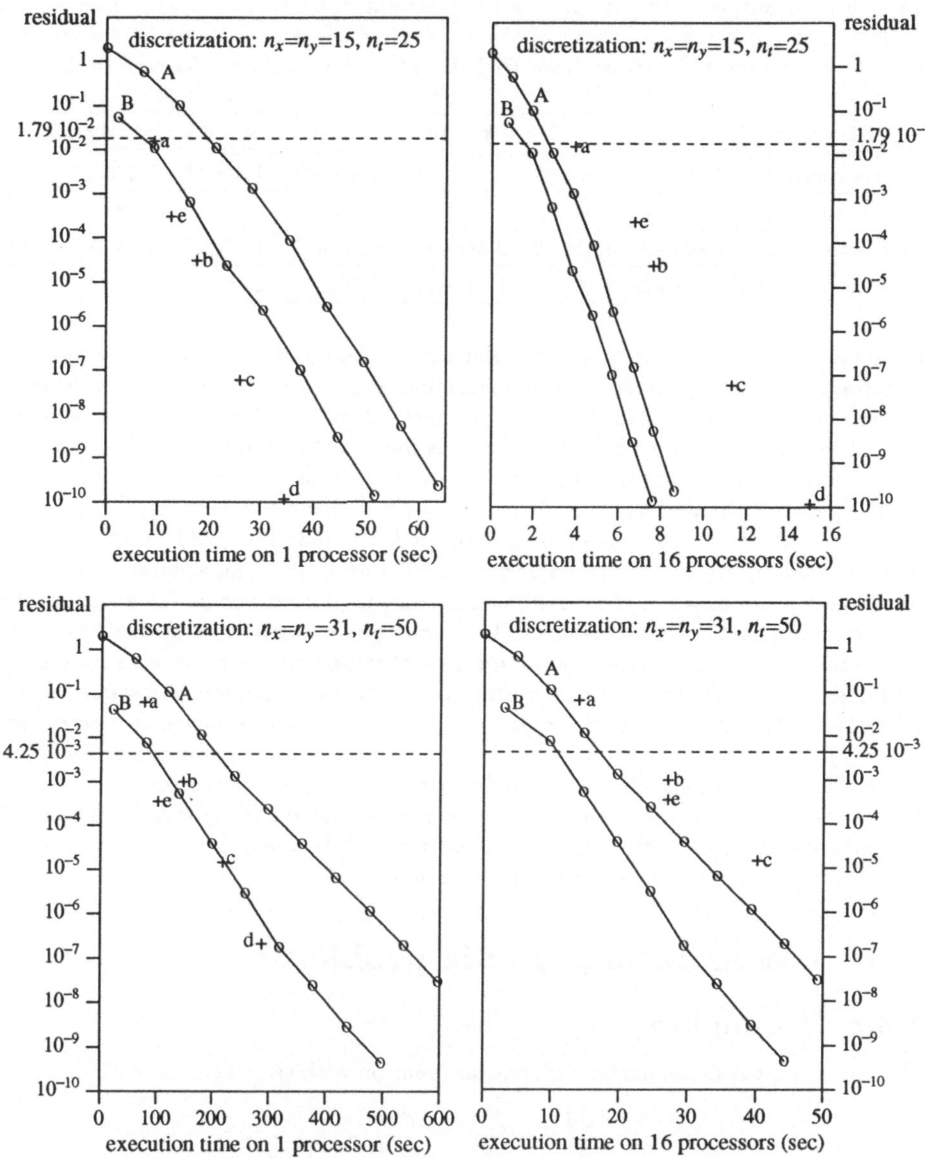

Figure 8.8: Example 4: Comparison of execution times on 1 and on 16 processors with multigrid waveform relaxation using V-cycles ("A": standard cycling, "B": full multigrid plus cycling) and with the Crank-Nicolson method with a fixed number of V-cycles ("a":1 cycle, "b":2 cycles, "c":3 cycles, "d":4 cycles, "e" 1 cycle with FMG).

Table 8.7: Example 4: the use of long windows and the effect of Newton linearization: evolution of the Euclidean norm of the residual of successive Gauss-Seidel waveform relaxation iterates $Y^{(k)}$ (Brusselator with $L=0.9$, $n_x=n_y=15$, $n_t=60$, $\tau=7/60$).

ODE solver	$k = 0$	$k = 8$	$k = 16$	$k = 24$	$k = 32$	$k = 40$	$k = 48$
exact solution	4.05e+2	8.43e+2	4.18e+2	1.98e+2	1.30e+2	2.27e-1	4.78e-5
1 Newton step	4.05e+2	4.51e+2	inf	inf	inf	inf	inf
2 Newton steps	4.05e+2	7.81e+2	2.59e+18	2.38e+12	1.82e+8	9.66e-1	4.96e-5
3 Newton steps	4.05e+2	8.38e+2	1.21e+12	8.66e+2	1.30e+1	2.29e-1	4.78e-5

behaviour a possible remedy is to restrict the length of the window. A second way is to refrain from using the Newton linearization at the differential equation level, but to use a standard time-stepping approach instead, with a few Newton steps in each time-step. The latter is a safe choice, but it is more costly. Indeed, the Newton process has to be iterated to convergence; otherwise the algebraic errors may accumulate.

In figure 8.8 the performance of the multigrid waveform relaxation and the Crank-Nicolson methods are compared, for mesh sizes $h=1/16$ and $h=1/32$. Time-integration is performed for $t \in [0,1]$ with 25 and with 50 time-steps. The accuracy, in this case the largest residual w.r.t. the discrete equations, is plotted versus the execution time on 1 processor and on 16 processors. The dashed line indicates the residual of the "true PDE solution", approximated by the solution obtained on the mesh with $h=1/64$ and $\tau=1/128$. On a single processor multigrid waveform relaxation turns out to be as efficient as the Crank-Nicolson method if the problem is to be solved to discretization accuracy (which is indicated by the dashed line). The method is somewhat less efficient if the discrete equations are to be solved to a high accuracy. The ranking is reversed on the 16 processor machine. Thanks to its better parallel efficiency waveform relaxation outperforms Crank-Nicolson by a factor of two. Note that this difference would be even more pronounced if vectorization were used.

8.6   Linear time-periodic problems

8.6.1   Example 5

We consider a parabolic partial differential equation with time-periodic right-hand side,

$$\frac{\partial u}{\partial t} = \frac{\partial^2 u}{\partial x^2} + \frac{\partial^2 u}{\partial y^2} + f \,, \tag{8.8}$$

defined on the unit square with four Dirichlet boundary conditions. The function f is chosen in such a way that the solution becomes 1-periodic and equal to,

$$u(t,x,y) = 1 + 100\,(x - x^2)^2(y - y^2)^2 \sin(2\pi t) \,. \tag{8.9}$$

The problem will be solved for two different mesh sizes, with equal spacing in spatial and in time direction, namely $h=\tau=1/64$ and $h=\tau=1/16$. The 16-processor timing results (without vectorization) are displayed in figure 8.9. Here, "error" signifies the largest error of the solution *at time-level t=0*. (Remember that standard periodic problem solvers, such as the shooting method and the multigrid method of the second kind, try to determine the value of the periodic solution at one time-level. They calculate one point on the periodic orbit.)

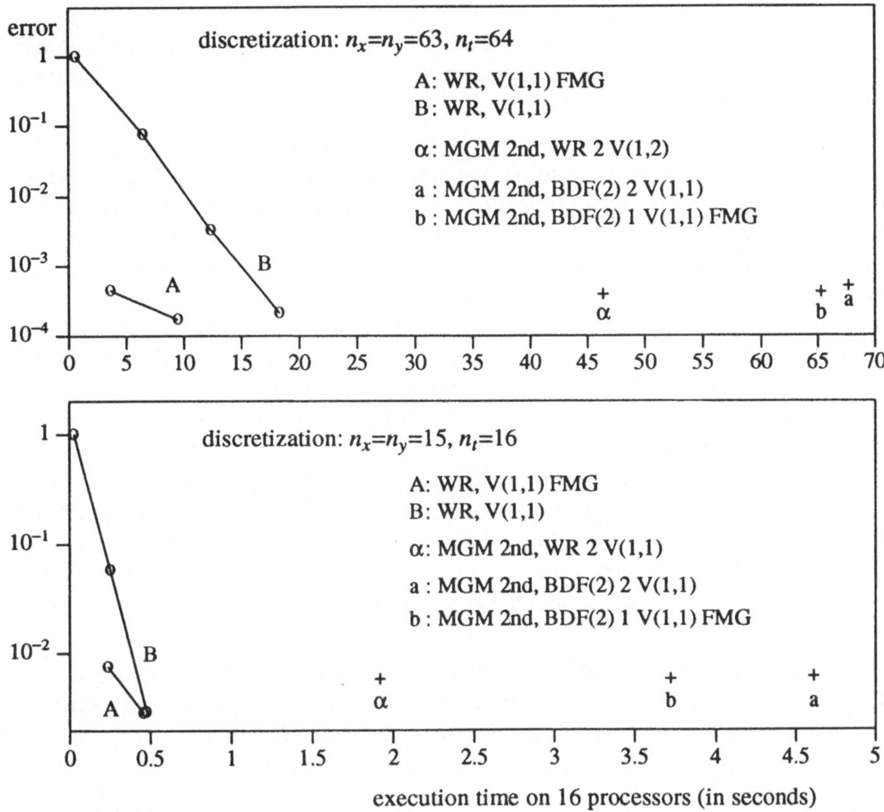

Figure 8.9: Example 5: comparison of the linear time-periodic multigrid waveform relaxation algorithm and multigrid method of the second kind (maximum norm of the error versus execution time).

Three methods are compared. The first method is a parallel implementation of the multigrid method of the second kind. The second order backward differentiation method, BDF(2), is used for time-integration because of its excellent smoothing properties, see [41]. The time-increments used for discretizing the initial boundary value problems on each grid level are chosen equal to the mesh size, which is determined by standard coarsening from the fine grid. The linear systems in each time-step of the BDF(2) scheme are solved by using standard multigrid with 2 V(1,1)-cycles or by full multigrid with 1 V(1,1)-cycle at grid each level. The multigrid method of the second kind is rapidly convergent, and one iteration applied to the initial approximation proves to be sufficient to obtain a solution with an error equal to the discretization error. (In this section we started the iteration with the initial approximation of $u(x, y, 0)$ chosen equal to zero.)

A related method is obtained when multigrid waveform relaxation is used as the initial boundary value problem solver inside the multigrid method of the second kind. The resulting algorithm is about 1.5 to 2 times as fast as the method with time-stepping as can be observed in figure 8.9. This was to be expected from the speedup values given in tables 8.1 and 8.2.

Finally, we consider the time-periodic multigrid waveform relaxation method (with starting iterate equal to the zero-function). Our implementation of this method proves to be faster than the implementation of the best standard algorithm by a factor of 7 to 10. This is due to the *lower arithmetic complexity* and the *much better parallel characteristics* of the waveform method. To verify the complexity estimates of both approaches determined in chapter 7 we report the one-processor times in table 8.8. The ratio of the execution times turns out to be close to 2.5, as theory had predicted. The same table shows that hardly any speedup is obtained with the multigrid method of the second kind. Yet, we applied several techniques to improve the parallel efficiency: agglomeration, unblocking message passing, communication-calculation overlap (see chapter 6). Whatever sophisticated techniques are used for speeding up the coarse grid operations, they remain a significant bottleneck. The multigrid method of the second kind visits these coarse grids very frequently, because of its "double multigrid" nature. It is basically a multigrid W-cycle, where in each smoothing step a time-stepping method is used which applies standard multigrid for solving the elliptic problems at each time-level. Consequently, the algorithm is not well-suited for implementation on a parallel computer.

We would also like to remark that vectorization will lead to an additional speedup in the case of the waveform relaxation algorithm. The performance difference on the 16-processor machine will then be in the range of 20 to 30, depending on the problem size.

Table 8.8: Example 5: execution time, in seconds, of two time-periodic solvers: time-periodic full multigrid waveform relaxation and 1 cycle of multigrid of the second kind.

method	$n_x = n_y = 15, n_t = 16$			$n_x = n_y = 31, n_t = 32$		
	1 proc.	16 proc.	Sp	1 proc.	16 proc.	Sp
Waveform Relaxation	1.77	0.46	3.9	14.33	1.78	8.1
Multigrid sec. kind	4.83	3.74	1.3	35.11	15.74	2.2

8.6.2 Example 6

We consider the diffusion equation on the unit square,

$$\frac{\partial u}{\partial t} = \frac{\partial^2 u}{\partial x^2} + \frac{\partial^2 u}{\partial y^2} \, , \tag{8.10}$$

with four Dirichlet boundary conditions, three of which are constant and homogeneous and one of which is a rapidly oscillating time-periodic function,

$$N: \quad u(t,x,1) = 0 \qquad\qquad E: \quad u(t,1,y) = 0$$
$$S: \quad u(t,x,0) = \sin(\pi x)\, \sin(10\pi t) \qquad W: \quad u(t,0,y) = 0$$

We calculate the stable periodic solution, the period of which equals $T=0.2$, on an equidistant grid with mesh size $h=1/32$ and with the time-interval, [0,0.2], being divided into 65 time-levels. The solution is plotted in figure 8.10 for different values of t. The graphs show the generation of a trailing wave, which is strongly damped and eventually fades away while receding backwards (i.e., northwards). Later on, a similar wave is generated, but with the opposite amplitude.

As in the previous example we shall compare three different algorithms for solving the above equation: multigrid of the second kind with BDF(2) time-stepping, multigrid of the second kind with waveform relaxation time-integration using the BDF(2) discretization, and time-periodic waveform relaxation.

The multigrid method of the second kind attains its very good theoretical convergence factor, which is of the order $O(h^2)$. This is evidenced by the numbers in table 8.9. Consequently, one iteration of the method applied to a suitably chosen starting approximation will often be satisfactory.

The successive iterates of the waveform method are illustrated in figure 8.11. They show the sinusoidal approximation at the grid point $(x,y)=(1/2,3/16)$, which, by a process of amplitude growing and phase shifting, bends its shape towards the shape of the solution of the semi-discrete problem.

The timing results on 1 and on 16 processors are presented in figure 8.12. (Here, the error is defined as the maximum difference at the grid points at time-level $t=0$, between the computed approximation and the solution of the problem obtained with $h=1/64$ and $\tau=1/128$.) The one-processor results are interesting in their own right as they again confirm the arithmetic complexities calculated in chapter 7. The periodic multigrid waveform method outperforms the "best" standard method, multigrid of the second kind, with a factor of 2.5. The timings on 16 processors illustrate the parallel characteristics. A very low parallel efficiency is obtained with the multigrid method of the second kind (although we have again applied various techniques to optimize its parallel performance). The multigrid method of the second kind combined with waveform relaxation achieves a speedup of about 6. As such it outperforms the parallel implementation of the standard method by a factor of 3. The periodic multigrid waveform method is faster than the multigrid method of the second kind by a factor of 12. This is due to its lower arithmetic complexity and its much better parallel characteristics. Again, we should note that vectorization will lead to an additional speedup, and this (only) in the case of waveform relaxation. From the numbers in

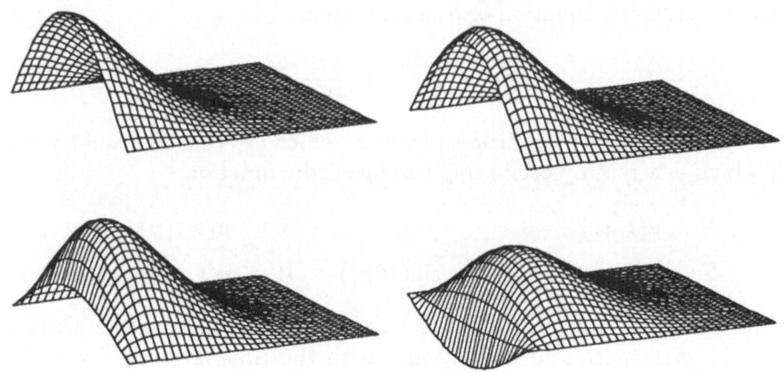

Figure 8.10: Example 6: solution $u(t_k, x, y)$ at $t_k = 0.08$, 0.088, 0.096 and 0.104.

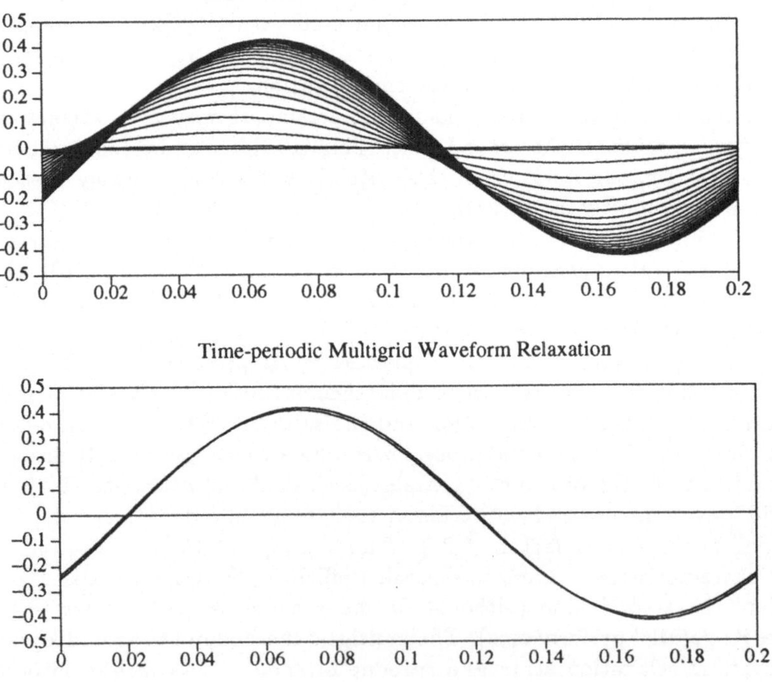

Figure 8.11: Example 6: successive iterates $u^{(k)}(t, 1/2, 3/16)$ with $k = 0, 5, 10, \ldots$ (upper picture) and $k = 0, 1, 2$ (lower picture).

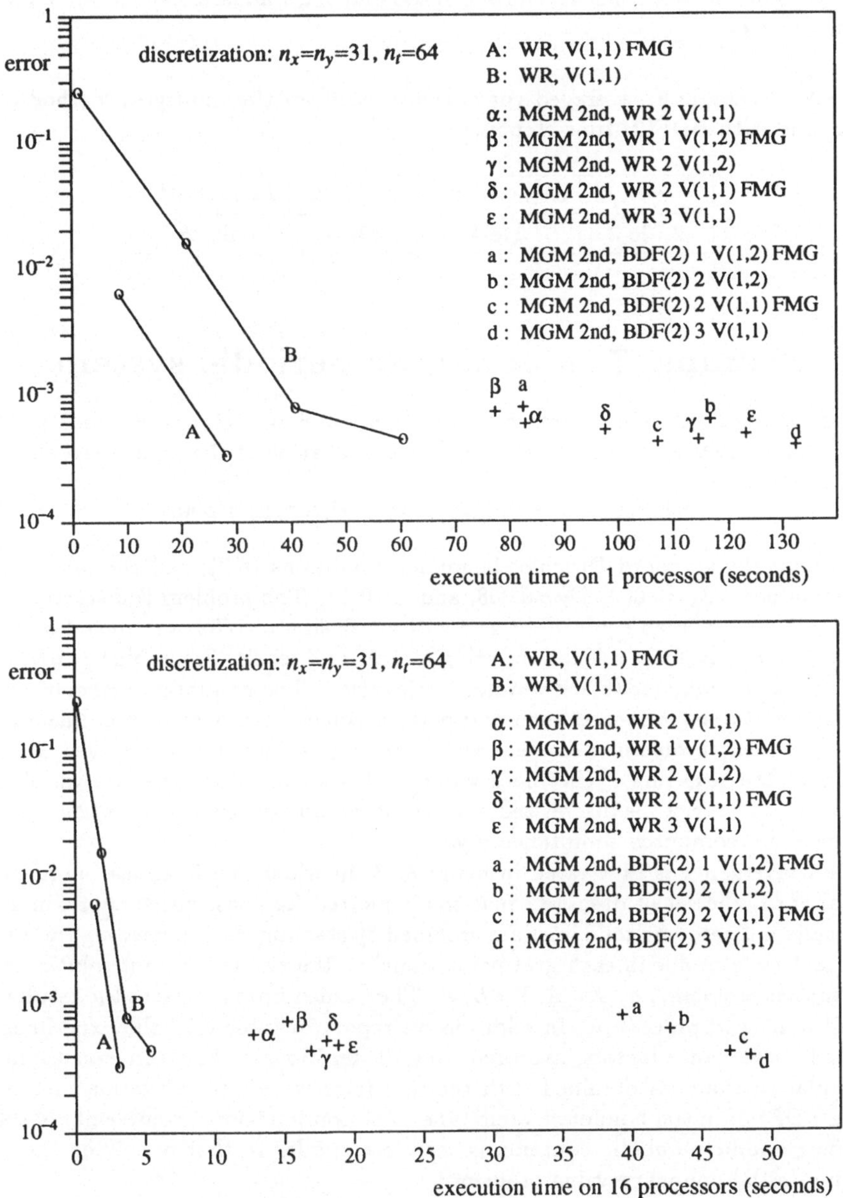

Figure 8.12: Example 6: comparison of the linear time-periodic multigrid waveform relaxation algorithm and multigrid method of the second kind.

table 8.1 we may expect a vector speedup of about 3. This would bring the ratio of the execution times between the waveform method and the standard sequential method up to a factor of 40.

Table 8.9: Example 6: averaged convergence factor of the multigrid method of the second kind with BDF(2) time-stepping.

$h=\tau=1/8$	$h=\tau=1/16$	$h=\tau=1/32$	$h=\tau=1/64$
2.72e-3	7.34e-4	1.87e-4	4.65e-5

8.7   Example 7: a nonlinear periodic system

In this final example we shall determine the stable periodic solution of the Brusselator model, equation (8.6), with a time-periodic concentration of the input reagents,

$$A(t, x, y) = 2 + \sin(4t) \quad \text{and} \quad B(t, x, y) = 5.45 . \tag{8.11}$$

We consider the standard Dirichlet boundary conditions (8.7), and the following parameter values: $D_X=0.004$, $D_Y=0.008$, and $L=0.15$. The problem is discretized with central differences on an equidistant grid with mesh size $h=1/32$, and solved with periodic multigrid waveform relaxation, using 5 multigrid levels. Smoothing is performed by blockwise periodic red/black waveform relaxation. The nonlinear system of two periodic differential equations at each grid point is solved with one Newton linearization step. The linear problem is discretized by the trapezoidal rule and solved by direct solution of the resulting cyclic block-bidiagonal matrix. The time-interval of interest $[0,\pi/2]$, one period, is discretized with constant time-increment $\tau=\pi/200$, i.e., 100 time-levels are computed simultaneously.

The convergence is illustrated in figure 8.13, in which the Euclidean norm of the residual w.r.t. the set of discrete equations is plotted for consecutive approximations. The results indicated by solid lines are obtained by starting the iterative algorithm with a constant time-profile in each grid point, equal to the stable constant solution of the unperturbed problem, i.e., $X=A$, $Y=B/A$. The dashed lines represent the results with the full multigrid procedure. In addition we report in table 8.10 the experimentally observed convergence factors, averaged over 10 iterations. These convergence factors are similar to the ones obtained with the multigrid waveform relaxation method for solving PDEs of initial boundary value type. The computational complexity of solving the time-periodic parabolic equation is therefore similar to that of solving the corresponding initial boundary value problem.

The profile of the stable periodic solution at the mid-point of the chemical reactor is plotted in the upper two diagrams of figure 8.14. The figure to the left shows the time-behaviour, whereas the figure to the right illustrates the limit cycle in the state diagram. This solution can also be determined by a straightforward time integration of the equations starting from some initial condition. This *dynamic simulation* process is

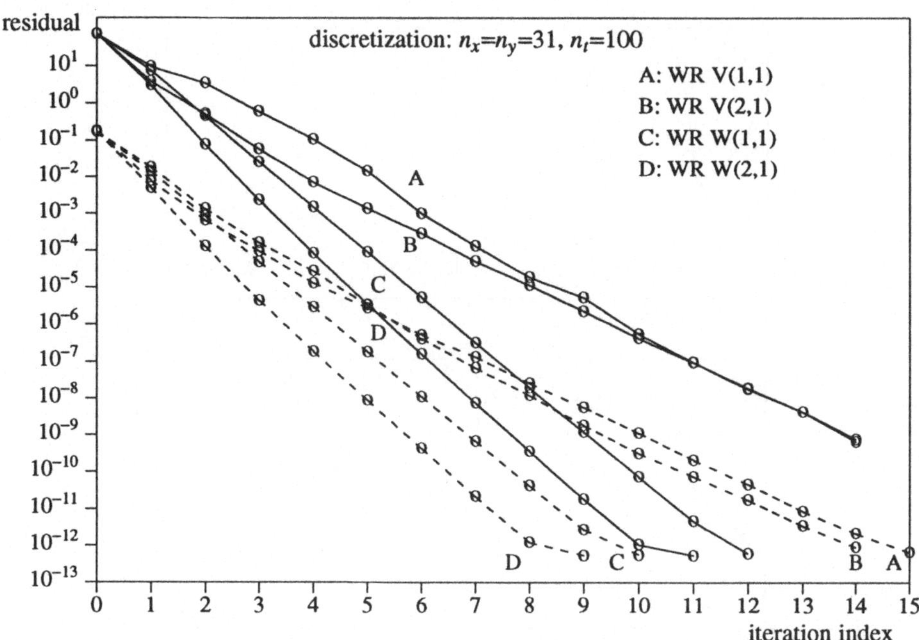

Figure 8.13: Example 7: convergence of the time-periodic multigrid waveform relaxation full approximation scheme (Brusselator with $L=0.15$, $h=1/32$, $\tau=\pi/200$).

illustrated in the lower two diagrams of figure 8.14. The periodic solution is obtained when the transient part of the solution has decayed. Integration over a long time-interval may be required when the periodic solution is not strongly attracting. In this example integration over a length of at least 5 periods is required, before the solution (graphically) remains on the limit cycle. Consequently, this *brute force* time-integration method is not competitive with the waveform relaxation method, as the latter requires only one time-integration.

Table 8.10: Averaged convergence factors of different time-periodic multigrid waveform relaxation cycles (periodic Brusselator).

V(1,1)	V(2,1)	W(1,1)	W(2,1)	F(1,1)	F(2,1)
0.163	0.163	0.063	0.041	0.064	0.041

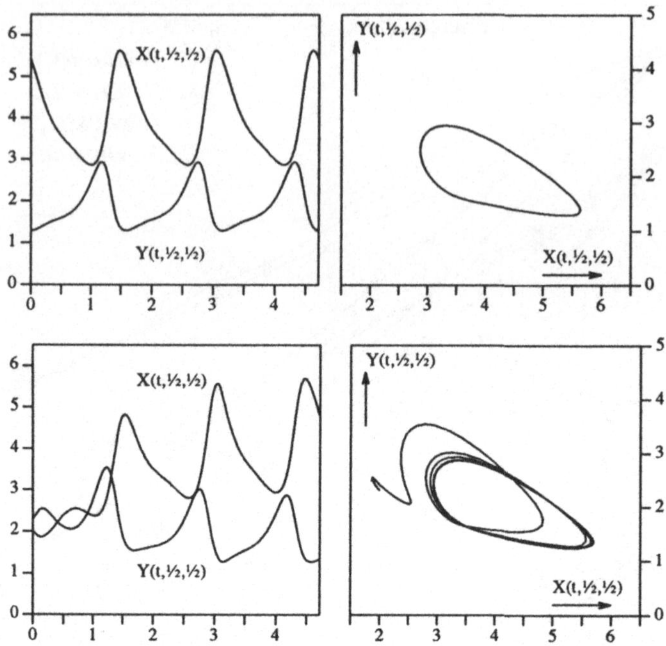

Figure 8.14: Example 7: comparison of time-periodic multigrid waveform relaxation (upper pictures) and dynamic simulation (lower pictures); evolution of two solution components in the time-domain (left) and in the phase-diagram (right).

8.8   Further remarks, limits of applicability

We have provided a number of case studies in which waveform relaxation compares favourably with standard time-stepping. The equations of our example problems are characterized by the fact that they are strongly parabolic. They have dominant diffusion terms and are more or less isotropic. For such problems both waveform multigrid and standard elliptic multigrid within a time-stepping code converge rapidly. Convergence factors are typically of magnitude 0.1, and one step of the nested iteration or full multigrid algorithm is sufficient to get a good approximation to the solution.

We do not claim, however, that waveform relaxation will *always* be superior to the classical techniques. We may expect that multigrid waveform relaxation for solving $\partial u/\partial t = \mathcal{L}u + f$ will perform unsatisfactorily whenever the corresponding elliptic multigrid method performs unsatisfactorily for solving $\mathcal{L}u + f = 0$. This follows immediately from corollary 3.4.4 in chapter 3, which states that the spectral radius of the waveform iteration operator is bounded from below by the spectral radius of the corresponding static iteration operator. *Anisotropy*, for instance, may cause a substantial performance degradation of the multigrid waveform relaxation algorithm (unless the red/black smoother is replaced by a more sophisticated one). *Large convection* terms in the differential equation have a similar effect. This was illustrated for model problems in section 3.5. We should note, however, that for such problems the convergence of the multigrid cycles in the time-stepping scheme is also affected.

Consider, e.g., the following problem, defined on the unit square, with Dirichlet boundary conditions and a given initial condition,

$$\frac{\partial u}{\partial t} = x\frac{\partial^2 u}{\partial x^2} + y\frac{\partial^2 u}{\partial y^2}. \tag{8.12}$$

The convergence of the basic multigrid cycle (with red/black smoothing) applied to the corresponding elliptic problem is strongly deteriorated because of the PDE coefficients, which become zero on some of the domain boundaries. The deterioration increases with decreasing fine grid mesh size. Following corollary 3.4.4 the same is true in the waveform relaxation case. This is illustrated in table 8.8, the second row of which shows averaged waveform V(1,1)-cycle convergence factors. (The time-interval [0,1] is large enough so that the convergence factors are independent of the time-increment.) The third, fourth and fifth rows show V(1,1)-cycle convergence factors of the multigrid cycles encountered in the Crank-Nicolson time-stepping scheme. They are smaller than those of the waveform method, because of the presence of the " $2/\tau\, u$ "- term in the elliptic operator, $2/\tau\, u - \mathcal{L}u$, which characterizes the problems at each time-level. Consequently, for certain combinations of h and τ (i.e., small h and very small τ), we may expect the time-stepping scheme to outdo waveform relaxation. For instance, when $h = 1/32$ and $\tau = 0.005$ the residual reduction caused by 5 WR V(1,1)-cycles is only slightly better than that of a single V(1,1)-cycle in the Crank-Nicolson method.

Table 8.11: Averaged V(1,1)-cycle convergence factors for waveform relaxation method ($t \in [0,1]$) and Crank-Nicolson time-stepping.

h	1/8	1/16	1/32	1/64	1/128
WR	0.23	0.56	0.78	0.90	0.95
CN ($\tau = 0.02$)	0.018	0.21	0.59	0.83	0.94
CN ($\tau = 0.01$)	0.011	0.10	0.47	0.78	0.92
CN ($\tau = 0.005$)	0.004	0.035	0.31	0.70	0.89

To conclude, for some classes of problems both time-stepping and waveform relaxation experience a performance degradation. This severest degradation occurs in the waveform algorithm, which may then be surpassed in performance by the time-stepping scheme. However, in this case the bad convergence factors indicate that the choice of multigrid operators is not optimal. In particular, one should look for more adequate smoothing operators, and, e.g., replace the simple red/black scheme by a smoother based on line-relaxation. This will improve the convergence characteristics of both approaches, the relative performance of which will then have to be re-evaluated.

**Twelve Ways to Fool the Masses
When Giving Performance Results on Parallel Computers**

David H. Bailey, Supercomputing Review, August 1991, pp. 54–55
(condensed version, article 2180 of comp.parallel, 7 Jan 92)

1. *Quote 32-bit performance results, not 64-bit results, or compare your 32-bit results with others' 64-bit results.*

2. *Present inner kernel performance figures as the performance of the entire application.*

3. *Quietly employ assembly code and other low-level language constructs, or compare your assembly-coded results with others' Fortran or C implementations.*

4. *Scale up the problem size with the number of processors, but don't clearly disclose this fact.*

5. *Quote performance results linearly projected to a full system.*

6. *Compare your results against scalar, unoptimized code on Crays.*

7. *Compare with an old code on an obsolete system.*

8. *Base MFLOPS operation counts on the parallel implementation instead of on the best sequential implementation.*

9. *Quote performance in terms of processor utilization, parallel speedups or MFLOPS per dollar (peak MFLOPS, not sustained).*

10. *Mutilate the algorithm used in the parallel implementation to match the architecture. In other words, employ algorithms that are numerically inefficient, compared to the best known serial or vector algorithms for this application, in order to exhibit artificially high MFLOPS rates.*

11. *Measure parallel run times on a dedicated system, but measure conventional run times in a busy environment.*

12. *If all else fails, show pretty pictures and animated videos, and don't talk about performance.*

Chapter 9

Concluding Remarks and Suggestions for Future Research

Mathematics is full of unanswered questions,
which far outnumber known theorems and results...
It is its nature to pose more problems than it can solve.
—I. Peterson, *"Searching for new mathematics"*
(SIAM Review, Vol. 33, No. 1,pp. 37-42, March 1991).

We started this book with the observation that time-marching methods for solving parabolic problems are inherently sequential. The computation proceeds from time-level to time-level, and the parallelism is limited to the parallelism inherent in the solvers applied in each time-step. It was further illustrated that standard solvers trade off numerical quality against parallel efficiency. Consequently, no time-stepping method proved to be really satisfactory for use on large-scale parallel processors.

We suggested to use the waveform relaxation method. This technique had shown to be very successful for solving certain very large systems of ordinary differential equations. A straightforward application of this method to the equations derived by the numerical method of lines did not result in a satisfactory algorithm because of poor convergence. Fortunately, the waveform algorithm could be accelerated by using the multigrid idea, and several *multigrid waveform relaxation* variants were developed. These methods proved to have good numerical performance, without sacrificing the favourable parallel characteristics of the waveform algorithm.

In the previous chapter we gathered a number of case studies in order to illustrate our experience with waveform relaxation methods. It was shown that for a certain class of problems, and the use of certain discretizations the new techniques surpass any of the standard numerical solution methods. This performance is obtained at the sole cost of having to provide a large memory to store the computed waveforms. We recall the three important reasons for the success of the waveform method:

- **a good sequential complexity:** For initial boundary value problems we were able to calculate solutions with algebraic error smaller than the discretization error at a similar cost as the cost involved when using standard time-stepping methods. Due to some implementation details (lower indexing overheads, etc.) and due to a slightly lower sequential complexity waveform relaxation turned out to be even slightly superior.

It was proven that with time-periodic waveform relaxation one can solve time-periodic problems at the cost of solving a single initial value problem (by the corresponding initial value waveform method). We compared the new time-periodic multigrid waveform method with standard time-periodic problem solvers and illustrated an actual performance improvement of at least a factor 2.5.

- **a low communication complexity:** The real advantage of using waveform relaxation techniques stems from their superior parallel characteristics. We theoretically verified that the algorithm strongly reduces the total number of messages required to solve a parabolic problem on a message-passing machine, possibly by orders of magnitude. The total message length remains unchanged (initial boundary value problems), or is substantially reduced (time-periodic problems).

- **vectorization in the time-direction:** For use with waveform relaxation methods, vectorization in the time-direction has shown to be very valuable, and attainable at a negligible implementation cost. It was also noted that this vectorization in the time-direction can be combined with vectorization in the spatial direction in order to increase the vector-lengths even further.

Many problems, questions and possible extensions to the methods discussed in this book require future attention. Some topics for future research are identified below.

More demanding elliptic operators

In the final section of the previous chapter, it was said that the multigrid waveform relaxation algorithm performs unsatisfactorily whenever the corresponding elliptic multigrid method performs unsatisfactorily for solving the elliptic problem $\mathcal{L}u + f = 0$. Such *elliptic* problems can often be solved efficiently with multigrid by using different smoothing operators, e.g., based on line relaxation or incomplete LU decomposition, by using other coarsening strategies, e.g., semi-coarsening, or by using non-trivial operator dependent intergrid transfer operators. Waveform equivalents of these multigrid operators still have to be studied.

The applicability of the waveform relaxation method is tied to the applicability of the numerical method of lines. As such, the method does not seem to be suited for time-dependent problems that require *dynamic grid adaptation*, i.e., for problems that require different spatial discretizations at different time-levels. As a possible remedy one could try to combine the waveform method with so-called *moving grid techniques*. It is unclear, however, whether this is going to be possible, and if so, whether it can be made efficient.

Three-dimensional domains

One of the main reasons for using waveform relaxation on message passing parallel computers is the excellent communication complexity. The method almost completely eliminates the cost associated with the "$t_{startup}$"-parameter (thanks to the enormous reduction in the number of messages). It slightly reduces the communication cost associated with the "t_{send}"-parameter (thanks to the better communication/calculation overlap). As such, the reduction in communication cost will only be significant when the message startup costs are dominant.

An analysis found in McBryan et al., [81], illustrates that the latter is often the case for two-dimensional problems solved on many present-day parallel architectures. However, it might be much less the case when three-dimensional problems are solved. With standard grid partitioning the message lengths in 3D-problem implementations are usually much longer than those encountered in 2D-problem implementations, and the influence of the startup cost is correspondingly reduced. We still have to analyse to what extent this will affect the overall performance of the waveform algorithm.

Variable and independent time-stepping

The use of variable and independent time-stepping to exploit *multirate behaviour* is conceptually straightforward, and, at first sight, it is expected to further improve the effectiveness of the multigrid waveform relaxation method. Many computational problems, however, must be resolved w.r.t. the data structures used to represent discrete functions and w.r.t. to the implementation of operators that manipulate such data structures. In addition, such an implementation will introduce additional problems related load balancing and message passing. It may also preclude efficient vectorization.

A second reason for using variable time-stepping is the use of *inaccurate iteration*, which was explained in section 2.6. The initial iterates are then calculated to low accuracy, and the error tolerance threshold used in the time-integrator is made smaller as more and more iterations are applied. The possible advantages of using this scheme remain to be investigated.

Autonomous time-periodic problems

Autonomous time-periodic problems still defy efficient computation. The modified shooting method presented in section 4.7 significantly reduces the number of time-integrations required by the standard shooting method. Yet, the algorithm still requires the solution of a significant number of initial boundary value problems in order to calculate the solution to a time-periodic problem. As such, its performance is far below that of the time-periodic multigrid waveform relaxation algorithm for solving *non-autonomous* time-periodic problems.

It is not clear whether a waveform relaxation algorithm can be found which calculates the periodic solution and the unknown period T with a cost comparable to the cost of solving *a single* initial boundary value problem.

Space-time concurrent multigrid waveform relaxation

The waveform relaxation method with spatial grid partitioning performs satisfactorily on medium scale parallel processors. On large-scale parallel processors (e.g., with a thousand processors) a loss of efficiency is caused by two reasons. First, the number

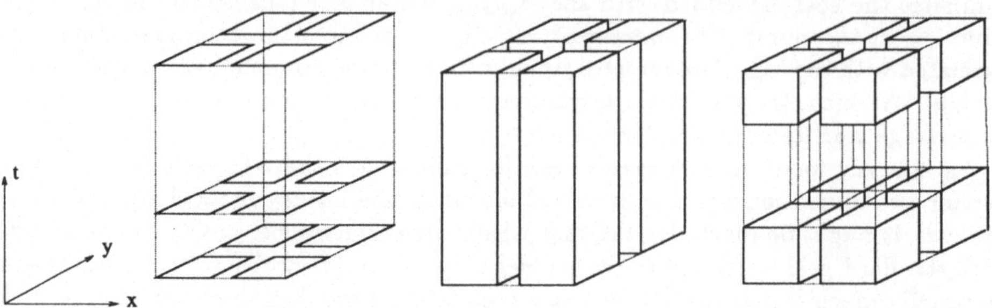

Figure 9.1:   Grid partitioning used for parallelizing parabolic solvers on a two-dimensional rectangular domain: time-stepping method (left), standard waveform relaxation (middle), space-time concurrent waveform relaxation (right).

of nodes may exceed the number of spatial grid points, e.g., on coarse spatial grids encountered in a multigrid hierarchy. The resulting loss of efficiency is much more severe on large multicomputers than it is on small ones. Second, even on relatively fine grids the number of spatial grid points may be too small to rely only on spatial grid partitioning, in which case good load balance is difficult to achieve.

The available computing power of large-scale parallel processors can be harnessed by using waveform relaxation methods that operate concurrently on different time levels. Each processor is then responsible for the calculations in a block of space-time grid points, see figure 9.1 . Most of the operations of the multigrid waveform relaxation algorithm can be parallelized straightforwardly in the time-direction (see, e.g., the discussion on the time-vectorization of the method in section 7.4). The only exception is a linear recurrence, arising in the core of the ODE solver, used to integrate the ODEs at every grid point. This step can be parallelized fairly efficiently by using a so-called *partitioning method*, a standard technique for vectorizing linear recurrence relations.

For implementation on massively parallel systems, it was suggested by Worley in [151] to use a pointwise space-time grid partitioning, and to calculate the linear recurrence relations by a (parallelizable) cyclic reduction method.

Comparison to parabolic multigrid with parallel smoothing

In section 3.7 we have pointed out the close connection between the discrete-time multigrid waveform relaxation method and the parabolic multigrid method introduced by Hackbusch in [43]. Hackbusch also proposes an algorithm, called *parabolic multigrid with parallel smoothing*, in which a smoothing operator is executed concurrently on different time-levels. This time-parallel variant is analysed theoretically by Burmeister, [13], and later on further developed in papers by Horton and coworkers, [58, 59, 60].

Time-parallel parabolic multigrid corresponds to a discrete-time waveform relaxation method in which the differential equations are solved inaccurately. Instead of solving the recurrence relations exactly, they are solved approximately by applying a Jacobi relaxation step to an approximation of the solution. Instead of calculating (7.3)

Table 9.1: Averaged V(1,1)-cycle convergence factors of multigrid waveform relaxation and parabolic multigrid with parallel smoothing. (model problem of section 3.2, $h=1/32$, $\tau=0.01$, $n_t=100$)

	BDF(1)	BDF(2)	C.-N.
waveform relaxation	0.10	0.10	0.11
parallel smoothing	0.36	0.45	0.75

as $x_n^{(\nu)} = a_n x_{n-1}^{(\nu)} + r_n$, one computes $x_n^{(\nu)} = a_n x_{n-1}^{(\nu-1)} + r_n$. This leads to an algorithm which is parallelizable in time, with one or more processors per time-level.

Some impressive results have been obtained with this method, e.g., for solving the Navier-Stokes equations on parallel machines, [59, 14]. The method parallelizes very well because of a low communication complexity. The processors on different time-levels exchange a few large messages, instead of many small messages as is the case in a standard spatial grid partitioning. Furthermore, when only one processor is assigned per time-level, there are no problems with coarse grids or processor idling, and perfect load-balance is assured. Yet, the method has a number of serious disadvantages. Most importantly, the convergence behaviour is strongly dependent on the time-discretization method and the discretization parameters h and τ. The convergence is also often worse than that of the waveform relaxation method. This shows, e.g., from the experimentally determined convergence factors displayed in table 9.1. Furthermore, the method does not allow local time-stepping or vectorization in the time-direction. It is also not clear whether the method can be used for solving time-periodic problems.

It would be interesting to study the relative merits of multigrid waveform relaxation and parabolic multigrid with parallel smoothing, and to determine for which applications and computer systems the methods are most effective.

Waveform Schwarz Iteration

Schwarz iteration is well-known as an iterative method for solving elliptic partial differential equations. The scheme is based on a decomposition of the spatial domain into several overlapping subdomains. When applied to a discretized elliptic problem it corresponds to a kind of overlapping-block Gauss-Seidel method. Its use as a smoother in a multigrid iteration is discussed in [44, section 15.3].

Overlapping-block waveform relaxation was studied by Jeltsch and Pohl in [61]. The application of this algorithm in the context of semi-discrete parabolic partial differential equations, and its behaviour as a smoother in a multigrid waveform algorithm remain to be studied. With use of the Schwarz iteration multigrid waveform relaxation could be implemented in a relatively straightforward way on certain block-structured domains (e.g. "L"-shapes, "T"-shapes, "H"-shapes, etc.).

This page intentionally left blank.
—IBM SPF/TSO Program Reference Manual, SH20-1975-2
p. iv, 6, 98

Appendix A

Discretization and Stencils

We illustrate the construction of the ordinary differential equations that are solved within the numerical method of lines. In particular, we detail the discretization of the spatial derivatives, the construction of five-point and nine-point stencils and the treatment of boundary conditions. The discussion is tailored towards the implementation that we used for obtaining the numerical results in previous chapters. We limit the discussion to linear parabolic partial differential equations. The treatment of nonlinear problems is qualitatively very similar. We consider the problem,

$$\frac{\partial u}{\partial t} = \mathcal{L}u + f \, , \tag{A.1}$$

defined on a two-dimensional rectangular domain, $\Omega = [a, b] \times [c, d]$, with second-order, variable coefficient elliptic partial differential operator,

$$\mathcal{L}u \equiv C_{xx}\frac{\partial^2 u}{\partial x^2} + C_{xy}\frac{\partial^2 u}{\partial x \partial y} + C_{yy}\frac{\partial^2 u}{\partial y^2} + C_x\frac{\partial u}{\partial x} + C_y\frac{\partial u}{\partial y} + Cu \, .$$

Discretization of the elliptic operator

In the finite differences method one places a regular mesh or grid over the domain. At each grid point one writes an ordinary differential equation which locally approximates the partial differential equation. Throughout the appendix we shall assume that the grid is rectangular with spacing h_x in the x-direction and h_y in the y-direction,

$$\Omega^{h_x, h_y} = \left\{ (x_i, y_j) \mid x_i = a + ih_x, \; h_x = \frac{b-a}{I} \text{ and } y_j = c + jh_y, \; h_y = \frac{d-c}{J} \right\}. \tag{A.2}$$

Let $u_{i,j}(t)$ (or $u_{i,j}$ for short) denote the unknown function which approximates $u(t, x_i, y_j)$. The use of *central differences* to approximate the spatial derivatives leads to the following second-order accurate approximations at each grid point (x_i, y_j),

$$\frac{\partial u}{\partial x} \simeq \frac{1}{2\,h_x}(u_{i+1,j} - u_{i-1,j}) \qquad \frac{\partial u}{\partial y} \simeq \frac{1}{2\,h_y}(u_{i,j+1} - u_{i,j-1})$$

$$\frac{\partial^2 u}{\partial x^2} \simeq \frac{1}{h_x^2}(u_{i+1,j} - 2\,u_{i,j} + u_{i-1,j}) \qquad \frac{\partial^2 u}{\partial y^2} \simeq \frac{1}{h_y^2}(u_{i,j+1} - 2\,u_{i,j} + u_{i,j-1})$$

$$\frac{\partial^2 u}{\partial x \partial y} \simeq \frac{1}{4\,h_x\,h_y}(u_{i+1,j+1} - u_{i+1,j-1} - u_{i-1,j+1} + u_{i-1,j-1})$$

When these finite difference approximations are used to replace the spatial derivatives in equation (A.1) at (x_i, y_j), the following ordinary differential equation results,

$$\frac{d}{dt}u_{i,j} = \alpha_{i,j}^c u_{i,j} + \alpha_{i,j}^n u_{i,j+1} + \alpha_{i,j}^e u_{i+1,j} + \alpha_{i,j}^s u_{i,j-1} + \alpha_{i,j}^w u_{i-1,j}$$
$$+\,\alpha_{i,j}^{ne} u_{i+1,j+1} + \alpha_{i,j}^{se} u_{i+1,j-1} + \alpha_{i,j}^{sw} u_{i-1,j-1} + \alpha_{i,j}^{nw} u_{i-1,j+1} + f_{i,j} \ .$$

Here, $f_{i,j}(t) = f(t, x_i, y_j)$. The coefficients, which are functions of t, are given by,

$$\alpha_{i,j}^c = C(t, x_i, y_j) - 2(C_{xx}(t, x_i, y_j)/h_x^2 + C_{yy}(t, x_i, y_j)/h_y^2)$$
$$\alpha_{i,j}^n = C_y(t, x_i, y_j)/2h_y + C_{yy}(t, x_i, y_j)/h_y^2$$
$$\alpha_{i,j}^e = C_x(t, x_i, y_j)/2h_x + C_{xx}(t, x_i, y_j)/h_x^2$$
$$\alpha_{i,j}^s = -C_y(t, x_i, y_j)/2h_y + C_{yy}(t, x_i, y_j)/h_y^2$$
$$\alpha_{i,j}^w = -C_x(t, x_i, y_j)/2h_x + C_{xx}(t, x_i, y_j)/h_x^2$$
$$\alpha_{i,j}^{ne} = C_{xy}(t, x_i, y_j)/4h_x h_y$$
$$\alpha_{i,j}^{se} = -C_{xy}(t, x_i, y_j)/4h_x h_y$$
$$\alpha_{i,j}^{sw} = C_{xy}(t, x_i, y_j)/4h_x h_y$$
$$\alpha_{i,j}^{nw} = -C_{xy}(t, x_i, y_j)/4h_x h_y$$

This ordinary differential equation may be rewritten by using *nine-point stencil* notation for the discretized elliptic operator,

$$\frac{d}{dt}u_{i,j} = \begin{bmatrix} \alpha_{i,j}^{nw} & \alpha_{i,j}^n & \alpha_{i,j}^{ne} \\ \alpha_{i,j}^w & \alpha_{i,j}^c & \alpha_{i,j}^e \\ \alpha_{i,j}^{sw} & \alpha_{i,j}^s & \alpha_{i,j}^{se} \end{bmatrix} u_{i,j} + f_{i,j} \ . \tag{A.3}$$

When there is no cross-derivative term in the elliptic operator, i.e., when $C_{xy} \equiv 0$, coefficients $\alpha_{i,j}^{ne}, \alpha_{i,j}^{se}, \alpha_{i,j}^{sw}$ and $\alpha_{i,j}^{nw}$ are equal to zero. In that case, the equation is somewhat shortened. Using *five-point stencil* (or *five-point star*) notation we get,

$$\frac{d}{dt}u_{i,j} = \begin{bmatrix} & \alpha_{i,j}^n & \\ \alpha_{i,j}^w & \alpha_{i,j}^c & \alpha_{i,j}^e \\ & \alpha_{i,j}^s & \end{bmatrix} u_{i,j} + f_{i,j} \ . \tag{A.4}$$

Treatment of boundary conditions

The incorporation of boundary conditions usually requires some special treatment of the equations at the grid points on the boundary or near the boundary. The choices that we have made in our implementation are discussed below. We assume that each side of the rectangular domain admits only one kind of boundary condition.

Dirichlet boundary conditions

Dirichlet conditions specify the solution on the boundary. They are of the form,

$$u(t, x, y) = g(t, x, y) , \quad (x, y) \in \partial\Omega .$$

There are no unknowns associated with grid points on these boundaries since their value is known exactly. They are eliminated from the system of equations by adapting the stencil coefficients and the right-hand sides of the ODEs at the points adjacent to the boundary. Consider, e.g., the domain in figure A.1 (upper picture) and assume an equation without cross-derivative term. Equation (A.4) at (x_i, y_j) becomes,

$$\frac{d}{dt} u_{i,j} = \begin{bmatrix} & \alpha_{i,j}^n & \\ \alpha_{i,j}^w & \alpha_{i,j}^c & 0 \\ & \alpha_{i,j}^s & \end{bmatrix} u_{i,j} + f_{i,j} + \alpha_{i,j}^e \, g_{i+1,j} . \tag{A.5}$$

Mixed boundary conditions

We consider mixed boundary conditions of the form,

$$\frac{\partial u}{\partial n}(t, x, y) + r(t, x, y) u(t, x, y) = s(t, x, y) , \quad (x, y) \in \Omega . \tag{A.6}$$

($\partial/\partial n$ denotes the outward normal derivative operator.) If $r(t, x, y) \equiv 0$, this condition is called a *Neumann* boundary condition. It explicitly specifies the boundary flux.

Consider the mesh in figure A.1 (middle picture) with a mixed condition on the boundary to the east. At grid point (x_i, y_j), equation (A.6) is discretized with central differences to obtain,

$$\frac{1}{2h_x}(u_{i+1,j} - u_{i-1,j}) + r_{i,j} u_{i,j} = s_{i,j} . \tag{A.7}$$

This equation relates the solution at three grid points, one of which is located outside the PDE domain. The contribution of this fictitious grid point is eliminated by combining (A.7) with the equations (A.3) or (A.4). For instance, in the case of a five-point stencil we get the following equation,

$$\frac{d}{dt} u_{i,j} = \begin{bmatrix} & \alpha_{i,j}^n & \\ \bar{\alpha}_{i,j}^w & \bar{\alpha}_{i,j}^c & 0 \\ & \alpha_{i,j}^s & \end{bmatrix} u_{i,j} + \bar{f}_{i,j} , \tag{A.8}$$

with $\quad \bar{f}_{i,j} = f_{i,j} + 2h_x s_{i,j} \alpha_{i,j}^e , \quad \bar{\alpha}_{i,j}^c = \alpha_{i,j}^c - 2h_x r_{i,j} \alpha_{i,j}^e , \quad \bar{\alpha}_{i,j}^w = \alpha_{i,j}^w + \alpha_{i,j}^e .$

Periodic boundary conditions

Periodicity along the horizontal or vertical coordinate direction implies the equality of the solution and of the first partial derivative of the solution on opposite borders. The location of the grid points and of the unknowns is illustrated in figure A.1 (lower picture). In this picture the variables at the eastern boundary are identified with the ones at the western boundary. The stencils at the adjacent grid points are adapted correspondingly. The equation at point (x_i, y_j) becomes,

$$\frac{d}{dt} u_{i,j} = \begin{bmatrix} & & & & \alpha^n_{i,j} & \\ \alpha^e_{i,j} & \cdots & \alpha^w_{i,j} & \alpha^c_{i,j} & 0 \\ & & & & \alpha^s_{i,j} & \end{bmatrix} u_{i,j} + f_{i,j} \, . \tag{A.9}$$

Its stencil extends all the way to the western boundary.

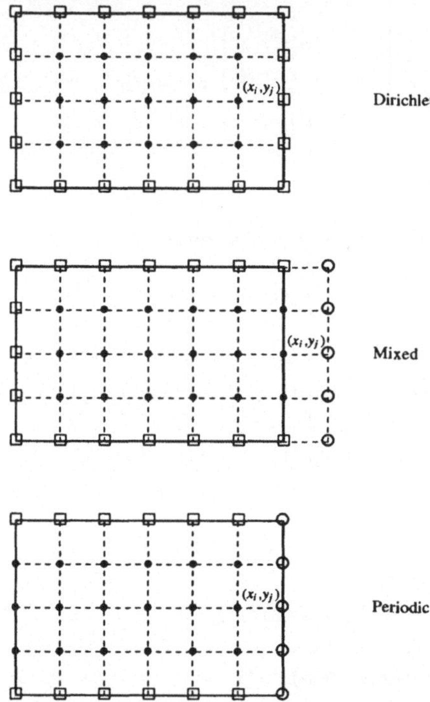

Figure A.1: The extent of the discrete grids in the case of different boundary conditions. The solid lines mark the boundary of domain Ω. "•": grid point with associated unknown. "□": grid point with known Dirichlet value. "o": fictitious grid point.

Boundary conditions and corner points

When using a nine-point stencil, special attention is needed at the corner points of the domain. Several combinations of boundary condition types are possible. The ones that require special action are depicted in figure A.2. We shall discuss in each of the three cases how coefficient $\alpha^{ne}_{i,j}$ can be eliminated from the stencil at grid point (x_i, y_j). The elimination of the other boundary points, away from the corner, proceeds as has been explained in the previous section.

When two Dirichlet boundaries meet, the value of $u_{i+1,j+1}$ is usually known. In the case of conflicting boundary conditions its value may, e.g., be chosen equal to the average of the two conflicting values. The coefficient $\alpha_{i,j}^{ne}$ is eliminated by adding the term $\alpha_{i,j}^{ne} u_{i+1,j+1}$ to the ODE right-hand side $f_{i,j}$.

Consider the case where a Dirichlet condition meets a mixed condition. Then, the mixed condition can be discretized at (x_i, y_{j+1}) which results in an algebraic equation in the unknowns $u_{i+1,j+1}$, $u_{i,j+1}$ and $u_{i-1,j+1}$. Since the latter two are known exactly by the Dirichlet condition, one can explicitly derive the value of $u_{i+1,j+1}$. This value is then easily eliminated.

In the mixed-mixed case the contribution of the fictitious corner grid point (x_{i+1}, y_{j+1}) is taken into account by considering the identity,

$$\frac{\partial u}{\partial n} = \frac{\partial u}{\partial x} n_x + \frac{\partial u}{\partial y} n_y . \tag{A.10}$$

Here, $\partial/\partial n$ denotes the directional derivative in the diagonal direction, from (x_{i-1}, y_{j-1}) to (x_{i+1}, y_{j+1}). (n_x, n_y) is the corresponding unit-length vector. The mixed boundary conditions can be used to replace $\frac{\partial u}{\partial x}$ and $\frac{\partial u}{\partial y}$ in (A.10). When discretized by using central differences, this equation becomes an algebraic equation in the unknowns $u_{i+1,j+1}, u_{i,j}$ and $u_{i-1,j-1}$. It can be used to eliminate $\alpha_{i,j}^{ne}$ from the stencil at grid point (x_i, y_j).

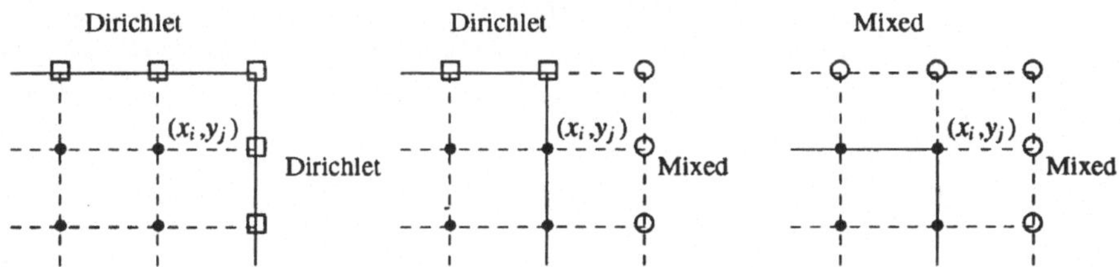

Figure A.2: Location of unknowns, known Dirichlet values and fictitious grid points near a corner of the domain. ("•", "□", "o": see figure A.1)

Number of equations

Consider again grid Ω^{h_x, h_y}, (A.2), with $I+1$ grid lines in x-direction and $J+1$ grid lines in the y-direction. The number of unknowns along a coordinate direction is determined as follows. For instance, along the x-direction it is equal to $I+1$ (mixed conditions on both sides), I (one mixed and one Dirichlet condition, or a periodicity condition), or $I-1$ (two Dirichlet conditions).

It is the man, not the method,
that solves the problem.
—H. Poincaré.

78% of all people who quote statistics
make them up.
—/bin/fortune.

Performance: A statement of the speed at which a computer system works.
Or rather, might work under certain circumstances.
Or was rumored to work last month.
—/bin/fortune.

$$P = NP \quad \text{if} \quad N = 1.$$
—/bin/fortune.

...many mathematicians still have the feeling
that using a computer is akin to cheating
and say that computation is merely an excuse
for not thinking harder...
—Ivars Peterson.

Bibliography

[1] M. Alef. Concepts for efficient multigrid implementation on SUPRENUM-like architectures. *Parallel Computing*, 17:1–16, 1991.

[2] G. Amdahl. Validity of the single processor approach to achieving large scale computing capabilities. In *Proc. AFIPS Comput. Conference*, volume 30, 1967.

[3] U. Asher, R. Mattheij, and R. Russell. *Numerical Solution of Boundary Value Problems for Ordinary Differential Equations*. Prentice Hall, Englewood Cliffs, 1988.

[4] M Bassett. Implementation of multigrid on a hypercube multiprocessor. In M. Heath, editor, *Proceedings of the First Conference on Hypercube Multiprocessors*, pages 211–220, Philadelphia, 1985. SIAM.

[5] P. Bastian, J. Burmeister, and G. Horton. Implementation of a parallel multigrid method for parabolic partial differential equations. In Hackbusch [45], pages 18–27.

[6] L. Beernaert, D. Roose, R. Struys, and H. Deconinck. A multigrid solver for the Euler equations on the iPSC/2 parallel computer. *Applied Numerical Mathematics*, 7:379–398, 1991.

[7] H. Blum, S. Lisky, and R. Rannacher. A domain splitting algorithm for parabolic problems. *Computing*, 49:11–23, 1992.

[8] L. Bomans and D. Roose. Benchmarking the iPSC/2 hypercube multiprocessor. *Concurrency: Practice and Experience*, 1(1):3–18, 1990.

[9] A. Brandt. Multi-level adaptive solutions to boundary-value problems. *Math. Comp.*, 31:333–390, 1977.

[10] A. Brandt. Multigrid solvers on parallel computers. In M. Schultz, editor, *Elliptic Problem Solvers*, pages 39–83, New York, 1981. Academic Press.

[11] A. Brandt and J. Greenwald. Parabolic multigrid revisited. In Hackbusch and Trottenberg [46], pages 143–154.

[12] L. Brochard. Efficiency of some parallel numerical algorithms on distributed systems. *Parallel Computing*, 12:21–44, 1989.

[13] J. Burmeister. *Paralleles lösen diskreter parabolischer Probleme mit Mehrgit-tertechniken*. Diplomarbeit, Universität Kiel, 1985.

[14] J. Burmeister and G. Horton. Time-parallel multigrid solution of the Navier-Stokes equations. In Hackbusch and Trottenberg [46], pages 155–166.

[15] R. Chamberlain. personal communication.

[16] R. Chamberlain. Gray codes, fast Fourier transforms and hypercubes. *Parallel Computing*, 6:225–233, 1988.

[17] D. Champeney. *A Handbook of Fourier Transforms*. Cambridge University Press, Cambridge, 1987.

[18] T. Chan and Y. Saad. Multigrid algorithms on the hypercube multiprocessor. *IEEE Trans. Comp.*, C-35:969–977, November 1986.

[19] T. Chan, Y. Saad, and M. Schultz. Solving elliptic partial differential equations on the hypercube multiprocessor. *Supercomputer*, (13):35–47, 1986.

[20] T. Chan and R. Schreiber. Parallel networks for multi-grid algorithms: Architecture and complexity. *SIAM J. Sci. Stat. Comput.*, 6(3):698–711, July 1985.

[21] T. Chan and R. Tuminaro. Implementation of multigrid algorithms on hypercubes. In Heath [49], pages 730–736.

[22] T. Chan and R. Tuminaro. Design and implementation of parallel multigrid algorithms. In McCormick [83], pages 101–115.

[23] T. Chan and R. Tuminaro. A survey of parallel multigrid algorithms. In A. Noor, editor, *Parallel Computations and Their Impact on Mechanics*, volume 86 of *AMD*, pages 155–170. The Americal Society of Mech. Engineers, 1988.

[24] R. Curtain and A. Pritchard. *Functional Analysis in Modern Applied Mathematics*. Academic Press, London, 1977.

[25] Ph. Davis. *Circulant Matrices*. John Wiley & Sons Inc., New York, 1979.

[26] C. Dawson, Q. Du, and T. Dupont. A finite difference domain decomposition algorithm for numerical solution of the heat equation. Technical Report TR90-24, Rice Univ., Houston, 1990.

[27] D. Dumlugol. *The segmented waveform relaxation method for mixed mode simulation of digital MOS VLSI circuits*. Phd-thesis, Katholieke Universiteit Leuven, Belgium, October 1986.

[28] D. Evans. The alternating group explicit method for the diffusion equation. *Appl. Math. Modeling*, 9:201–206, 1985.

[29] D. Evans, G. Joubert, and F. Peters, editors. *Parallel Computing 89*, Amsterdam, 1990. North Holland.

[30] M. Flynn. Very high-speed computing systems. In *Proc. IEEE, Volume 54*, pages 1901–1909, 1966.

[31] G. Fox, M. Johnson, G. Lyzenga, S. Otto, and J. Salmon. *Solving Problems on Concurrent Processors, Volume I.* Prentice-Hall, Englewood Cliffs, 1988.

[32] P. Frederickson and O. McBryan. Parallel superconvergent multigrid. In McCormick [83], pages 195–210.

[33] E. Gallopoulos and Y. Saad. On the parallel solution of parabolic equations. In *Proc. 1989 ACM Int'l. Conference on Supercomputing, Herakleion, Greece*, pages 17–28, 1989.

[34] E. Gallopoulos and Y. Saad. Efficient solution of parabolic equations by Krylov approximation methods. *SIAM J. Sci. Stat. Comput.*, 13(5):1236–1264, 1992.

[35] D. Gannon and J. van Rosendale. On the structure of parallelism in a highly concurrent PDE solver. *J. of Parallel and Distributed Computing*, 3:106–135, 1986.

[36] C. Gear. *Numerical initial value problems in ordinary differential equations.* Prentice-Hall, Englewood Cliffs, 1971.

[37] C. Gear and F. Juang. The speed of waveform methods for ODEs. In R. Spigler, editor, *Applied and Industrial Mathematics*, pages 37–48, Boston, 1991. Kluwer.

[38] E. Gendler. *Multigrid methods for time-dependent parabolic equations.* Ms. Sc. thesis, The Weizmann Institute, Rehovot, Israel, August 1986.

[39] I. Gohberg and S. Goldberg. *Basic Operator Theory.* Birkhäuser, Boston, 1981.

[40] J. Gustafson, G. Montry, and R. Benner. Development of parallel methods for a 1024-processor hypercube. *SIAM J. Sci. Stat. Comput.*, 9(4):609–638, July 1988.

[41] W. Hackbusch. Fast numerical solution of time-periodic parabolic problems by a multigrid method. *SIAM J. Sci. Stat. Comput.*, 2(2):198–206, June 1981.

[42] W. Hackbusch. A multi-grid method applied to a boundary value problem with variable coefficients in a rectangle. Bericht nr. 2/83, Institut für Informatik und praktische Mathematik, Kiel, 1983.

[43] W. Hackbusch. Parabolic multigrid methods. In R. Glowinski and J.-R. Lions, editors, *Computing Methods in Applied Sciences and Engineering VI*, pages 189–197, Amsterdam, 1984. North Holland.

[44] W. Hackbusch. *Multi-Grid Methods and Applications.* Springer Verlag, Berlin, 1985.

[45] W. Hackbusch, editor. *Parallel Algorithms for PDEs (Proceedings of the 6th GAMM Seminar Kiel, January 19-21, 1990)*, Wiesbaden, 1990. Vieweg Verlag.

[46] W. Hackbusch and U. Trottenberg, editors. *Multigrid methods III (Proceedings of the third European Multigrid Conference, Bonn, 1990)*, number 98 in ISNM, Basel, 1991. Birkhaüser Verlag.

[47] W. Hairer and G. Wanner. *Solving Ordinary Differential Equations II.* Springer-Verlag, Berlin, 1991.

[48] B. Hassard, N. Kazarinoff, and Y. Wan. *Theory and Applications of Hopf Bifurcation.* Cambridge University Press, Cambridge, 1981.

[49] M. Heath, editor. *Hypercube Multiprocessors 1987*, Philadelphia, 1987. SIAM.

[50] D. Hebert. Random-walk simulation of diffusion-reaction-convection systems. In Rodrigue [104], pages 175–180.

[51] D. Heller. A survey of parallel algorithms in numerical linear algebra. *SIAM Review*, 20(4):740–777, October 1978.

[52] P. Hemker and H. Schippers. Multiple grid methods for the solution of Fredholm integral equations of the second kind. *Math. Comput.*, 36(153):215–232, January 1981.

[53] R. Hempel. The SUPRENUM communications subroutine library for grid-oriented problems. Technical Report ANL-87-23, Argonne Nat. Lab., Argonne, Illinois, 1987.

[54] R. Hempel and A. Schüller. Experiments with parallel multigrid algorithms using the SUPRENUM communications subroutine library. Gmd-studien nr. 141, G.M.D., Sankt Augustin 1, April 1988.

[55] C. Ho and S. Johnsson. Optimizing tridiagonal solvers for alternating direction methods on boolean cube multiprocessors. *SIAM J. Sci. Stat. Comput.*, 11(3):563–592, 1990.

[56] R. Hockney and C. Jesshope. *Parallel Computers.* Adam Hilger, Bristol, 1981.

[57] M. Holodniok, P. Knedlík, and M. Kubíček. Continuation of periodic solutions in parabolic partial differential equations. Number 79 in International Series of Numerical Mathematics, pages 122—130, Basel, 1987. Birkhäuser Verlag.

[58] G. Horton. *Ein zeitparalleles Lösungsverfahren für die Navier-Stokes-Gleichungen.* Phd-thesis, Universität Erlangen-Nünberg, 1991.

[59] G. Horton. Time-parallel multigrid solution of the Navier-Stokes equations. In C. Brebbia, editor, *Applications of Supercomputers in Engineering.* Elsevier, August 1991.

[60] G. Horton and R. Knirsch. A time-parallel multigrid-extrapolation method for parabolic partial differential equations. *Parallel Computing*, 18:21–29, 1992.

[61] R. Jeltsch and B. Pohl. Waveform relaxation with overlapping splittings. Research report no. 91-02, E.T.H. Zürich, Seminar für Angewandte Mathematik, February 1991.

[62] S. Johnsson. Communication efficient basic linear algebra computations on hypercube architectures. *J. of Par. Distr. Computing*, 4:133–172, 1987.

[63] S. Johnsson. Solving tridiagonal systems on ensemble architectures. *SIAM J. Sci. Stat. Comput.*, 8(3):354–392, 1987.

[64] S. Johnsson and C. Ho. Algorithms for matrix transposition on boolean n-cube configured ensemble architectures. *SIAM J. Matrix Anal. Appl.*, 9(3):419–454, 1988.

[65] S. Johnsson, Y. Saad, and M. Schultz. Alternating direction methods on multiprocessors. *SIAM J. Sci. Stat. Comput.*, 8(5):686–700, 1987.

[66] F. Juang. *Waveform Methods for Ordinary Differential Equations*. Phd-thesis, University of Illinois at Urbana-Champaign, Dept of Computer Science, January 1990.

[67] L. Kantorovich and G. Akilov. *Functional Analysis*. Pergamon Press, Oxford, sec. edition, 1982.

[68] O. Kolp. Optimal color relaxations for parallel systems. In Evans et al. [29], pages 121–126.

[69] A. Krechel, H. Plum, and K. Stüben. Parallel solution of tridiagonal linear systems. In F. André and J. Verjus, editors, *Hypercube and distributed computers*, pages 49–64, Amsterdam, 1989. North-Holland.

[70] R. Kress. *Linear Integral Equations*, volume 82 of *Applied Mathematical Sciences*. Springer Verlag, Berlin, 1989.

[71] J. Lambert. *Computational Methods in Ordinary Differential Equations*. John Wiley & Sons, Chichester, 1973.

[72] L. Lapidus and G. Pinder. *Numerical solution of partial differential equations in science and engineering*. John Wiley and Sons, New York, 1982.

[73] J. Lawson and S. Morris. The extrapolation of first order methods for parabolic partial differential equations I. *SIAM J. Num. Anal.*, 15(6):1212–1224, 1978.

[74] E. Lelarasmee. *The Waveform Relaxation Method for the Time Domain Analysis of Large Scale Nonlinear Systems*. Phd-thesis, University of California, Berkeley, 1982.

[75] E. Lelarasmee, A. Ruehli, and A. Sangiovanni-Vincentelli. The Waveform Relaxation Method for Time-Domain Analysis of Large Scale Integrated Circuits. *IEEE Trans. on CAD of IC and Sys.*, 1(3):131–145, July 1982.

[76] M. Lemke. Experiments with a vectorized multigrid Poisson solver on the CDC 205, CRAY X-MP and Fujitsu VP200. Arbeitspapiere der GMD 179, G.M.D., St. Augustin, November 1985.

[77] D. Lim and R. Thanakij. A survey of ADI implementations on hypercubes. In Heath [49], pages 674–679.

[78] Ch. Lubich and A. Ostermann. Multigrid dynamic iteration for parabolic equations. *BIT*, 27:216–234, 1987.

[79] J. Massera and J. Schäffer. *Linear Differential Equations and Function Spaces*. Academic Press, New York, 1966.

[80] S. Mattison. *CONCISE: a concurrent circuit simulation program*. Phd-thesis, Lund Institute of Technology, Dept of Applied Electronics, Lund, Sweden, 1986.

[81] O. McBryan, P. Frederickson, J. Linden, A. Schüller, K. Solchenbach, K. Stüben, C. Thole, and U. Trottenberg. Multigrid methods on parallel computers — a survey of recent developments. *IMPACT of Computing in Science and Engineering*, 3:1–75, 1991.

[82] O. McBryan and E. Van de Velde. Hypercube algorithms and implementations. *SIAM J. Sci. Stat. Comput.*, 8:s227–s287, 1987.

[83] S. McCormick, editor. *Multigrid Methods: Theory, Applications, and Supercomputing*, New York, 1988. Marcel Dekker.

[84] U. Miekkala and O. Nevanlinna. Convergence of dynamic iteration methods for initial value problems. *SIAM J. Sci. Stat. Comput.*, 8(4):459–482, July 1987.

[85] U. Miekkala and O. Nevanlinna. Sets of convergence and stability regions. *BIT*, 27:554–584, 1987.

[86] A. Mitchell and D. Griffiths. *The finite difference method in partial differential equations*. John Wiley and Sons, New York, 1980.

[87] V. Naik and S. Ta'asan. Performance studies of the multigrid algorithms implemented on hypercube multiprocessor systems. In Heath [49], pages 720–729.

[88] O. Nevanlinna. Remarks on Picard-Lindelöf iteration, PART I. *BIT*, 29:328–346, 1989.

[89] O. Nevanlinna. Remarks on Picard Lindelöf iteration, PART II. *BIT*, 29:535–562, 1989.

[90] O. Nevanlinna. Linear acceleration of Picard-Lindelöf iteration. *Numer. Math.*, 57:147–156, 1990.

[91] O. Nevanlinna. Power bounded prolongations and Picard-Lindelöf iteration. *Numer. Math.*, 58:479–501, 1990.

[92] A. Newton and A. Sangiovanni-Vincentelli. Relaxation-based electrical simulation. *SIAM J. Sci. Stat. Comput.*, 4(3):485–524, September 1983.

[93] P. Odent. *Electrical-level simulation of VLSI MOS circuits using multi-processor systems*. Phd-thesis, Katholieke Universiteit Leuven, Belgium, January 1990.

[94] J. Ortega and R. Voigt. Solution of partial differential equations on vector and parallel computers. *SIAM Review*, 27:149–240, 1985.

[95] M. Osborne. A note on the numerical solution of a periodic parabolic problem. *Numer. Math.*, 7:155–158, 1965.

[96] M. Osborne. The numerical solution of a periodic parabolic problem subject to a nonlinear boundary condition II. *Numer. Math.*, 12:280–287, 1968.

[97] M. Ozisik. *Heat Conduction*. Wiley, New York, 1980.

[98] T. Parker and L. Chua. *Practical Numerical Algorithms for Chaotic Systems*. Springer Verlag, New York, 1989.

[99] L. Peterson and S. Mattison. Partitioning tradeoffs for waveform relaxation in transient analysis circuit simulation. In D. Walker and Q. Stout, editors, *Proc. of the Fifth Distributed Memory Computing Conference*, pages 612–621, Los Alamitos, CA., 1990. IEEE.

[100] A. Poularikas and S. Seely. *Elements of Signals and Systems*. PWS-Kent Publishing Company, Boston, 1988.

[101] M. Quinn. *Designing Efficient Algorithms for Parallel Computers*. McGraw-Hill, New York, 1987.

[102] S. Raman, L. Patnaik, and R. Mall. Parallel implementation of circuit simulation. *Int. J. High Speed Computing*, 2(4):351–373, 1990.

[103] M. Rees and K. Morton. Moving point and particle methods for convection diffusion equations. In Rodrigue [104], pages 170–174.

[104] G. Rodrigue, editor. *Parallel Processing for Scientific Computing*, Philadelphia, 1989. Proceedings of the Third SIAM Conference on Parallel Processing for Scientific Computing, Los Angeles, December 1-4, 1987, SIAM.

[105] G. Rodrigue and D. Wolitzer. Preconditioned time-differencing for the parallel solution of the heat equation. In J. Dongarra, P. Messina, D. Sorensen, and R. Voigt, editors, *Proceedings of the Fourth SIAM conference on Parallel Processing for Scientific Computing*, pages 268–271, Philadelphia, 1990. SIAM.

[106] D. Roose and S. Vandewalle. Efficient parallel computation of periodic solutions of parabolic partial differential equations. In R. Seydel, F. Schneider, A. Küpper, and H. Troger, editors, *Bifurcations and Chaos: Analysis, Algorithms, Applications*, pages 307–317, Berlin, 1991. Birkhäuser Verlag.

[107] N. Rouche and J. Mawhin. *Ordinary Differential Equations: Stability and Periodic Solutions*. Pitman Publishing Ltd, London, 1980.

[108] Y. Saad and M. Schultz. Topological properties of hypercubes. Technical Report YALEU/DCS/RR-389, Dept. of Computer Science, Yale University, June 1985.

[109] F. Saied, C. Ho, S. Johnsson, and M. Schultz. Solving Schrödinger's equation on the Intel iPSC by the alternating direction method. In Heath [49], pages 680–691.

[110] J. Saltz and V. Naik. Towards developing robust algorithms for solving partial differential equations on MIMD machines. *Parallel Computing*, 6:19–44, 1988.

[111] A. Sameh and D. Kuck. On stable parallel linear system solvers. *J. ACM*, 25:81–91, 1978.

[112] H. Schippers. Analytical and numerical results for the non-stationary rotating disk flow. *J. Eng. Math.*, 13(2):173–191, April 1979.

[113] H. Schippers. Application of multigrid methods for integral equations to two problems from fluid dynamics. *J. of Comp. Physics*, 48(3):441–461, December 1982.

[114] J. Scroggs. Parallel processing of a domain decomposition method. In Rodrigue [104], pages 164–168.

[115] S. Seidel, M. Lee, and S. Fotedar. Concurrent bidirectional communication on the Intel iPSC/860 and iPSC/2. In Q. Stout and M. Wolfe, editors, *Proc. of the Sixth Distributed Memory Computing Conference*, pages 283–286, Los Alamitos, CA., 1991. IEEE.

[116] C. Seitz. The cosmic cube. *Comm. ACM*, 28(1):22–33, 1985.

[117] R. Seydel. *From Equilibrium to Chaos. Practical Bifurcation and Stability Analysis*. Elsevier, New York, 1988.

[118] R. Skeel. Waveform iteration and the shifted Picard splitting. *SIAM J. Sci. Stat. Comput.*, 10(4):756–776, July 1989.

[119] A. Skjellum. *Concurrent dynamic simulation: multicomputers algorithms research applied to ordinary differential-algebraic process systems in chemical engineering*. Phd-thesis, California Institute of Technology, May 1990.

[120] K. Solchenbach. *Einsatz Schneller Elliptischer Löser zur Lösung nichtlinearer parabolischer Anfangsrandwertaufgaben*. Diplomarbeit, G.M.D., Bonn, 1980.

[121] K. Solchenbach. Grid applications on distributed memory architectures: implementation and evaluation. *Parallel Computing*, 7:341–356, 1988.

[122] B. Sommeijer. Increasing the real stability boundary of explicit methods. *Computers Math. Applic.*, 19(6):27–49, 1990.

[123] B. Sommeijer and P. Van der Houwen. ALGORITHM 621. Software with low storage requirements for two-dimensional, nonlinear, parabolic differential equations. *ACM Trans. on Math. Software*, 10(4):378–396, 1984.

[124] M. Steuerwalt. The existence, computation, and number of solutions of periodic parabolic problems. *SIAM J. Num. Anal.*, 16(3):402–420, June 1979.

[125] K. Stüben and U. Trottenberg. Multigrid methods: fundamental algorithms, model problem analysis and applications. In W. Hackbusch and U. Trottenberg, editors, *Multigrid Methods*, number 960 in Lecture Notes in Mathematics, pages 1–176, Berlin, 1982. Springer Verlag.

[126] G. Tee. An application of p-cyclic matrices, for solving periodic parabolic problems. *Numer. Math.*, 6:142–159, 1964.

[127] C. Thole. Experiments with multigrid methods on the CalTech hypercube. GMD-Studien 103, G.M.D., St. Augustin, 1985.

[128] R. Van Driessche. *Parallelle algoritmen voor parabolische partiële differentiaalvergelijkingen.* Ms. Sc. thesis, Department of Computer Science, Katholieke Universiteit Leuven, Belgium, June 1989. (in Dutch).

[129] S. Vandewalle. Activiteitenverslag 1985-1986. Interim Report N.F.W.O. (in Dutch), October 1986.

[130] S. Vandewalle. Waveform relaxation methods for solving parabolic partial differential equations. In D. Walker and Q. Stout, editors, *Proceedings of the Fifth Distributed Memory Computing Conference*, pages 575–584, Los Alamitos, CA., 1990. IEEE.

[131] S. Vandewalle, J. De Keyser, and R. Piessens. The numerical solution of elliptic partial differential equations on a hypercube multiprocessor. In J. Devreese and P. Van Camp, editors, *Scientific Computing on Supercomputers*, pages 69–97, New York, 1989. Plenum Press.

[132] S. Vandewalle and R. Piessens. A comparison of parallel multigrid strategies. In F. André and J. Verjus, editors, *Hypercube and Distributed Computers*, pages 65–79, Amsterdam, 1989. North Holland.

[133] S. Vandewalle and R. Piessens. A comparison of the Crank-Nicolson and waveform relaxation multigrid methods on the Intel hypercube. In J. Mandel, S. McCormick, J. Dendy, C. Farhat, G. Lonsdale, S. Parter, J. Ruge, and K. Stüben, editors, *Proceedings of the Fourth Copper Mountain Conference on Multigrid Methods*, pages 417–434, Philadelphia, 1990. SIAM.

[134] S. Vandewalle and R. Piessens. A parallel and vectorizable algorithm for solving parabolic partial differential equations. In Hackbusch [45], pages 216–227.

[135] S. Vandewalle and R. Piessens. Multigrid waveform relaxation for solving parabolic partial differential equations. In Hackbusch and Trottenberg [46], pages 377–388.

[136] S. Vandewalle and R. Piessens. Numerical experiments with nonlinear multigrid waveform relaxation on a parallel processor. *Applied Numerical Mathematics*, 8(2):149–161, 1991.

[137] S. Vandewalle and R. Piessens. Efficient parallel algorithms for solving initial-boundary value and time-periodic parabolic partial differential equations. *SIAM J. Sci. Stat. Comput.*, 13(6):1330–1346, November 1992.

[138] S. Vandewalle and R. Piessens. On the existence of convergent and non-convergent time-periodic dynamic iteration splittings. Technical Report TW 169, Department of Computer Science, Katholieke Universiteit Leuven, April 1992.

[139] S. Vandewalle and R. Piessens. On dynamic iteration methods for solving time-periodic differential equations. *SIAM J. Num. Anal.*, 30(1):286–303, February 1993.

[140] S. Vandewalle and D. Roose. The parallel waveform relaxation multigrid method. In Rodrigue [104], pages 152–156.

[141] S. Vandewalle, D. Roose, and R. Piessens. A comparison of two parallel multigrid methods for the numerical solution of parabolic partial differential equations. In *Proceedings of the Fourth Conference on Hypercubes, Concurrent Computers and Applications*, pages 1287–1290, Los Altos, California, 1990. Golden Gate Enterprises.

[142] S. Vandewalle, R. Van Driessche, and R. Piessens. The implementation of parabolic partial differential equation solvers on a hypercube multiprocessor. In Evans et al. [29], pages 61–66.

[143] S. Vandewalle, R. Van Driessche, and R. Piessens. The parallel implementation of standard parabolic marching schemes. Technical Report TW 125, Department of Computer Science, Katholieke Universiteit Leuven, December 1990.

[144] S. Vandewalle, R. Van Driessche, and R. Piessens. The parallel performance of standard parabolic marching schemes. *Int. J. High Speed Computing*, 3(1):1–29, 1991.

[145] R. Varga. *Martrix Iterative Analysis*. Prentice-Hall, Englewood Cliffs, 1965.

[146] H. Wang. A parallel method for tridiagonal equations. *ACM Trans. on Math. Software*, 7:170–183, 1981.

[147] J. White and A. Sangiovanni-Vincentelli. *Relaxation Techniques for the Simulation of VLSI Circuits*. Kluwer Academic Publishers, Boston, 1987.

[148] J. White, A. Sangiovanni-Vincentelli, F. Odeh, and A. Ruehli. Waveform relaxation: Theory and practice. *Trans. of the Soc. for Comp. Simulation*, 2(1):95–133, 1985.

[149] D. Womble. A time-stepping algorithm for parallel computers. *SIAM J. Sci. Stat. Comput.*, 11(5):824–837, 1990.

[150] D. Womble and B. Young. A model and implementation of multigrid for massively parallel computers. *Int. J. High Speed Computing*, 2(3):239–255, 1990.

[151] P. Worley. Parallelizing across time when solving time-dependent partial differential equations. ORNL/TM-11963, Oak Ridge National Laboratory, September 1991.

[152] E. Xia. Parallel Waveform-Relaxation-Newton for circuit simulation. Csrd rpt. no. 772, University of Illinois, Center for Supercomputing Research and Development, May 1988.

[153] A. Zaanen. *Linear Analysis: measure and integral, Banach and Hilbert space, linear integral equations.* North-Holland, Amsterdam, 1964.

[154] P. Zabreyko, A. Koshelev, M. Krasnosel'skii, S. Mikhlin, L. Rakovshchik, and V. Stet'senko. *Integral Equations - a reference text.* Noordhoff, Leiden, 1975.

> *Classical is a book*
> *which everybody praises*
> *but nobody reads.*
> —Mark Twain.

*In theory, there is no difference between theory
and practice. But, in practice, there is.*
—/bin/fortune.

*Our present position on the use of advanced programming techniques is that,
although they are essential for the expert,
the use of "simplified" pseudo-codes by the novice is dangerous,
for he will produce too many solutions by improper methods to incorrect problems.*
—M. Newman, J. Todd, "Automatic Computers",
in Survey of Numerical Analysis, J. Todd (ed), McGraw-Hill, New York, 1962.

*Rayleigh asserted the irrelevance of formal proofs of existence theorems,
"since they tell us only what we knew before"*
—F. Hossfeld, "Nonlinear dynamics: a challenge on high-speed computation"
in Parallel Computing 83, M. Feilmeier et al. (eds), North-Holland, Amsterdam, 1984.

*Scientists are bogged down with day-to-day problems and find it tiresome when forced
repeatedly to be ingenious in overcoming inadequacies in the computer system or the
support software. When they are responsible for the physics as well as the numerics and the
computer system's idiosyncrasies, they become less effective physicists or engineers.
These extraneous duties gobble up their scarcest resource—time.*
—J. Hyman, "Future directions in large scale scientific computing"
in Large scale scientific computing, S. Parter (ed), Academic Press, Orlando, 1984

*A conclusion is simply the place
where someone got tired of thinking.*
—/bin/fortune.

Index

accommodative scheme, 208
accuracy, 197
    order of, 33
ADI, 138, 172, 175–182
agglomeration, 146, 150–154, 167–170
Amdahl's law, 133, 193
anisotropic problem, 77, 80, 222
autonomous problem, 84, 116–122, 227

backward differentiation formula, 42
Banach space, 30, 35, 36, 100
BDF method, 42, 138, 175–182
binary representation, 127
block-circulant matrix, 102
Brusselator, 120, 210, 220

central differences, 231
characteristic polynomial, 41, 107
coarse grid
    approximation, 60, 79
    correction, 59, 65, 80, 111, 114
    problem, 146
coarsening
    semi-, 81, 226
    standard, 60, 62
colouring, 47, 146, 148
communication, 125
    parameters, 131
    schemes, 142–165
compactness, 96
consistent grid, 142
continuous-time
    analysis, 38–40, 52, 68–70, 90–99, 112–113
    iteration, 25, 28
contraction, 30, 85
    mapping principle, 30
    strict, 32
convection-diffusion problem, 78, 222

convergence, 30, 36
    set of, 108
    factor, 39
        averaged, 55
        iteration, 55
        time-level, 56
convolution property, 97
Crank-Nicolson method, 42, 138, 175–182
    arithmetic complexity, 186
    communication complexity, 191
    stability region, 107
cyclic reduction, 159, 194
    balanced, 160–163
    timing results, 170–172

deagglomeration, 150–154
defect, 60, 64, 65, 111, 114
    equation, 60, 64, 65, 79
    parallel computation, 147
discrete
    Fourier series, 100
    solvability condition, 100
discrete-time
    analysis, 40–45, 53, 70–72, 100–109, 114–116
    iteration, 28
distributed memory, 125
divergence, 37
    of SOR WR, 56
DuFort-Frankel method, 138, 175–182
dynamic
    iteration, 24, 38
    simulation, 85, 220

efficiency, 132
eigenvalue, 36
error, 174
    actual, 199

algebraic, 198, 199
    discretization, 199
Euler method, 175–182
    backward, 107, 138
    forward, 137
explicit
    method, 18, 137
    update, 142–144
        timing results, 165–166
extended subdomain, 142

F-cycle, 61, 62
Fourier series, 94, 95, 100
full approximation scheme, 78–80, 208
full multigrid, 61, 62, 63,
    waveform relaxation, 66, 67, 111

Gauss-Seidel splitting, 37
Gauss-Seidel WR, 25–27
    accuracy increase, 34
    blockwise, 29, 46
    convergence
        general, 31–33
        linear, 40
    hierarchical, 46
    iteration function, 29
    time-periodic, 88–90
        convergence, 98
Gauss-Seidel WR for PDEs, 51–57
    convergence, 53
    lexicographic, 51
    red/black, 52, 64
    time-periodic, 110, 115–117
graphical convergence check, 108-109
Gray code, 128
    embedding, 129–130
grid, 17, 50, 137, 231
    partitioning, 18, 141, 190, 228
        strip or square, 172

heat equation, 51, 178, 204, 207
Heun method, 137, 175–182
hypercube, 126–132

implicit method, 18, 138
infinite loop, *see* infinite loop
initial boundary value problem, 49

iPSC/2, 130–132

Jacobi
    matrix, 40
    splitting, 37
Jacobi WR, 26, 29
    accuracy increase, 34
    blockwise, 29
    convergence
        general, 31–33
        linear, 40
    iteration function, 29
    time-periodic, 88, 92
        convergence, 98
Jacobi WR for PDEs, 51–55
    convergence, 53
    time-periodic, 110, 115, 116
JOR splitting, 37

Lebesgue, 35
line hopscotch method, 139, 172, 175–
    182
linear multistep method, 40–42, 100

maximum principle, 39
mesh, 50
MIMD, 125
model problem, 51
    anisotropic, 77
    convection-diffusion, 78
multi-rate, 24, 26, 45, 46, 227
multigrid, 59–63, 139
    cycle, 61, 73
    nonlinear, 79
    of the second kind, 86–87
        arithmetic complexity, 189–190
        communication complexity, 191–193
        timing results, 215–220
        with WR, 216
    on space-time grid, 80–82
    parallel implementation, 144–154
    procedure, 63
    timing results, 166–170
multigrid WR, 21, 64–67
    arithmetic complexity, 184–187
    communication complexity, 191–193

convergence
  continuous-time, 68–70
  discrete-time, 70–72
  experimental results, 73–78
  full approximation scheme, 79–80
  limits of applicability, 222–223
  model problem analysis, 70, 72
  nonlinear, 78–80
  operator, 68, 71
  operators, 64–65
  procedure, 66
  time-periodic, *see* time-periodic multigrid WR
  vectorization, 193

N-periodic, 100
nested iteration, *see* full multigrid
node, 126
non-critical
  matrix, 91
  splitting, 93
norm
  ∞-norm, 35
  exponentially scaled, 31
  maximum, 30
  of operator, 36
  p-norm, 35
numerical method of lines, 17, 49, 137

operator
  bounded, 36
  compact, 96
  convolution, 38
  discrete convolution, 43
  discrete periodic convolution, 105
  Fredholm, 95
  periodic convolution, 94
ordering of equations, 34, 46
overlap
  area, 142, 143
  of computation and communication, 19, 132, 144
  of messages, 143

parabolic multigrid, 20
  with parallel smoothing, 20, 228

  with sequential smoothing, 82
parabolic PDE, 17, 49, 83, 136
parallel
  computer
    classification, 124–126
    model, 140
  computing, 123–134
  efficiency, 132
  speedup, 132–134
period, 83
Picard-Lindelöf iteration, 24
prolongation, 60, 65
  parallel implementation, 148, 149

recursive doubling, 157–158
  timing results, 170–172
red/black, 52, 139, 148
residual, 147, 198
  equation, 118
restriction, 60
  full-weighting, 64, 185
  half-weighting, 65
  injection, 65
  parallel implementation, 148, 149
Richardson splitting, 37
root locus curve, 41

semi-implicit method, 138
shared memory, 125
shifted interpolation, 119
shooting, 26, 86, 118–122
SIMD, 124
smoothing, 59, 64, 65, 79, 111, 114
  parallel implementation, 147
  x-line, 81
SOR splitting, 37
SOR WR, 37
  convergence, 40, 41
  time-periodic, 92
SOR WR for PDEs, 52
  convergence 53, 55
  divergence, 56, 58
  time-periodic, 110
space
  of p'th power integrable functions, 35, 90

of p-summable sequences, 36, 100
of continuous functions, 30, 90
of essentially bounded functions, 35
space-time grid, 80–82, 85, 184, 228
spectral
    picture, 107–109
    radius, 36
spectrum, 36, 91
    of block circulant matrix, 102
    of compact operator, 96
    of convolution operator, 68
speedup, 132–134
splitting, 37
    convergent, 99
    non-critical, 93
stability region, 41–42
startup-time, 131, 141
static iteration, 24, 38
stencil, 64
    five-point, 232
    nine-point, 232
strict contractivity, 32
subcube, 128
    induction, 129
substructured Gaussian elimination, 155–157
    timing results, 170–172
successive approximation, 36
superlinear
    convergence, 38
    speedup, 133, 166
symbol, 39

T-periodicity condition, 83
TGET, 163–165
    timing results, 170–172
Thomas algorithm, 140, 154
time-marching, *see* time-stepping
time-periodic multigrid WR, 21, 110–117
    arithmetic complexity, 187–190
    communication complexity, 191
    convergence
        continuous-time, 112–113
        discrete-time, 114–115
    full approximation scheme, 111

numerical example, 115–117
    operator, 112, 114
time-periodic PDE, 83
    standard solvers, 85–87
time-periodic WR, 88–89
    convergence
        continuous-time, 90–99
        discrete-time, 100–109
    Gauss-Seidel, 88-90, 98
    Jacobi, 88, 92, 98
    operator
        continuous-time, 94–95
        discrete-time, 105
    SOR, 92
time-periodic WR for PDEs, 110–116
time-stepping, 18, 50
    acceleration of, 19
    arithmetic complexity, 186–187
    as a multigrid method, 82
    communication complexity, 191
    numerical examples, 174–182
    parallel implementation, 135–182, 228
transposition by reflection, 163, 164
trapezoidal rule, 42
    stability region, 107
tridiagonal system, 140
    parallel solution, 154–165
    timing results, 170–173
two-grid method
    linear, 59–60
    nonlinear, 79
    time-periodic WR, 111, 114
    WR, 65, 79–80
two-point boundary value problem, 86, 90, 118

update right, 18, 141

V-cycle, 61, 62
vectorization, 193

W-cycle, 61, 62
Wang's algorithm, 155
    timing results, 170–172
waveform
    continuous, 25

discrete, 43
vectorization, 193
waveform iteration, 28
classification of methods, 28–30
linear, 37
operator, 28
waveform Newton method, 29
accuracy increase, 33
waveform Picard method, 28, 30
waveform quasi-Newton method, 29
waveform relaxation, 20, 23–48
acceleration techniques, 45–48
basic ideas, 25–28
block-Jacobi point-Gauss-Seidel, 47
convergence
linear, 35–45
nonlinear, 30–35
for linear systems, 37
for PDEs, *see* WR for PDEs
Gauss-Seidel, *see* Gauss-Seidel WR
hierarchical Gauss-Seidel, 46
Jacobi, *see* Jacobi WR
multigrid, *see* multigrid WR
Newton method, 29, 90
operator
continuous-time, 38
discrete-time, 43
overlapping block, 46, 229
parallel implementation, 47
shooting, 120
SOR, *see* SOR WR
time-periodic, *see* time-periodic WR
waveform Schwarz method, 229
waveform vectorization, 193
window, 19, 46, 208, 214
WR for PDEs, 49–82
convergence
continuous-time, 52–53
discrete-time, 53
Gauss-Seidel, *see* Gauss-Seidel WR
for PDEs
Jacobi, *see* Jacobi WR for PDEs
multigrid, *see* multigrid WR
numerical experiments, 54–58
parallel implementation, 190–192, 228

SOR, *see* SOR WR for PDEs
standard methods, 51–59
time-periodic, 110–116

Wittum
Filternde Zerlegungen

Schnelle Löser für große Gleichungssysteme

Das Lösen großer Gleichungssysteme ist ein zentraler Bestandteil vieler Aufgaben in Naturwissenschaft und Technik. Insbesondere die numerische Lösung partieller Differentialgleichungen erfordert die schnelle Lösung großer Gleichungssysteme. In vielen Fällen ist der Gleichungslöser der zeit- und speicheraufwendigste Teil einer Simulation und begrenzt damit deren Umfang.
Das Buch stellt eine neue Klasse von schnellen Lösern für große Gleichungssysteme vor, die filternden Zerlegungen, und ordnet sie vergleichend in die bestehende Landschaft von schnellen Lösern ein. Filternde Zerlegungen sind eine Verallgemeinerung unvollständiger Block-Zerlegungen. Sie ermöglichen die Konstruktion eines Verfahrens mit Mehrgittereffizienz, das jedoch nur ein Gitter benötigt. In verschiedenen Anwendungen wird der Vorteil dieses Vorgehens deutlich.

Aus dem Inhalt:

Historischer Überblick – Mehrgitterverfahren – Filternde Zerlegungen – Glätter-Korrektor-Verfahren – Modellanalyse – Konvergenztheorie – Anwendungen – Symmetrische Probleme – Unsymmetrische Probleme – Nichtlineare Probleme.

Von Prof. Dr.
Gabriel Wittum,
Universität Heidelberg

1992. 176 Seiten.
16,2 x 23,5 cm.
Kart. DM 29,–.
ISBN 3-519-02715-1

(Teubner Skripten zur Numerik)

Preisänderungen vorbehalten.

B. G. Teubner Stuttgart